2nd edition
Precalculus
Bridge to College Math

Meijun Zhu

Everyone Academy
Boston, USA

Precalculus
Bridge to College Math
2nd edition

Everyone Academy
Boston, USA

www.everyone-math.com

Copyright @ Meijun Zhu 2025
Cover design by Lei Cai 2025

All rights reserved.

No part of this work covered by the copyright
herein may be reproduced or distributed in any form or by any means,
except as permitted by U.S. copyright law, without the prior written
permission of the copyright owner.

For product information, permission to use materials from this book,
contact us at: everyone2academy@gmail.com

ISBN: 979-8-9987409-0-9

Contents

Preface   **1**
    0.1   About this book 1
    0.2   Why do we need Calculus 2

1   Numbers and Operations   **1**
    1.1   Rational numbers and decimal numbers 1
    1.2   Algebraic expressions, polynomials 27
    1.3   Properties of equality 36
    1.4   Properties of inequality 52
    1.5   Real numbers 59
    1.6   Complex numbers 71
    1.7   Chapter review and exercises 79

2   Function   **86**
    2.1   Sets 86
    2.2   Counting 94
    2.3   Concepts of functions 101
    2.4   Chapter review and exercises 116

3   Elementary Functions   **120**
    3.1   Polynomials 120
    3.2   Power functions 136
    3.3   Exponential functions 140
    3.4   Logarithmic functions 145
    3.5   Extra reading: Fundamental Theorem of Algebra for quadratic equations   149
    3.6   Chapter review and exercises 153

4   Trigonometric functions   **158**
    4.1   Definition and graph 158
    4.2   Trigonometric identities 165
    4.3   Trigonometric functions in triangles 184
    4.4   Extra reading: Euler formula* 187

| | 4.5 | Chapter review and exercises . . . . | 189 |

5 Inverse trigonometric functions   194
5.1 Definitions 194
5.2 Basic properties 201
5.3 Chapter review and exercises 205

6 Geometric properties of functions   207
6.1 Two-dimension and three-dimension coordinate systems 207
6.2 Graphs of functions in Cartesian coordinate system 211
6.3 Polar coordinate system 219
6.4 Operations on vectors 225
6.5 Equations for lines and planes 239
6.6 Parametric equations and vector functions 246
6.7 Chapter review and exercises 250

7 Sequence and series   256
7.1 Sequence 256
7.2 Limit 268
7.3 Series 288
7.4 Chapter review and exercises 293

8 Differentiation and Integration   297
8.1 Tangent and instantaneous speed— concept of differential 297
8.2 Derivative and applications 300
8.3 Anti-derivative—indefinite integral 313
8.4 Definite integral 317
8.5 Chapter review and exercises 325

9 Solutions to even numbered questions   330

Preface

0.1 About this book

The main goal of this book is to help college-bound students prepare for future challenges during their study of advanced mathematics.

In the first three chapters, we list the most important properties of number sets and the elementary functions, with the exception of trigonometric functions, which will be covered in Chapters 4 and 5.

In order to prepare students for college mathematics courses, we shall answer some questions left behind while we study elementary mathematics. Why do we have the Fundamental Assumption of Elementary Math? What on the earth, is a repeating decimal? And so on. We introduce some basic knowledge of limits, continuity, and series in Chapters 7 and Chapter 8. At the end of the book, we introduce derivatives and integrals for single variable functions. These concepts are often used in the study of other modern scientific subjects.

The corresponding mathematical level for each chapter is listed as follows:

Chapter 1. Numbers and Operations (Math Level: Algebra I)

Chapter 2. Functions (Math Level: Algebra II)

Chapter 3. Elementary Functions (Math Level: Algebra II)

Chapter 4. Trigonometric Functions (Math Level: Precalculus)

Chapter 5. Inverse Trigonometric Functions (Math Level: Precalculus)

Chapter 6. Geometric properties of functions (Math Level: Precalculus)

Chapter 7. Sequence and Series (Math Level: Calculus)

Chapter 8. Differential and Integral (Math Level: Calculus)

Chapter 9. Solutions to even numbered questions

I am indebted to everyone in my family for their immeasurable contributions to my lifelong education goal. I would like to thank my two children for helping me realize that the potential of students depends also on trust. I would like to thank my wife for proofreading the book and helping make our household conducive to learning.

I would also like to express my gratitude to some of my former current middle and high school students—Lilian Dai, Bryan Joo, Hannah Joo, Jenny Zhu, and Claire Zou—for pointing out numerous typos and errors in earlier drafts of this book. The cover of this book was designed by Professor Lei Cai, who also provided invaluable suggestions for editing the second edition. In this updated version, we corrected many typos; We also included solutions to even-numbered exercises, and added an index.

0.2 Why do we need Calculus

Here we try to illustrate some motivations for studying differentiation and integration after we complete the study of "Introductory Algebra" and "Introductory Geometry and Proofs".

We first recall some old questions left to be answered.

1. Repeating (Recurring) decimals

Although we introduced the concept of a reciprocal of a nonzero integer, we did not show what a reciprocal number looks like. A naive arithmetic calculation leads us to the following confusing "long division":

$$
\begin{array}{r}
0.3333 \\
3\,\overline{)\,1.0000} \\
-9 \\
\hline
10 \\
-090 \\
\hline
10 \\
-9 \\
\hline
1
\end{array}
$$

We thus have the following notation:
$$\frac{1}{3} = 0.\dot{3} = 0.33\cdots$$
and call the right-hand side a "repeating decimal number".

On a closer inspection, the first step in the above calculation is correct:
$$\frac{1}{3} = 0.3 + \frac{0.1}{3}.$$
The next step is also okay:
$$\frac{1}{3} = 0.3 + 0.03 + \frac{0.01}{3}.$$

But who can guarantee that the computation can go on? Even if it can go on, what is the real meaning of this procedure? This is where the concept of an infinite series comes in.

2. Fundamental Assumption of Elementary Math

In Chapter 11 of "Introductory Algebra, 2nd edition", we introduced the *Fundamental Assumption of Elementary Math*[1]: *The elementary operation rules that hold for all rational numbers are true for all real numbers.* Using this assumption we extend many rules that hold for rational numbers to all real numbers. A typical example is the definition of the area of a rectangle. In this book, we will use continuity properties for elementary functions to verify that the Fundamental Assumption holds.

3. Approximation by polynomials

In the study of Algebra and Geometry, we have already encountered some basic questions. For example, how is the value of $\sqrt{2.3}$ calculated? How is the value for π approximated?

Here is a seemingly unrelated question: how does one find the tangent line for parabola $y = x^2$ at point $(1, 1)$? Essentially, we need to systematically understand all elementary functions, and then use this understanding to solve more complicated problems.

As an analogy: if we want to understand the structure of an onion, we must peel off its outer skin layer by layer. Peeling off the outer skin is analogous to calculating the tangent line of a function. Once we know how to find the tangent line of a function at one point, we can then approximate the function near that point. This idea enables us to compute $\sqrt{2.3}$, π, and the value of e (via the Taylor series). It also helps us to better understand complex numbers. In particular, we will introduce how to derive the amazing Euler formula

$$e^{ix} = \cos x + i \sin x.$$

4. Motivation from Physics

The study of motion also relies on calculus. The concept of derivatives enables us to understand instantaneous speed. The integration enables us to derive the distance formula from the speed function easily. In fact, differentiation and integration can also be used to derive Kepler's three famous theorems describing the motion of planets in our solar system. Kepler's theorem usually will be covered in a calculus course.

---

[1]The introduction of this assumption is inspired by H. Wu's book: *Teaching School Mathematics: Pre-Algebra*, AMS, 2016.

Chapter 1 Numbers and Operations

> **Introduction**
>
> - Place value and decimal number system
> - Operations and operation rules
> - Opposite and negative number
> - Reciprocal and fraction
> - Power operation
> - Fundamental Theorem of Arithmetic (FTA)
> - Monomial
> - Polynomial
> - Operations on polynomials
> - Factorizations for polynomials
> - Equations and the properties of equalities
> - Inequalities and the properties of inequalities
> - Number line
> - Absolute value
> - Radical expressions
> - Rational and irrational numbers
> - Fundamental Assumption of Elementary Math
> - General rule for power operations
> - Exponent and the operation
> - Modulus and argument for complex number
> - Euler formula

1.1 Rational numbers and decimal numbers

1.1.1 Natural numbers, operations, decimal numbers, and positive exponents

In practice, people start to recognize the numbers "1, 2, 3, ..." while counting sheep, money, etc. The introduction of the number 10 is revolutionary in a way that gets hidden by its Roman numeral equivalent X. In particular, we learn that there is a "not very natural" number 0 in the number 10, and we get introduced to the tens place! Similarly, we then can introduce the numbers one hundred (100), one thousand (1000), and the hundreds place, the thousands place, etc.

1.1 Rational numbers and decimal numbers

> **Definition 1.1. Natural numbers**
>
> *We call 0, 1, 2, ... natural numbers. Natural numbers are ordered. From 0 to 1 to the next number, natural numbers get larger and larger. We use $\mathbb{N}$ to represent the set of all natural numbers.*

While counting natural numbers, we naturally introduce the first operation, "addition", which we notate as "+". For example, after the number "2" is the number "3". We can represent this as

$$2 + 1 = 3.$$

The fourth number after number "5" is number "9". So we can represent this as

$$5 + 4 = 9$$

While adding the same numbers, people discover the second operation "multiplication", which we notate using "×". For instance, if we wish to calculate three "10"s added up, we can also use "multiplication":

$$10 + 10 + 10 = 3 \times 10.$$

Since multiplication is more efficient than the addition of multiple duplicate numbers, memorizing the 9×9 multiplication table can be a useful skill.

With multiplication, we can further explore these "special numbers" in the decimal number system: 10, 100, 1000, etc.

We first observe that

$$100 = 10 \times 10, \ 1000 = 10 \times 10 \times 10, \ 10000 = 10 \times 10 \times 10 \times 10, ..., \text{etc.}$$

For convenience, we introduce "exponentiation"

$$10 \times 10 = 10^2, \ 10 \times 10 \times 10 = 10^3, \ 10 \times 10 \times 10 \times 10 = 10^4.$$

More generally, we define

> **Definition 1.2. Exponentiation**
>
> *For any given two positive natural numbers b and n, we define*
>
> $$b^n = \underbrace{b \times \cdots \times b}_{n \text{ copies}}. \qquad (1.1)$$
>
> *We call b the base for the exponentiation, and n the exponent or the power. We read b^n: b to the power of n, or b to the n-th power.*

Note, all of these numbers 10 (ten), 10^2 (hundred), 10^3 (thousand) have the same base of "10". We can then represent the number 500 as

$$500 = 5 \times 10^2.$$

Once we understand "addition", "multiplication" and "exponentiation", we can understand the structure of any natural numbers. For example, we know

$$3506 = 3 \times 10^3 + 5 \times 10^2 + 6.$$

It is clear, that the expression of the number on the left-hand side is simpler, but the expression on the right-hand side (the way to place value) is easier to recognize and easier to do operations on. We often refer to the right-hand expression of a natural number as the expression of a number using the decimal system (it is also called the standard form for a decimal number). Here is another example:

$$2000000009 = 2 \times 10^9 + 9.$$

In the left-hand expression, the number 2 has to occupy the billions place (we carry out this by filling in eight "0" between the number 2 and the number 9). On the contrary, the right-hand expression is simpler and easier to recognize if one knows that a billion is "10^9".

There are other number systems for natural numbers. A popularly used one is the binary system. For example, the binary number 1010_2 converted into the decimal system becomes [1]

$$1010_2 = 1 \times 2^3 + 1 \times 2^1 = 10.$$

A base 5 number, for example, 23_5 converted to decimal (base 10) is

$$23_5 = 2 \times 5^1 + 3 \times 1 = 28.$$

In daily life, people observe the following operation rules for "addition" and "multiplication".

> **Proposition 1.1. Commutative rule**
> *In an addition or a multiplication, if we switch the position of two numbers, we get the same result:*
> *1. $A + B = B + A$*
> *2. $A \times B = B \times A$*

Note We suggest students use numbers in the previous and subsequent rules to verify the correctness of these rules.

The following figure (fig 1.1) illustrates the commutative rule.

The commutative rule can also help us to count. For instance, counting six of one

---

[1] The subscript 2 indicates it is in the base 2 number system. We usually drop the subscript 10 for decimal numbers, since it is the commonly used number system.

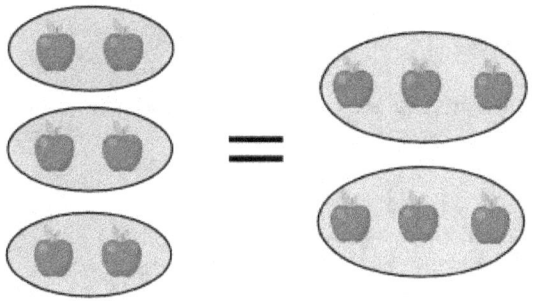

Figure 1.1: **Commutative rule for multiplication:** $3 \times 2 = 2 \times 3$

hundred will be faster than counting one hundred of 6, even though the results are the same.

> **Proposition 1.2. Associative rule**
> For any given three numbers A, B and C, the following holds:
> 1. $A + B + C = (A + B) + C = A + (B + C)$
> 2. $A \times B \times C = (A \times B) \times C = A \times (B \times C)$

The following figure (fig 1.2) illustrates the associative rule.

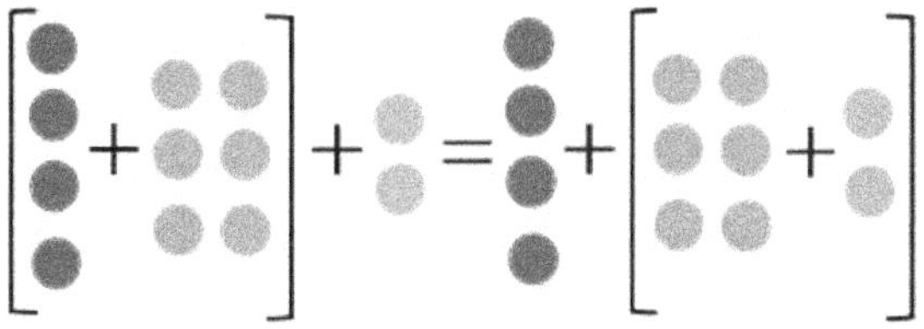

Figure 1.2: **Associative rule for addition:** $(4 + 6) + 2 = 4 + (6 + 2)$

Often, when we use the commutative rule and the associative rule together, we can quickly and easily obtain the result.

Example 1.1 Calculate $37 + 68 + 63$.

Solution:

$$37 + 68 + 63 = 37 + 63 + 68 \quad \text{(Commutative rule)}$$
$$= (37 + 63) + 68 \quad \text{(Associative rule)}$$
$$= 100 + 68$$
$$= 168.$$

Exercise 1.1

1. $73 + 68 + 27 =$
2. $5 \times 37 \times 2 =$

When the calculation involves both addition and multiplication, we have the following rule:

> **Proposition 1.3. Distributive rule**
> For any given three numbers A, B, C, the following holds:
> $$A \times (B + C) = A \times B + A \times C$$

We shall use the following example to illustrate the distributive rule.

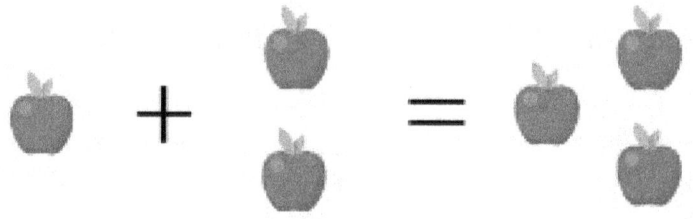

one apple + two apples = three apples

Figure 1.3: **Distributive rule:** $1a + 2a = (1+2)a = 3a$

Suppose that you have one apple and your friend has two. Together, you and your friend have three apples. This can be expressed in an algebraic formula:

$$1 \text{ apple} + 2 \text{ apples} = 3 \text{ apples}$$

It is kind of inconvenient to write "apple" three times. So we prefer to use a letter to represent "apple". How about "a" for apple, "b" for "banana"? So we have

$$1a + 2a = (1+2)a = 3a.$$

In the above formula, we keep "a" unchanged, but only add numbers together (we call this procedure "combining like terms" in Algebra) — this is just the distributive rule!

Example 1.2 Compute $37 \times 68 + 63 \times 68$.

Solution:

$$37 \times 68 + 63 \times 68 = (37 + 63) \times 68 \quad \text{(Distributive rule)}$$
$$= 100 \times 68$$
$$= 6800.$$

□

Basic philosophy for algebraic operations: Even though people discovered the above three rules from operations on natural numbers, we want these three rules to always hold for other numbers (integers, rational numbers, irrational numbers, and finally complex numbers!) until we discuss operations on vectors.

Using these three operation rules, we shall explain why we can use a vertical form to carry out calculations[2].

Example 1.3 Calculate

$$1389 + 356.$$

Solution: We first use the traditional horizontal form to carry out the computation.

$$1389 + 356 = (1 \times 1000 + 3 \times 100 + 8 \times 10 + 9) + (3 \times 100 + 5 \times 10 + 6)$$

(Decimal number system)

$$= 1 \times 1000 + 3 \times 100 + 8 \times 10 + 9 + 3 \times 100 + 5 \times 10 + 6$$

(Associative rule)

$$= 1 \times 1000 + (3 \times 100 + 3 \times 100) + (8 \times 10 + 5 \times 10) + (9 + 6)$$

(Commutative+Associative rules)

$$= 1 \times 1000 + (6 \times 100) + (13 \times 10) + (15)$$

(Operations inside the parentheses first)

$$= 1 \times 1000 + 6 \times 100 + 1 \times 100 + 3 \times 10 + 1 \times 10 + 5$$

(Carrying numbers)

$$= 1 \times 1000 + 7 \times 100 + 4 \times 10 + 5$$

(Associative rule)

$$= 1745.$$

(Decimal number system)

□

We recommend students use the vertical form to carry out the computation.

---

[2]The vertical form is a two-dimensional picture, which is easier to understand and remember

Example 1.4 Use the vertical form to compute $1389 + 356$.

Solution:

Step 1. We add the numbers in the ones position first. These are 9 and 6. $9+6 = 15$, 15 is bigger than 10, so we put number 5 in the ones position and carry number 1 to tens position for the next step (the addition of numbers at the tens position).

$$\begin{array}{r} 1\ 3\ {}^18\ 9 \\ +\quad 3\ 5\ 6 \\ \hline 5 \end{array}$$

Step 2. We add the numbers in the tens position. $1 + 8 + 5 = 14$ (do not forget we carry 1 into tens position from the first step!). The result is bigger than 10, so we put 4 in the tens position and carry 1 to the hundreds position for the next step (the addition of numbers at hundreds position).

$$\begin{array}{r} 1\ {}^13\ {}^18\ 9 \\ +\quad 3\ 5\ 6 \\ \hline 4\ 5 \end{array}$$

Step 3. We add the numbers in the hundreds position. $1 + 3 + 3 = 7$ (do not forget we carry 1 into hundreds position from the second step!). The result is less than 10, so we put 7 in the hundreds position. In the thousands position, the number is just 1. We thus complete the addition.

$$\begin{array}{r} 1\ {}^13\ {}^18\ 9 \\ +\quad 3\ 5\ 6 \\ \hline = 1\ 7\ 4\ 5 \end{array}$$

□

Exercise 1.2 Compute (1) $3087 + 2876$; (2) $3095 + 308$.

Introduction of binary system: The decimal system consists of ten natural numbers $0, 1, 2, \ldots 8, 9$ and the exponent number 10^n with 10 as the base. Similarly, we can use only two numbers 0, 1, and the exponent number 2^n with 2 as the base to represent all natural numbers. For example

$$3 = 2 + 1 = 1 \times 2 + 1 = 11_2.$$

We use the subscript "2" to indicate that number 11_2 is represented in the base 2 number system — often, we call it the binary system. Another example:

$$10101_2 = 1 \times 2^4 + 0 \times 2^3 + 1 \times 2^2 + 0 \times 2^1 + 1 = 16 + 4 + 1 = 21,$$

where 21 is the number in the decimal system. As we mentioned before: since the decimal system is used in daily life, we often omit the subscribe "10": $21_{10} = 21$. The above equality verifies that 21 in the decimal system is the same as 10101_2 in the binary system.

✎ **Exercise 1.3** Use the definition for binary numbers (not converting into decimal numbers) to compute
$$10101_2 + 11_2.$$
In particular, understand "carrying number in the binary system".

We now study the multiplication for decimal numbers.

Example 1.5 Compute 38×42.

Solution:

$$\begin{aligned}
38 \times 42 &= (30 + 8) \times (40 + 2) &&\text{(Decimal numbers)} \\
&= (30 + 8) \times 40 + (30 + 8) \times 2 &&\text{(Distributive rule)} \\
&= 30 \times 40 + 8 \times 40 + 30 \times 2 + 8 \times 2 &&\text{(Distributive rule)} \\
&= 1200 + 320 + 60 + 16 &&\text{(Calculation)} \\
&= 1596. &&\text{(Addition)}
\end{aligned}$$

□

Next, we examine how to use the vertical form to carry out multiplication.

Example 1.6 Compute 38×42.

Solution:

Using the **distributive rule**, we have $38 \times 42 = 38 \times (40 + 2) = 38 \times 2 + 38 \times 40$.

Step 1: Calculate $38 \times 2 = 76$:

$$\begin{array}{r} 3\ 8 \\ \times\ 2 \\ \hline 7\ 6 \end{array}$$

Step 2: We continue to compute $38 \times 40 = 1520$. We align the numbers in the same places: the tens place in 1520 is "2", we align it with "7" — the number in tens place of the number 76 obtained in the first step. We also can omit "0" in the number

1520 since it's implied (what happens if you leave it there?):

$$
\begin{array}{r}
3\ 8 \\
\times\quad 4\ 2 \\
\hline
7\ 6 \\
1\ 5\ 2
\end{array}
$$

Step 3: In Step 1 and Step 2, we have calculated 38×2 and 38×40. Now we add the results together, that is, add 76 to 1520 to get the final result.

$$
\begin{array}{r}
3\ 8 \\
\times\quad 4\ 2 \\
\hline
7\ 6 \\
+\ 1\ 5\ 2 \\
\hline
1\ 5\ 9\ 6.
\end{array}
$$

$\square$

✎ **Exercise 1.4** Compute 381×21.

✎ **Exercise 1.5** Without converting the binary numbers to decimal numbers, directly compute

$$101_2 \times 101_2.$$

1.1.2 Opposite numbers—introduction to negative numbers

In arithmetic, we naturally introduce the subtraction operation as the inverse operation of addition[3]. For example, from $3 + 7 = 10$, we can get

$$10 - 3 = 7,$$

where we call 10 the minuend, 3 the subtrahend, and the result 7 the difference.

When the minuend is more than the subtrahend, subtraction is easy. However, when the minuend is less than the subtrahend, we have trouble carrying out the operation. In fact, before we learn about negative numbers, we can not understand the outcome of the subtraction $3 - 7$. Even worse, the commutative rule does not hold for subtraction.

To avoid confusion, we will not initially introduce the subtraction operation. Instead, we will introduce the opposite number and subsequently, the definition for the negative number will follow. Then, we can use addition and the opposite number to replace subtraction in arithmetic.

---

[3] We often call operations that do not involve negative numbers "arithmetic operations".

1.1 Rational numbers and decimal numbers

> **Definition 1.3. Opposite number**
>
> *The opposite number of a given number is a new number such that the addition of the new number and the original number is zero.*
>
> *For a given number A, we write its opposite number as $-A$.*[a]
>
> ---
> [a] We encourage the use of hat notation to represent the opposite of a number: the opposite of A is denoted as $\hat{A}$. This notation offers certain advantages. See, for example, Section 2.3.1.

If A is a positive number[4], then we call its opposite number $-A$ a negative number. Number 0 is special: its opposite is itself. Natural numbers and their opposites together form the integer set $\mathbb{Z}$.

Exercise 1.6 Find the opposite of the following numbers: 3, 4, -4, -3.

We can now define how to subtract a number.

> **Definition 1.4. Subtraction**
>
> *Subtracting a number B from another number A is defined as adding the opposite of B to A.*
>
> $$A - B = A + (-B)$$

Note *Note we add parenthesis for $-B$. Without it, the above calculation will be $A + -B$, which may cause confusion.*

Now the "subtraction" becomes "addition", so all three operation rules hold. For example

$$A + (-B) = (-B) + A$$

Often, we just write $(-B) + A$ as $-B + A$. There is no ambiguity here.

In arithmetic, we know how to subtract a smaller number from a larger number (for example, $12 - 5 = 7$). With the help of a negative number, we can carry out subtracting a larger number from a smaller number. For example, $5 - 12$, which is $5 + (-12)$.

Example 1.7

$$\begin{aligned} 5 - 12 &= -(-5 + 12) && \text{(The opposite of the opposite is itself)} \\ &= -(12 - 5) && \text{(Commutative rule)} \\ &= -7. \end{aligned}$$

□

---
[4] Before we extend the number set, here positive numbers refer to nonzero natural numbers

For any given natural number n, its opposite is $-n$. From the definition of the opposite number, we can verify

$$\underbrace{(-1) + \cdots + (-1)}_{n \text{ times}} = -n.$$

That is

$$-n = n \times (-1)$$
$$= (-1) \times n. \qquad \text{(Commutative rule)}$$

So for any nonzero natural number n, $-n = (-1) \times n$. **If we wish for this property to also hold for a negative n, we shall define**

$$(-1) \times (-1) = -(-1) = 1.$$

This identity makes sense[5], since the opposite of -1 is 1. This identity also has a significant impact on multiplication: multiplication is no longer a replacement for fast addition[6], it has logical implication: namely, the opposite of the opposite is itself.

Using the associative rule, we thus have:

$$-a = (-1) \times a$$

holds for all integers. In particular, the following distributive rule holds

$$(-1) \times (a + b) = -(a + b)$$
$$= -a - b \qquad \text{(Definition)}$$
$$= (-1) \times a + (-1) \times b.$$

We thus agree that all three rules hold for all integers. **From now on, we agree that all three rules shall hold for all numbers.**

Example 1.8 Compute

$$1389 - 356.$$

Solution:

---

[5] If we agree that operation rules hold for both natural numbers and negative numbers, we then can prove the above identity. See the challenging problem 9 at the end of this section.

[6] or, we say that multiplication can no longer be replaced by addition.

$$1389 - 356 = (1 \times 1000 + 3 \times 100 + 8 \times 10 + 9) - (3 \times 100 + 5 \times 10 + 6)$$

(Decimal numbers)

$$= 1 \times 1000 + 3 \times 100 + 8 \times 10 + 9 - 3 \times 100 - 5 \times 10 - 6$$

(Associative rule)

$$= 1 \times 1000 + (3 \times 100 - 3 \times 100) + (8 \times 10 - 5 \times 10) + (9 - 6)$$

(Commutative and associative rules)

$$= 1 \times 1000 + (0 \times 100) + (3 \times 10) + (3)$$

(Operations inside the parentheses first)

$$= 1033.$$

(Decimal number)

We can also use the vertical form to carry out the subtraction

Example 1.9 Use the vertical form to carry out the subtraction $1329 - 356$.

Solution: Step 1: First, we do subtraction for numbers in the ones place: $9 - 6 = 3$.

$$\begin{array}{r} 1\ 3\ 2\ 9 \\ -\ \ \ 3\ 5\ 6 \\ \hline 3 \end{array}$$

Step 2: Second, we do subtraction for numbers in the tens place. In the tens place, the minuend "2" is less than the subtrahend "5". So we borrow number 1 from the hundreds place of the minuend (now the number at the hundreds place for the minuend is 2. We will use this for the next step). Thus we have 12 in the tens place for the minuend. Now we do subtraction: $12 - 5 = 7$.

$$\begin{array}{r} 1\ \ 3_2\ \ ^12\ \ 9 \\ -\ \ \ \ \ 3\ \ 5\ \ 6 \\ \hline 7\ \ 3 \end{array}$$

Step 3: Next, we do subtraction for numbers in the hundreds place. Again, in the hundreds place, the minuend (2 is left after borrowing 1 in the last step) is less than the subtrahend "3". So we borrow 1 from the thousands place of the minuend (now the number at thousands place for the minuend is 0). Thus we have 12 in the hundreds place for the minuend. Now we do subtraction: $12 - 3 = 9$.

$$\begin{array}{r} 1_0 \ {}^13_2 \ {}^12 \ 9 \\ - \quad\quad 3 \ \ 5 \ \ 6 \\ \hline \quad\quad 9 \ \ 7 \ \ 3 \end{array}$$

Step 4: Finally, we do subtraction for numbers in the thousands place. The number at thousands place for the minuend is 0. We do subtraction: $0 - 0 = 0$.

$$\begin{array}{r} 1_0 \ {}^13_2 \ {}^12 \ 9 \\ - \quad\quad 3 \ \ 5 \ \ 6 \\ \hline 0 \ \ 9 \ \ 7 \ \ 3 \end{array}$$

The final answer is: $1329 - 356 = 973$.

$\square$

To further help readers understand "borrowing" numbers from the upper places better, we give an example to illustrate how to do subtraction in the base-5 number system.

Example 1.10 Calculate $413_5 - 32_5$.

Solutions: **Step 1:** First, we do subtraction for numbers in the ones place. $3 - 2 = 1$.

$$\begin{array}{r} 4 \ \ 1 \ \ 3 \\ - \quad 3 \ \ 2 \\ \hline 1 \end{array}$$

Step 2: Second, we do subtraction for numbers in the tens place. In tens place, the minuend "1" is less than the subtrahend "3". So we borrow 1 from the hundreds place of the minuend (now the number in the hundreds place for the minuend is 3. We will use this for the next step). In the base-5 system, the borrowed number is actually $100_5 = 1 \times 5^2 + 0 \times 5^1 + 0 \times 1 = 5 \times 5^1 + 0 \times 1$. Thus we have $5 + 1 = 6$ in the tens place for the minuend. Now we do subtraction: $6 - 3 = 3$.

$$\begin{array}{r} 4_3 \ \ {}^11 \ \ 3 \\ - \quad\quad 3 \ \ 2 \\ \hline 3 \ \ 1 \end{array}$$

Step 3: Next, we do subtraction for numbers in hundreds place. In the hundreds place, the minuend (3 is left after borrowing 1 in the last step) is larger than the subtrahend (0). So we do subtraction: $3 - 0 = 3$.

$$\begin{array}{r} 4_3\ {}^11\ \ 3 \\ -\ \ \ 3\ \ 2 \\ \hline 3\ \ 3\ \ 1 \end{array}$$

The final answer is: $413_5 - 32_5 = 331_5$. □

Exercise 1.7 Compute (1) $354 - 213$; (2) $3124 - 456$.

Exercise 1.8 Compute (1) $356_7 - 234_7$; (2) $312_5 - 243_5$.

In Exercise 1.4, we observe that $381 = 400 - 20 + 1 = 4 \times 10^2 - 2 \times 10 + 1$. With the help of a formula[7], we can mentally get the result: 8001. We can also use the vertical calculation:

$$\begin{array}{r} 4\ -2\ \ 1 \\ \times \ \ 2\ \ 1 \\ \hline 4\ -2\ \ 1 \\ +\ 8\ -4\ \ 2 \\ \hline 8\ \ 0\ \ 0\ \ 1. \end{array}$$

Exercise 1.9 Compute

$$(10000 - 1000 + 100 - 10 + 1) \times (10 + 1).$$

1.1.3 Reciprocal, fraction and negative power

We shall discuss another operation that we learned in arithmetic— division. Division is in fact, the inverse operation of multiplication. For example,

$$43 \times 47 = 2021.$$

Knowing this, we also know that

$$2021 \div 43 = 47, \quad \text{and} \quad 2021 \div 47 = 43.$$

We call 2021 the dividend and 43 and 47 the divisors in the above calculation. We also say that the number 2021 is divisible by 43 and 47. In real life, we can say that 2021 apples can be evenly shared by 43 people or 47 people. The result of division is called the quotient.

---

[7] Addition of cubes formula:
$$(x^2 - x + 1)(x + 1) = x^3 + 1.$$

We will learn the operations on polynomials shortly.

We can also use the vertical form to carry out division. Recall the example that we covered in "Introductory Algebra".

Example 1.11 Compute: $27001 \div 31$.

Solution:
$$270001 \div 31 = 871.$$

Step 1: Since the leading number 27 in 27001 is less than 31, we try the leading three numbers $270 \div 31$. Note that $9 \times 31 = 271 > 270$, and $8 \times 31 = 248 < 270$. We shall use 8.

$$\begin{array}{r} 8 \\ 31 \overline{)\, 2\,7\,0\,0\,1} \\ -\,2\,4\,8 \\ \hline \end{array}$$

Step 2: $270 - 248 = 22$. We align 22 under 48. We then bring the second "0" in 27001 down to have 220. If we wrote each number and operation completely, the previous step represents $27001 - 24800 = 2201$. However, we only address the remainder "1" in the ones place in the next step.

$$\begin{array}{r} 8 \\ 31 \overline{)\, 2\,7\,0\,0\,1} \\ -\,2\,4\,8 \\ \hline 2\,2\,0 \end{array}$$

Step 3: We repeat the procedure in steps 1-2. We shall try 7 since 7 is the largest integer that multiplied by 31 is less than 220. We then calculate $220 - 7 \times 31 = 220 - 217 = 3$, and align 3 under 7. We then bring down the last number 1 in 27001.

$$\begin{array}{r} 8\,7 \\ 31 \overline{)\, 2\,7\,0\,0\,1} \\ -\,2\,4\,8 \\ \hline 2\,2\,0 \\ -\,2\,1\,7 \\ \hline 3\,1 \end{array}$$

Step 4: Similar to the above step, since $31 = 1 \times 31$, we put 1 in the quotient. Since

$31 - 31 = 0$, the computation is completed.

```
            8 7 1
     ┌─────────────
  31 ) 2 7 0 0 1
      - 2 4 8
        ─────
          2 2 0
        - 2 1 7
          ─────
              3 1
            - 3 1
              ───
                0.
```

$\square$

To discuss more general divisions, we need to expand the number set again.

> **Definition 1.5. Reciprocal**
>
> If A is a non-zero number, we define its reciprocal as $\frac{1}{A}$, which has the property: $A \times \frac{1}{A} = \frac{1}{A} \times A = 1$.

With the notation of reciprocal, we then can see that a number m is divided by another number n is in fact number m multiple the reciprocal of number n.

> **Definition 1.6. Division**
>
> A number m divided by a non-zero number n is the same as the number m multiplying the reciprocal of n:
> $$m \div n = m \times \frac{1}{n}.$$

From the definition of division, We define a fraction.

> **Definition 1.7. Fraction**
>
> The fraction number $\frac{m}{n}$ (where $n \neq 0$) is defined as:
> $$\frac{m}{n} = m \times \frac{1}{n} = m \div n.$$
> We call m the numerator of the fraction, n the denominator of the fraction. We use $\mathbb{Q}$ to represent the set of all fraction numbers whose numerators are integers, and whose denominators are nonzero integers.

So from Definition 1.7 we have formula

$$\frac{m}{n} = m \times \frac{1}{n}. \tag{1.2}$$

Once we understand fractions, we can learn the operations involving fractions. For two nonzero number n and b, we can verify that $\frac{1}{n} \times \frac{1}{b}$ is in fact, the reciprocal of nb, thus we have the first important formula

$$\frac{1}{n} \times \frac{1}{b} = \frac{1}{n \times b} = \frac{1}{nb}. \tag{1.3}$$

From this and formula (1.2), we have

$$\frac{m}{n} \times \frac{a}{b} = \frac{ma}{bn}, \quad \text{and} \quad \frac{l}{n} + \frac{m}{n} = \frac{l+m}{n}. \tag{1.4}$$

Exercise 1.10 Using formula (1.4) and the definition of the reciprocal number to prove: the reciprocal of the fraction $\frac{m}{n}$ is $\frac{n}{m}$.

Now we can add two fractions

$$\frac{m}{n} \quad \text{and} \quad \frac{a}{b}$$

together. First, we observe that for any non-zero number k,

$$k \div k = \frac{k}{k} = 1.$$

So

$$1 = \frac{b}{b}, \quad 1 = \frac{n}{n}.$$

Then we have

$$\frac{m}{n} = \frac{m}{n} \times 1 = \frac{m}{n} \times \frac{b}{b} = \frac{mb}{nb},$$

and

$$\frac{a}{b} = \frac{a}{b} \times 1 = \frac{a}{b} \times \frac{n}{n} = \frac{an}{bn}.$$

Since the above two fractions

$$\frac{mb}{nb} \quad \text{and} \quad \frac{an}{bn}$$

have the same denominator nb, we can use formula (1.4) to obtain:

formula

$$\frac{m}{n} + \frac{a}{b} = \frac{mb + an}{nb}. \tag{1.5}$$

Example 1.12 Calculate and simplify the result:

$$\frac{1}{4} + \frac{1}{6}.$$

Solution: According to formula (1.5), we have

$$\frac{1}{4} + \frac{1}{6} = \frac{6+4}{24} = \frac{10}{24}.$$

We can then simplify the result

$$\frac{10}{24} = \frac{2 \times 5}{2 \times 12} = \frac{2}{2} \times \frac{5}{12} = \frac{5}{12}.$$

So, after simplification, we have
$$\frac{1}{4} + \frac{1}{6} = \frac{5}{12}.$$

□

We will discuss simplifying fractions in the next subsection (Section 1.1.4).

To end this subsection, we introduce the zero exponent and negative exponents. For any non-zero number a, we define
$$a^0 = 1, \quad \text{and} \quad a^{-1} = \frac{1}{a}.$$
Thus, we can use a^{-1} to represent the reciprocal for a non-zero number a. In particular, we can represent the decimal number 0.1 as
$$0.1 = 10^{-1} = \frac{1}{10}.$$
In the next subsection, we will also show
$$0.01 = 10^{-2}, \quad 0.001 = 10^{-3}, \quad \text{and} \quad 0.0001 = 10^{-4}, \quad \text{etc.}$$

Now, we can represent any decimal number in the decimal number system.

Example 1.13 Number 3201.04 can be formally represented by
$$3201.04 = 3 \times 10^3 + 2 \times 10^2 + 1 \times 10^0 + 4 \times 10^{-2}.$$

1.1.4 Integer exponents and the exponentiation

For a natural number n, we defined b^n in Definition 1.2. For a general number p, it is hard to understand the expression b^p (usually we give the definition in a calculus course). If we instead limit p to a negative number or zero, we can define exponents with the help of reciprocal numbers

Definition 1.8. Negative exponent

For any give nonzero number a and a positive natural number n, define
$$a^{-n} = \underbrace{\frac{1}{a} \times \cdots \times \frac{1}{a}}_{n} = \frac{1}{a^n}. \tag{1.6}$$

That is, a^{-n} is the reciprocal of a^n.

Definition 1.9. Zero exponent

For any non-zero number a,
$$a^0 = 1. \tag{1.7}$$

For any two natural numbers m, n and any nonzero number a, we have

$$a^m \times a^n = \underbrace{a \times \cdots \times a}_{\text{m copies}} \times \underbrace{a \times \cdots \times a}_{\text{n copies}} = \underbrace{a \times \cdots \times a}_{\text{m+n copies}} = a^{m+n}.$$

From Definition 1.8 and 1.9, we obtain

$$a^m \times a^{-n} = a^m \times \frac{1}{a^n} = \begin{cases} a^{m-n}, & \text{if } m \geq n \\ \dfrac{1}{a^{n-m}}, & \text{if } m < n \end{cases}$$

$$= a^{m-n}.$$

We thus obtain the following two rules for exponentiation.

> **Proposition 1.4. The same base operation rule**
>
> *For any two integers m, n and any non-zero number a, we have*
>
> $$a^m \times a^n = a^{m+n}$$
>
> *and*
>
> $$(a^m)^n = a^{mn}.$$

Exercise 1.11 Derive from the definitions the following result

$$3^4 \div (3^{-1}) = 3^5.$$

Exercise 1.12 Derive from the definitions the following result

$$(3^4)^{-2} = 3^{-8}.$$

> **Proposition 1.5. The same exponent rule**
>
> *For any two non-zero number a, b and an integer m, we have*
>
> $$(a \cdot b)^m = a^m \cdot b^m.$$

With the above three exponential operation rules, we can simplify fractions.

Example 1.14 Simplify

$$\frac{15^3}{25^2}$$

Solution

$$\frac{15^3}{25^2} = \frac{(3 \cdot 5)^3}{(5^2)^2} = \frac{3^3 \cdot 5^3}{5^4}$$

$$= \frac{3^3}{5} = \frac{27}{5}.$$

□

Exponentiation plays a fundamental role in algebraic operations. We list a few applications below.

1.1.4.1 Scientific notation and percentage

Very often, we write a positive number as a number between 1 and 10 (10 is excluded) multiplied by 10 to a certain power.

> **Definition 1.10. Scientific notation**
>
> $$m = a \times 10^k,$$
>
> where a is a number in the interval $[1, 10)$. This expression of m is called the scientific notation of m.

Example 1.15

$$13{,}000 = 1.3 \times 10^4, \qquad 130.5 = 1.305 \times 10^2.$$

Recall one type of special fraction — the percentage, which is a fraction with 100 as the denominator.

Example 1.16

$$\frac{25}{100} = 25\%, \qquad \frac{3}{4} = \frac{3}{4} \times \frac{25}{25} = \frac{75}{100} = 75\%.$$

Exercise 1.13 Convert the following numbers into percentage numbers: $\frac{1}{4}$; 0.375.

1.1.4.2 Factorization for natural numbers

While simplifying a fraction, it is very helpful to write the numerator and the denominator as products of smaller natural numbers.

> **Definition 1.11. Factor**
>
> If a nonzero natural number m is divisible by another nonzero natural number l, then we call l a factor of m and m is a multiple of l.

Example 1.17

$$1000 = 10 \times 100 = 10 \times 10 \times 10,$$

so 10 is a factor of 1000.

Example 1.18

$$459 = 3 \times 153 = 3 \times 3 \times 51 = 3 \times 3 \times 3 \times 17,$$

so both 3 and 17 are factors of 459.

If we write a natural number as a product of its factors, we call such a product a "factorization" of the natural number. For example, 4×6 is a factorization of 24. 2×12 is another factorization of 24. We may wonder: how far can we factorize a given natural number? To answer this, we need to recall the definition of a prime number.

> **Definition 1.12. Prime number**
>
> If a natural number p (greater than 1) has only 1 and p as its factors, we call this number a prime number. A natural number that is larger than 1 and is not a prime number is called a composite number. A composite number has another factor that is not 1 and the number itself.

Example 1.19 Numbers 7, 13, 17 are all prime numbers.

Example 1.20
$$2021 = 43 \times 47,$$

so 2021 is a composite number.

Exercise 1.14 Find all prime numbers in the following list:
$$12, \ 27, \ 34, \ 51, \ 151, \ 267.$$

Roughly speaking, prime numbers are relatively simple numbers as compared to composite numbers. A natural question arises: are there infinitely many prime numbers? This was answered by Euclid around 300 BC (in his book "Elements") via a contradiction argument.

> **Theorem 1.1. Infinitely many prime numbers**
>
> There are infinitely many prime numbers.

Proof. We assume that there are only a finite number of prime numbers (opposite assumption), and list them all as $p_1, p_2, ..., p_n$.

Let P be the product of all the prime numbers in the list: $P = p_1 p_2 ... p_n$, and let $q = P + 1$.

q can not be prime, since q is bigger than all prime numbers in the list.

So q is not prime, thus q is divisible by some prime factor p. Since p is in the list (which includes all prime numbers), P is divisible by p (since P is the product of every number in the list). However, $P + 1 = q$ is also divisible by p, thus the difference between the two numbers $q - P = 1$ is divisible by p. Impossible! We thus derive a contradictory result. The contradiction indicates the assumption is false. We complete the proof of the theorem.

Euclid expanded on this and went on to establish the following Fundamental Theorem of Arithmetic (FTA): indexFundamental Theorem of Arithmetic (FTA)

Theorem 1.2. Fundamental Theorem of Arithmetic

Every positive integer that is greater than 1 is either a prime number itself or can be represented as the product of prime numbers (ignoring the order). That is, for any natural number m that is larger than 1, it can be represented as:

$$m = p_1 \times \cdots \times p_k,$$

where $p_1, \cdots, p_k$ are all prime numbers (there may be duplicates), or uniquely as

$$m = q_1^{l_1} \times \cdots \times q_n^{l_n} \tag{1.8}$$

where $q_1 < q_2 < \cdots < q_n$ are different prime numbers and l_1 to l_n are positive natural numbers.

The proof of this Fundamental Theorem of Arithmetic usually will be given in an "Abstract Algebra" course for math major students in college. We call expression (1.8) the factorization of the number m.

Example 1.21 Factorize 720.

Solution:

$$720 = 2 \times 2 \times 2 \times 2 \times 3 \times 3 \times 5$$
$$= 2^4 \times 3^2 \times 5$$

Once we can do factorization for a natural number, we then can figure out how many different factors it has.

Exercise 1.15 Factorize 350, then count totally how many different prime number factors it has, and how many different factors it has.

In Chapter 2, we will discuss the general method for counting different combinations, which will make it easier to count how many factors a number has.

We now introduce some concepts that are often used while simplifying fractions.

Definition 1.13. Common divisor, Common multiplier

For any given two natural numbers A and B, which are greater than 1: If a natural number $f > 1$ is a factor of A and a factor of B, we call f a common factor of A and B. If a natural number $g > 1$ is a multiple of A, as well as a multiple of B, we call g a common denominator (or common multiple) of A and B.

> If p is the largest common divisor of A and B, we call p the greatest common divisor (GCD) of A and B, denoted as (A, B); If q is the smallest common multiplier of A and B, we call q the least common multiplier (LCM) of A and B.

Exercise 1.16 Find the greatest common divisor (GCD) and the least common multiplier (LCM) of 54 and 207.

Definition 1.14. Classification of fractions

For a given fraction $\frac{p}{q}$, if $(p, q) = 1$, we call this fraction the simplest fraction (fraction in the simplest form).

For a simplest fraction $\frac{p}{q}$, if $p \geq q$, we call it an improper fraction; If $p < q$, we call it a proper fraction

For a given proper fraction $\frac{p}{q}$ and a nonzero natural number r, we write $r\frac{p}{q} = r + \frac{p}{q}$, and call it a mixed fraction[a].

For a given fraction $\frac{p}{q}$, if one of p and q is a fraction, we call this fraction a complex fraction.

---
[a]Here, r, p, q are given numbers. For example, $3\frac{1}{4}$. Sometimes, people use $\overline{r\frac{p}{q}}$ to represent $r + \frac{p}{q}$, to avoid confusion with $r \cdot \frac{p}{q}$.

Example 1.22 Find the greatest common divisor (GCD) and the least common multiplier (LCM) of 54 and 207.

Solution:
$$54 = 2 \times 27 = 2 \times 3^3;$$
and
$$72 = 8 \times 9 = 2^3 \times 3^2.$$

So the least common multiplier is $2^3 \times 3^3 = 216$, and the greatest common divisor is $2 \times 3^2 = 18$.

□

Example 1.23 Simplify: (1) $\frac{54}{72}$; (2) $\frac{2^3 3^2 5^2}{2^2 3^3 5^3}$; (3) $\frac{2^3 3^2 5^{-2}}{2^2 3^3 5^{-3}}$.

Solution: (1)
$$\frac{54}{72} = \frac{2 \times 3^3}{2^3 \times 3^2} \qquad \text{(Factorization)}$$
$$= \frac{3}{2^2} \qquad \text{(Eliminating common factor)}$$

(2)
$$\frac{2^3 3^2 5^2}{2^2 3^3 5^3} = \frac{2}{3 \times 5}$$ (Eliminating common factor)
$$= \frac{2}{15}.$$ (In the simplest form)

(3)
$$\frac{2^3 3^2 5^{-2}}{2^2 3^3 5^{-3}} = \frac{2}{3 \times 5^{-1}}$$ (Eliminating common factor)
$$= \frac{2 \times 5}{3}$$ (Exponential operation)
$$= \frac{10}{3}$$ (Simplest form, but an improper fraction)
$$= 3\frac{1}{3}$$ (Changing to a mixed fraction)

$\square$

Sometimes we can use simple observation to convert an improper fraction into a mixed fraction. For example, $\frac{11}{7}$. We observe $\frac{11}{7} = \frac{7+4}{7} = 1 + \frac{4}{7} = 1\frac{4}{7}$. For a more complicated improper fraction, we can use long division to convert it into a mixed number. For example, we look at $\frac{203}{13}$. Doing the long division

$$\begin{array}{r} 1\ \ 5 \\ 13 \overline{)\ 2\ 0\ 3} \\ -\ 1\ 3 \\ \hline 7\ 2 \\ -\ 6\ 5 \\ \hline 7, \end{array}$$

we conclude
$$\frac{203}{13} = 15\frac{7}{13}.$$

1.1.5 Exercises after class

1. **Basic skills**

   (1) $38 - 49 + 20$; (2) $123 - 29 + 37$;

   (3) 38×23; (4) 38×33;

   (5) $61 \times 2 - 62$; (6) $62 - 2 \times 61$;

   (7) $(32 - 21) \times 4$; (8) $48 \div 14 \times 7$;

   (9) $54321 \div 3$; (10) $54522 \div 3$;

   (11) $57 \times 9 \div 19$; (12) $(32 + 23) \div 5$;

   (13) $\frac{1}{7} + \frac{1}{6}$; (14) $\frac{2}{7} \times \frac{5}{6}$;

   (15) $(5 - \frac{1}{3}) \div 7$; (16) $1 + \frac{1}{3} \times (1 + \frac{1}{3})$;

   (17) $78 \times 98 \times \frac{1}{7} \div \frac{39}{4}$; (18) $\frac{1}{16} + \frac{1}{8} + \frac{1}{4} + \frac{1}{2}$;

   (19) $(5 + \frac{1}{3}) \times (3 + \frac{1}{4})$; (20) $(1 + \frac{1}{2} + \frac{1}{3}) \times (\frac{5}{11} + \frac{1}{2})$.

2. **Use operation rules to mentally calculate**

   (1) $38 - 49 + 12$; (2) $123 - 69 + 67$;

   (3) $8 \times 25 \times 125$; (4) $25 \times 33 \times 8$;

   (5) $60 \div 3 \div 2$; (6) $60 \div (3 \div 2)$;

   (7) $(32 + 21) \times 73 + (21 + 32) \times 27$; (8) $48 \div 7 \times 14$;

   (9) $2 \times (58 + 73) - 120$; (10) $29 \times 47 - 47 \times 13 - 16 \times 27$;

   (11) $27 \times (21 - 4 \times 7) + 36 \times (21 - 2 \times 7)$;

   (12) $$\frac{23 \times 13 - 13 \times 14}{15} + \frac{46 + 11 \times 7}{15}.$$

3. **Use vertical form to compute**

   (1) 421×19; (2) $(400 + 20 + 1) \times (20 - 1)$;

   (3) 381×21; (4) $(400 - 20 + 1) \times (20 + 1)$;

   (5) $8001 \div 21$; (6) $(8000 - 1) \div (20 - 1)$;

4. **Operation in different base number systems**

   (1) $101_2 + 10_2$; (2) $213_5 + 121_5$;

   (3) $101_2 + 11_2$; (4) $323_5 + 32_5$;

   (5) $213_5 \times 10_5$; (6)* $215_5 \times 5_{10}$;

   (7) $312_5 - 133_5$; (8) $312_7 \times 23_7$.

5. **Find all prime number factors in the following numbers**

   (1) 1000

   (2) 2610

   (3) 1024

(4) 1001

(5) 2009

(6) 2021

6. **Counting the number of different factors**

    (1) Count: how many different factors does $2^3 5^4$ have?

    (2) Count: how many different factors does $2^3 3^2 5^4$ have?

    (3) Count: how many different factors does $4^2 7^3$ have?

    (4) Count: how many different factors that are divisible by 10 does $2^3 5^4$ have?

    (5) Count: how many different factors that are divisible by 20 does $2^3 5^4$ have?

7. **Convert improper fractions to mixed fractions or mixed fractions to improper fractions**

    (1) $\frac{12}{5}$; (2) $\frac{23}{11}$;

    (3) $\frac{100}{9}$; (4) $1\frac{3}{10}$;

    (5) $3\frac{3}{13}$.

8. **Calculation**

    (1) $2\frac{4}{13} - 1\frac{1}{13}$;

    (2) $2\frac{4}{13} \div 1\frac{1}{13}$;

    (3) $9 \times (4\frac{2}{9} - 2\frac{5}{9})$;

    (4) $24 \div (6 \div 9 + 2)$;

    (5) $\frac{2^3 5^{-2} 3^4}{3^3 5^{-3} 2^2}$;

    (6) $\frac{10^2 6^{-1}}{2^2 \div 5^{-2} \div 12}$.

9. **Challenging problem**

    (1) Use the vertical form to compute $1235_7 - 432_7$.

    (2) Calculate without converting to decimal numbers: $2402_5 \div 21_5$.

    (3) Use operation rules (Commutative rule, associative rule and distributive rule) to prove
    $$(-1) \times (-1) + (-1) = 0.$$

    This leads to show that $(-1) \times (-1)$ is the opposite of -1, thus it is 1: $-(-1) = (-1) \times (-1) = 1$.

1.2 Algebraic expressions, polynomials

An algebraic expression is an expression involving variables, usually represented by letters. For example, to represent "twice of a number", we use the letter x to represent the number, then twice of x is two copies of x, or simply: $2 \times x$. We can simplify the notation to $2 \times x = 2x$.

Example 1.24 Let n be a positive natural number. If we use $C_n, C_{n-1}, \cdots, C_1, C_0$ to each represent a natural number between 0 to 9, and assume that $C_n \neq 0$, then

$$C_n \times 10^n + C_{n-1} \times 10^{n-1} + \cdots + C_1 \times 10 + C_0$$

represents a $n + 1$ digit natural number. For example, $3 \times 10^2 + 8 \times 10 + 3 = 383_{10}$ is a three digit number. Since people usually use decimal numbers, we often omit the subscript 10, and just write 383 for 383_{10}. Note that here, $C_n C_{n-1} \cdots C_1 C_0$ is used to represent a decimal number, not the multiplication of $C_n \times C_{n-1} \times \cdots \times C_1 \times C_0$.

□

Example 1.25 Let n be a positive natural number. If we use $C_n, C_{n-1}, \cdots, C_1, C_0$ to represent any natural number between 0 to 4, and assume that $C_n \neq 0$, then

$$C_n \times 5^n + C_{n-1} \times 5^{n-1} + \cdots + C_1 \times 5 + C_0$$

represents a $n + 1$ digit natural number in the base 5 number system. For example, $3 \times 5^2 + 2 \times 5 + 3 = 323_5$ is a three-digit number in the base 5 number system.

□

Inspired by different number systems, we now introduce polynomial expressions.

1.2.1 Polynomials: Definition

> **Definition 1.15. Monomial**
> If n is a natural number, and c is any nonzero number, we call the algebraic expression
> $$cx^n$$
> a monomial. We usually call c the coefficient, and n the degree of the monomial.

> **Definition 1.16. Polynomials**
> A polynomial is a summation of monomials. If we arrange the summation according to the degree of each monomial, a polynomial can be written as
> $$c_n x^n + c_{n-1} x^{n-1} + \cdots + c_1 x + c_0.$$

> We then call the largest degree among the monomials the degree of the polynomial. ♣

We observe that the structures of polynomials are very similar to that of decimal numbers, or binary numbers, etc. So we can view polynomials as an extension of different number-based systems. In Section 8.2, we will show how we can use polynomials to approximate any "good" function.

1.2.2 Operation on polynomials

1.2.2.1 Operation on monomials

Why don't polynomials have multiple terms with the same degree? The reason is simple: terms with the same degree can be combined using the distributive law.

$$Ax^n + Bx^n = (A+B)x^n.$$

Example 1.26 Simplify:
$$3x^2 - x^2 + 10x^2 = (3 - 2 + 10) \cdot x^2 = 11x^2.$$

□

Monomials with the same degree are referred to as "like terms," and the process of combining them is called "combining like terms." Unlike terms, however, cannot be combined. For instance, one apple plus one banana remains an apple and a banana—it cannot turn into two "apple bananas."

Using the commutative law and Exponent rules, we can multiply monomials:

$$cx^n \times dx^m = c \times d \times x^n \times x^m = cdx^{m+n}.$$

Example 1.27 Simplify
$$5x^2 \times 2x^5 = 5 \times 2 \times x^2 \times x^5 = 10x^7.$$

□

Example 1.28 Simplify
$$5x^5 \div 2x^2 = 5x^5 \times \frac{1}{2x^2} = \frac{5}{2}x^3.$$

□

1.2.2.2 Operation on polynomials

With the help of operation rules (including Exponent rules), we can carry out operations on polynomials.

When adding two polynomials, combine like terms.

Example 1.29
$$(5x^3 - 4x^2 + 7) + (x^5 + 4x^2 - 9) = 5x^3 - 4x^2 + 7 + x^5 + 4x^2 - 9$$
$$= x^5 + 5x^3 - 4x^2 + 4x^2 + 7 - 9$$
$$= x^5 + 5x^3 + (-4x^2 + 4x^2) + (7 - 9)$$
$$= x^5 + 5x^3 - 2.$$

□

Using the distributive rule, we can compute the multiplication of one monomial and one polynomial.

Example 1.30
$$5x^2 \cdot (2x^3 + 3x) = 5x^2 \cdot (2x^3) + 5x^2 \cdot (3x) = 10x^5 + 15x^3.$$

□

We can also compute the multiplication of one polynomial and another polynomial.

Example 1.31
$$(x+1)(x+1) = (x+1) \cdot x + (x+1) \cdot 1$$
$$= x \cdot x + 1 \cdot x + x + 1$$
$$= x^2 + x + x + 1$$
$$= x^2 + 2x + 1.$$

□

Exercise 1.17 Calculate:

(1).
$$(x+a)(x+a) =$$

(2).
$$(x-a)(x-a) =$$

(3).
$$(x-a)(x+a) =$$

The above three products are called special products or formulas: (1) is called the square of the sum of two numbers; (2) is called the square of the difference; and (3) is called the difference of two squares. We will list these important formulas with other formulas later in this section.

Similar to operations on decimal numbers, we can also use the vertical form to carry out operations on polynomials.

Example 1.32 Compute $(3x + 8) \times (4x + 2)$.

Solution: **Step 1**, compute $(3x+8) \times 2 = 6x+16$. Note we do not need to carry numbers to other places while doing multiplication for polynomials; we can view polynomials as part of an infinitely large number-based system.

$$\begin{array}{r} 3 \quad 8 \\ \times \quad 4 \quad 2 \\ \hline 6 \quad 16 \end{array}$$

(or, with more details,

$$\begin{array}{r} 3x \quad 8 \\ \times \quad 4x \quad 2 \\ \hline 6x \quad 16 \end{array}$$

if we do not want to omit x.)

Step 2, we continue to compute $(3x + 8) \times (4x) = 12x^2 + 32x$. Note, we shift positions

$$\begin{array}{r} 3 \quad 8 \\ \times \quad 4 \quad 2 \\ \hline 6 \quad 16 \\ 12 \quad 32 \quad \end{array}$$

(or, with more details,

$$\begin{array}{r} 3x \quad 8 \\ \times \quad 4x \quad 2 \\ \hline 6x \quad 16 \\ 12x^2 \quad 32x \quad \end{array}$$

if we do not want to omit x.

Step 3, finally, we add them together

$$\begin{array}{r} 38 \\ \times 42 \\ \hline 616 \\ +1232 \\ \hline 123816 \end{array}$$

Let us see what we have obtained: at the constant place (analogous to the "ones place" while doing calculations for decimal numbers), we have 16; For the x place, or the "linear term", we have $38x$; In the x^2 place, we have $12x^2$. So the final answer is $12x^2 + 38x + 16$.

(or, with more details,

$$\begin{array}{r} 3x8 \\ \times 4x2 \\ \hline 6x16 \\ +12x^232x \\ \hline 12x^238x16 \end{array}$$

if we do not want to omit x.)

□

Similarly, we can perform polynomial division using the vertical form, commonly referred to as "long division."

Example 1.33 Compute: $(x^3 + 1) \div (x + 1)$.

Solution: First, we write the divisor into a full expression of a polynomial by adding missing terms with coefficient 0: $x^3 + 1 = x^3 + 0x^2 + 0x + 1$.

Step 1, let us try x^2 (since $x^2 \cdot x = x^3$, which matches the leading term x^3):

$$\begin{array}{r} x^2 \\ x+1\overline{)x^3 +0x^2 +0x +1} \\ -(x^3 +x^2) \\ \hline -x^2 +0x. \end{array}$$

Step 2, we then try $-x$ (since $-x \cdot x = -x^2$, the leading term in the remainder):

$$
\begin{array}{r}
x^2 \quad -x \\
x+1 \overline{)\; x^3 \quad +0x^2 \quad +0x \quad +1} \\
-(x^3 \quad +x^2) \\
\overline{ -x^2 \quad +0x } \\
-(-x^2 \quad -1x) \\
\overline{ x \quad +1}
\end{array}
$$

Step 3, we try 1 ($1 \cdot x = x$, x is the leading term for the remainder):

$$
\begin{array}{r}
x^2 \quad -x \quad +1 \\
x+1 \overline{)\; x^3 \quad +0x^2 \quad +0x \quad +1} \\
-(x^3 \quad +x^2) \\
\overline{ -x^2 \quad +0x } \\
-(-x^2 \quad -1x) \\
\overline{ x \quad +1} \\
-(x \quad +1) \\
\overline{ 0.}
\end{array}
$$

Therefore, we obtain

$$(x^3 + 1) \div (x + 1) = x^2 - x + 1.$$

$\square$

The divisibility of a natural number by another natural number depends on their factors and is therefore closely related to the factorization of numbers.

Similarly, whether a polynomial is divisible by another polynomial also depends on their factors, tying it to the factorization of polynomials. In the next chapter, we will explore topics such as simplification, factorization, and solving polynomial equations in detail.

1.2.2.3 A few formulas and their applications

Since we will see the following products often, we list them as formulas.

1.2 Algebraic expressions, polynomials

> **Proposition 1.6. Special product formulas**
>
> *(1) (Perfect square formula-1)*
> $$(a+b)^2 = a^2 + 2ab + b^2$$
>
> *(2) (Perfect square formula-2)*
> $$(a-b)^2 = a^2 - 2ab + b^2$$
>
> *(3) (Difference of two squares)*
> $$(a-b)(a+b) = a^2 - b^2$$
>
> *(4) (Difference of two cubes)*
> $$(a-b)(a^2 + ab + b^2) = a^3 - b^3$$
>
> *(5) (Sum of two cubes)*
> $$(a+b)(a^2 - ab + b^2) = a^3 + b^3$$

The best way to help students remember these formulas is to encourage them to derive these formulas. The following "computation tricks" may be used to encourage students to do similar calculations.

Exercise 1.18 (**Panel discussion**): Compute 25×25, 45×45, 65×65, etc. Find and discuss the pattern.

Exercise 1.19 (**Panel discussion**): Compute 22×28, 43×47, 64×66, etc. Find and discuss the pattern.

Example 1.34 (**Discussion about Mersenne prime numbers**) Why is $2^p - 1$ a prime number probably only when $p > 1$ is a prime number? We will show that $2^m - 1$ can not be a prime number if m is a composite number, and $2^m + 1$ can not be a prime number if $m > 1$ is an odd number.

Solution: Assume $m = kl$, where both k and l are two natural number larger than 2. Then
$$2^m - 1 = (2^k)^l - 1$$
$$= (2^k - 1)[(2^k)^{l-1} + (2^k)^{l-2} + \cdots + (2^k) + 1].$$

So $2^m - 1$ is not a prime number if m is a composite number.

Assume $m = 2k + 1$, where k is a positive natural number. Then
$$2^m + 1 = 2^{2k+1} + 1$$
$$= (2+1)[2^{2k} - 2^{2k-1} + 2^{2k-2} - 2^{2k-2} + \cdots + 2^2 - 2 + 1].$$

So $2^{2k+1} + 1$ is not a prime number.

1.2.3 Exercises after class

1. **Basic skills on computation: computing and simplifying**
   (1) $3x^2 + (x - 2x^2)$;   (2) $3x^2 - (x - 2x^2)$;
   (3) $6x^2 \times \frac{5x^3}{3}$;   (4) $3x^4 \div (\frac{3}{7}x^2)$;
   (5) $3x^2 \times (2x^3 - 3)$;   (6) $(3x^4 + 5x^2) \div x^2$;
   (7) $(x - 2) \times (x^2 + 2)$;   (8) $-(2 - x^2) - 3x(2 - x)$.

2. **More basic skills on computation: computing and simplifying**
   (1) $(x + 1)(x + 2)$;   (2) $(x^2 + 1)(x^2 + 2)$;
   (3) $(x - 1)(x - 2)$;   (4) $(x^2 - 1)(x^2 - 2)$;
   (5) $(x - 8)(x + 6)$;   (6) $(x^2 + 6)(x^2 - 7)$;
   (7) $(x + 8)(x - 6)$;   (8) $(x^2 - 6)(x^2 + 7)$;
   (9) $(2x - 1)(x + 3)$;   (10) $(x^2 + 3)(2x^2 - 1)$;
   (11) $(2x + 1)(x - 3)$;   (12) $(x^2 - 3)(2x^2 + 1)$;
   (13) $(x - 1)(x^2 + x + 1)$;   (14) $(x^2 - 1)(x^4 + x^2 + 1)$;
   (15) $(x + 1)(x^2 - x + 1)$;   (16) $(x^2 + 1)(x^4 - x^2 + 1)$;
   (17) $(x + 2)(x^2 - 2x + 4)$;   (18) $(x^2 + a)(x^4 - ax^2 + a^2)$.

3. **Comparing: Using vertical form to carry out the operations on decimal numbers and polynomials**
   (1) 25×25;   (2) $(x + 5)^2$;
   (3) $(400 - 20 + 1) \times (20 + 1)$;   (4) $(x^2 - x + 1)(x + 1)$;
   (5) $(64 \times 10^3 - 1) \div (40 - 1)$;   (6) $(64x^3 - 1) \div (4x - 1)$;
   (7) $1034 \div 11$;   (8) $(x^3 + 3x + 4) \div (x + 1)$;
   (9) $8064 \div 21$;   (10) $(8x^3 + 6x + 4) \div (2x + 1)$.

4. **Proof**: Assume a, b, c, d, e, f, g are one-digit natural numbers (between 0 to 9). We introduce the following notation:
   $$\overline{1abcde} = 1 \times 10^5 + a \times 10^4 + b \times 10^3 + c \times 10^2 + d \times 10 + e \times 10^0.$$
   For example, $\overline{1ab} = 1 \times 10^2 + a \times 10 + b = 100 + 10a + b$.
   Prove:
   (1) $\overline{1ab1ab}$ is always divisible by 7;
   (2) $\overline{1ab1ab}$ and $\overline{1abcd1abcd}$ are always divisible by 11.

1.3 Properties of equality

With the help of algebraic expressions, we can convert a real problem into a mathematical problem. Usually, the identity relation in a real problem can be converted into equality in mathematics.

1.3.1 Using equality to write equations

In "Introductory Algebra", we have the following question:

Find a number where $\frac{1}{3}$ of the number is 2 more than $\frac{1}{4}$ of the number minus 2.

Many students get confused: the question involves too many relations! However, do not panic. We can find the number by going backward. The key step is to write down an algebraic equation that represents the identity.

Example 1.35 Find a number where $\frac{1}{3}$ of the number is 2 more than $\frac{1}{4}$ of the number minus 2.

Solution: Using x to represent the number, we write down the corresponding relations, converting English into Mathematics:

$$\frac{1}{3} \text{ of the number} \Leftrightarrow \frac{1}{3}x;$$

$$\text{The number minus 2} \Leftrightarrow x - 2;$$

$$\frac{1}{4} \text{ of the new number} \Leftrightarrow \frac{1}{4} \cdot (x - 2).$$

Finally, we convert the identity relation into the following equation[8]:

$$\frac{1}{3}x = 2 + \frac{1}{4} \cdot (x - 2). \tag{1.9}$$

Once we find the number which satisfies equation (1.9), we are done.

Hereby, we learn the procedure to set up the equation. Next, we learn how to solve equations like (1.9). We will come back later to find the solution.

□

Exercise 1.20 Let m represent any integer. Write down algebraic expressions representing all even numbers and all odd numbers.

---

[8] We call the equality with an unknown variable x an "equation".

1.3 Properties of equality

✎ **Exercise 1.21** Assume Ann's age is x. If her dad is one year older than three times her age, what is the algebraic expression for her dad's age? If we know that her age and her dad's age together are 49, how can we represent that information using an equation?

1.3.2 Solve equations

After learning how to classify equations, it is natural to explore the corresponding techniques for solving each type. However, it is important to recognize that there are countless equations that humans cannot solve, even with advanced mathematical tools. As a result, understanding the underlying principles behind solving equations is far more valuable than merely memorizing specific techniques.

By focusing on these principles, we develop a deeper comprehension of how solutions are derived, allowing us to tackle a wider range of problems and adapt to new challenges. To guide this process, we rely on a set of fundamental equivalence rules that serve as the foundation for solving equations. These rules ensure that the transformations applied to equations maintain their integrity while leading us closer to the solution.

> **Proposition 1.7. Equivalent equations**
>
> (1) (**Addition invariance**): For any number C,
> $$A = B \Longleftrightarrow A + C = B + C.$$
>
> (2) (**Multiplication invariance**): For any $C \neq 0$,
> $$A = B \Longleftrightarrow AC = BC.$$
>
> (3) (**Symmetric property**):
> $$A = B \Longleftrightarrow B = A.$$

☝ **Note** *In fact, we can derive the symmetric property from invariant properties.*

Proof. If $A = B$, then

$$A = B \Longleftrightarrow 0 = B - A \qquad \text{(Addition invariance)}$$
$$\Longleftrightarrow -B = -A \qquad \text{(Addition invariance)}$$
$$\Longleftrightarrow B = A \qquad \text{(Multiplication invariance).}$$

□

We will use the above three rules to solve equations.

1.3 Properties of equality

Example 1.36 Solve the following equation (In other words, find x such that it satisfies the following equation):
$$5x - 12 = 13.$$

Solution: First, using Addition invariance, we add 12 on both sides of the equation and obtain an equivalent equation:
$$5x - 12 + 12 = 13 + 12.$$
Using the associative rule, we simplify the above equation:
$$5x = 25.$$
Using Multiplication invariance we multiply $\frac{1}{5}$ on both sides and obtain
$$5x \cdot \frac{1}{5} = 25 \cdot \frac{1}{5}.$$
We simplify the above and obtain:
$$x = 5.$$

□

We give another example.

Example 1.37 Solve equation:
$$\frac{5}{x} - 12 = 13.$$

Solution:
$$\frac{5}{x} - 12 = 13 \iff \frac{5}{x} - 12 + 12 = 13 + 12 \quad \text{(Addition invariance)}$$
$$\iff \frac{5}{x} = 25 \quad \text{(Simplifying)}$$
$$\iff \frac{1}{x} = 5 \quad \text{(Multiplication invariance)}$$
$$\iff x = \frac{1}{5}. \quad \text{(Definition of reciprocal)}$$

□

Note In example 1.36, if we let $y = \frac{1}{x}$, we obtain a new equation for y:
$$5y - 12 = 13.$$
It is the same equation in example 1.35. Thus we immediately know that $y = 5$. From this, we conclude $x = \frac{1}{5}$. This approach to solving an equation is often called the "substitution" method.

Exercise 1.22 Solve equation: (1) $2x + 3 = 7$; (2) $\frac{2}{x} + 3 = 7$.

Next, we shall complete the solution to (1.9). We first simplify the right side of the

equation:
$$\frac{1}{3}x = 2 + \frac{1}{4}x - \frac{2}{4} = \frac{1}{4}x + 2 - \frac{1}{2} = \frac{1}{4}x + \frac{3}{2}.$$

Using Addition invariance, we have
$$\frac{1}{3}x - \frac{1}{4}x = \frac{3}{2}.$$

Combining like terms, we obtain:
$$(\frac{1}{3} - \frac{1}{4})x = \frac{3}{2}.$$

That is
$$\frac{1}{12}x = \frac{3}{2}.$$

Using Multiplication invariance, we obtain
$$x = \frac{3}{2} \cdot 12 = 18.$$

So, we conclude that this number is 18.

□

The following transitive property for equality is also important.

> **Proposition 1.8. Transitive property**
>
> **Transitive property:** If $A = B$ and $B = C$, then
> $$A = C.$$

Note *Mathematical derivation (logical derivation): Using Addition invariance and transitive property, we can obtain equivalent equations by multiplying a positive natural number on both sides of the equation (the proof of invariant under integer multiplication). For example, from*
$$A = B$$
we can derive
$$2 \times A = A + B = 2 \times B.$$

1.3.3 Classification of equations

There are various equations in real life. There are equations that are used to calculate economic growth and equations to describe the trajectory of objects. In mathematics, there are algebraic equations as we just saw, and differential equations that are used to determine option prices. Here we focus on algebraic equations, which only involve algebraic (polynomial and exponential) expressions and operations.

> **Definition 1.17. First order equation for one variable**
>
> Assume x is the unknown variable, other letters represent constants (we often call these other letters "parameters"). We call the following equation
>
> $$ax + b = 0$$
>
> the first-order equation for one variable. One variable refers to variable x; First order means that the terms involving x are at most first-degree monomials.

Example 1.38 Which of the following equations are first-order equations for one variable (assuming x is variable)?

(1) $\frac{x}{2} - 2 = 5$;

(2) $\frac{x-2}{3} = 1$;

(3) $ax = 2$;

(4) $\frac{2}{x} - 1 = \frac{1}{3}$.

Solution: (1), (2), and (3) are first-order equations for one variable, but (4) is not. (However, later, we will show that (4) can be converted to a first-order equation for one variable via "substitution").

□

Again, in a first-order, single variable equation (here x is the designated variable)

$$ax + b = 0$$

we call other letters (here, a and b) parameters.

Example 1.39 Solve the following equation for x variable (a is a parameter):

$$ax = 2.$$

Solution: Whether the parameter a is zero or not is crucial for us to solve the equation.

(1) If $a \neq 0$, then it has a reciprocal $\frac{1}{a}$. Using Multiplication invariance we multiply $\frac{1}{a}$ on both sides and obtain

$$x = \frac{2}{a}.$$

(2) If $a = 0$, then for whatever x, the left side of the equation is 0, which can not be the same as the right side. So the equation has no solution!

□

Example 1.40 Solve the following equation for x variable:

$$ax = 0.$$

Solution: Again, we discuss two cases.

(1) If $a \neq 0$, then it has a reciprocal: $\frac{1}{a}$. Using Multiplication invariance we multiply $\frac{1}{a}$ on both sides and obtain
$$x = 0 \times \frac{1}{a} = 0.$$

(2) If $a = 0$, then for whatever x, the left side of the equation is 0, which is the same as the right side of the equation. So any arbitrary number is a solution to the equation.

□

Example 1.41 Set up the equation and solve the word problem: Find a number such that half of the difference between this number and 4 is 3. There shall be two such numbers.

Solution: Let x be this number.

If x is greater than 4, then
$$\frac{1}{2}(x - 4) = 3.$$

Multiplying both sides by 2, we have
$$x - 4 = 6.$$

Adding 4 to both sides, we have $x = 10$.

If x is less than 4, then
$$\frac{1}{2}(4 - x) = 3.$$

Multiplying both sides by 2, we have
$$4 - x = 6.$$

Adding x to both sides, we have
$$4 = 6 + x.$$

Now, adding -6 to both sides, we have $-2 = x$, that is: $x = -2$.

Answer: this number is either 10, or -2.

□

First-order, single variable equations can usually be solved via the properties of the equation, as we saw from the previous examples. Some other equations, though they are not first-order, can be converted into first-order equations. For example, in Example 1.36, we viewed $\frac{1}{x}$ as a variable and solved for that new variable first. Then we find x from its reciprocal. The following example demonstrates how to use this "substitution" method to find a solution.

Example 1.42 Use the "substitution method" to convert the following equation into a

first-order equation[9], then solve the equation for x:
$$\frac{3}{x} - 1 = \frac{1}{2}.$$

Solution: Let $y = \frac{1}{x}$. Then the original equation becomes
$$3y - 1 = \frac{1}{2}.$$

This is a first-order equation. Solving this first-order equation, we have $y = \frac{1}{2}$. That is: $\frac{1}{x} = \frac{1}{2}$. It follows that $x = 2$.

□

Exercise 1.23 Use the substitution method to convert the following equation into a first-order equation, then solve the equation for x.
$$\frac{1}{x+2} - \frac{1}{2} = -\frac{1}{3}.$$

Next, we will study some relatively complicated equations. The methods to solve these equations were given in our first two books ("Introductory Algebra" and "Introductory Geometry and Proofs").

Assume that x, y are two variables, and other letters are constants. We call the following equation
$$c_{11}x + c_{12}y = b_1$$

a first-order two-variable equation. Here, the two variables are x and y. First order means the terms involving x, y are at most first-degree polynomials. In the Cartesian coordinate system, a first-order, two-variable equation represents a straight line. That is, every pair of the solution (x, y) is a point on a straight line.

Example 1.43 Check which of the following pair is a solution to equation $x - 2y = 3$.

(1) $(1, 2)$,

(2) $(3, 0)$,

(3) $(-1, -1)$,

(4) $(2, -1)$.

Solution: (1) We first check $(1, 2)$:
$$1 - 2 \times 2 = -3 \neq 3.$$

This pair does not satisfy the equation.

(2) For $(3, 0)$:
$$3 - 2 \times 0 = 3.$$

It satisfies the equation, thus it is a solution.

---

[9]This is only for demonstration. In practice, it is not necessary to write down this "first-order equation"

(3) For $(-1, -1)$:
$$-1 - 2 \times (-1) = 1 \neq 1.$$

This pair does not satisfy the equation.

(4) For $(2, -1)$:
$$2 - 2 \times (-1) = 4 \neq 3.$$

The pair does not satisfy the equation.

□

Example 1.44 Check: which point is on the line represented by equation $x - 2y = 3$:
  (1) $(1, 2)$,
  (2) $(3, 0)$,
  (3) $(-1, -1)$,
  (4) $(2, -1)$.

Solution: We know from the previous example that point $(3, 0)$ satisfies the equation and thus is on the line. Other points are not on the line.

□

If two variables satisfy two first-order two-variable equations simultaneously, these two equations combined together are called a system.

> **Definition 1.18. 2 by 2 linear system**
> *Assume x, y are two variables, and other letters are constants. We call the following*
> $$\begin{cases} c_{11}x + c_{12}y = b_1 \\ c_{21}x + c_{22}y = b_2 \end{cases} \quad (1.10)$$
> *a "2×2 (read as 2 by 2) linear system".*

Recall: in "Introductory Algebra", we learned two methods to solve a 2×2 system: substitution method and elimination method.

Example 1.45 Solve the following 2×2 system:
$$\begin{cases} x + y = 5 \\ 4x + 2y = 8. \end{cases}$$

Solution: Method 1: We first recall the substitution method.

We solve for y in the second equation $4x + 2y = 8$:
$$y = 4 - 2x. \quad (1.11)$$

1.3 Properties of equality

We then substitute $4 - 2x$ for y in the first equation $x + y = 5$, and obtain:

$$x + (4 - 2x) = 5$$
$$\Longleftrightarrow 4 - x = 5$$
$$\Longleftrightarrow x = -1.$$

After we obtain $x = -1$, we then use (1.11) to find y:

$$y = 4 - 2(-1) = 6.$$

We obtain the answer:

$$\begin{cases} x &= -1 \\ y &= 6. \end{cases}$$

Method 2: We then recall the elimination method.

We first observe that $x + y = 5$ is equivalent to $2x + 2y = 10$, and both equations $2x + 2y = 10$ and $4x + 2y = 8$ contain term $2y$. So we can subtract one equation from another equation (left side minus left side, right side minus right side), and obtain:

$$2x + 2y - (4x + 2y) = 10 - 8$$
$$\Longleftrightarrow -2x = 2$$
$$\Longleftrightarrow x = -1.$$

Once we obtain $x = -1$, we then bring it back to one of the original equations to find $y = 5 - x = 6$. Thus obtain the answer:

$$\begin{cases} x &= -1 \\ y &= 6. \end{cases}$$

$\square$

Note *The substitution method is based on the following principle: Suppose $A + B = C$, and $B = D$, then $A + D = C$. The elimination method, meanwhile, is based on the principle: If $A = B$, and $C = D$, then $A + C = B + D$. Can you prove the above two statements from the properties of equality?*

In our second book "Introductory Geometry and Proofs", we also show that we can write a 2×2 system in a matrix form, and then use matrices to derive a general formula.

Let

$$A = \begin{bmatrix} c_{11} & c_{12} \\ c_{21} & c_{22} \end{bmatrix}, \quad \vec{X} = \begin{bmatrix} x \\ y \end{bmatrix}, \quad \vec{V} = \begin{bmatrix} b_1 \\ b_2 \end{bmatrix}.$$

Then, 2×2 system (1.10) can be written as

$$A\vec{X} = \vec{V}. \tag{1.12}$$

We usually call A the coefficient matrix, $\vec{X}$ the unknown vector, $\vec{V}$ the given vector.

If the coefficient matrix A has an inverse matrix A^{-1}, we then can multiply the above equation on both sides from the left to derive the solution[10]:

$$\vec{X} = A^{-1}\vec{V}.$$

So, in order to solve (1.10), we first need to compute the determinant of the coefficient matrix $|A|$. If $|A| \neq 0$, then we use

$$A^{-1} = \frac{1}{c_{11}c_{22} - c_{12}c_{21}} \cdot \begin{bmatrix} c_{22} & -c_{12} \\ -c_{21} & c_{11} \end{bmatrix} = \begin{bmatrix} \frac{c_{22}}{c_{11}c_{22}-c_{12}c_{21}} & \frac{c_{12}}{c_{11}c_{22}-c_{12}c_{21}} \\ \frac{c_{21}}{c_{11}c_{22}-c_{12}c_{21}} & \frac{c_{11}}{c_{11}c_{22}-c_{12}c_{21}} \end{bmatrix}$$

to compute the inverse matrix and derive the solution.

It looks like solving a system via matrices is much more complicated than the previous two methods. However, we can easily standardize the matrix method. With the help of a computer, the matrix method can be extended to 3×3, 4×4 and even more complex systems.

Example 1.46 Use matrices to solve

$$\begin{cases} x + y = 5 \\ 4x + 2y = 8 \end{cases}$$

Solution: First,

$$A = \begin{bmatrix} 1 & 1 \\ 4 & 2 \end{bmatrix}, \quad \vec{V} = \begin{bmatrix} 5 \\ 8 \end{bmatrix}.$$

We compute

$$A^{-1} = \frac{1}{|A|} \begin{bmatrix} 1 & 1 \\ 4 & 2 \end{bmatrix} = \begin{bmatrix} -1 & \frac{1}{2} \\ 2 & -\frac{1}{2} \end{bmatrix},$$

So

$$\vec{X} = A^{-1}\vec{V} = \begin{bmatrix} -1 \\ 6 \end{bmatrix}.$$

□

Exercise 1.24 Use matrices to solve

$$\begin{cases} x + y = 1 \\ 3x + 4y = 0. \end{cases}$$

Generally, we define the following polynomial equations

Definition 1.19. Quadratic equations, higher order equations

Assume that x is the variable and other letters are constants and $a \neq 0$. We call

---

[10]**Warning**: Commutative rule does not hold for matrix multiplication.

> the following equation
> $$ax^2 + bx + c = 0$$
> a second-order, one-variable equation or quadratic equation. One variable refers to the x variable, and second order means the highest degree of the terms involving x is the second-degree.
>
> Assume that x is a variable, n is a natural number larger than 2, and other letters are constants. We call the following equation
> $$x^n + C_{n-1}x^{n-1} + \cdots + C_1 x + C_0 = 0$$
> a one variable, n-th order equation. One variable refers to the x variable, and n-order means the highest degree of the terms involving x is n-th degree.

1.3.4 Factorization and solving quadratic equations in rational number set

Factorizing a natural number can help us to understand the number better, help us to simplify fractions, find the number of different factors, etc. Factorizing a polynomial can similarly help us to dissolve a complex equation into a few simpler equations[11].

Let us see how to solve a quadratic equation.

Example 1.47 Solve
$$(x-2)(x-3) = 0.$$

Let's analyze this first: Logically, from
$$A \times B = 0$$
we obtain
$$\text{either } A = 0 \quad \text{or} \quad B = 0.$$

(Can you explain the reason?)

Solution:
$$\text{Either } x - 2 = 0 \quad \text{or} \quad x - 3 = 0,$$
That is
$$\text{either } x = 2 \quad \text{or} \quad x = 3.$$

---

[11] This philosophy works for most 4th-order or below polynomial equations. Later, in college, some of you may learn Galois theory, in which you will find out that in general a 5th-order or above polynomial equation may not be able to dissolve into simpler equations via factorization.

We say this equation has two solutions (often, we also say this equation has two roots): $x = 2$ or $x = 3$.

□

Exercise 1.25 Solve equation:
$$(x - 1)(y + 2) = 0.$$

Now, suppose we are asked to solve
$$x^2 - 5x + 6 = 0.$$
We may argue: if we know
$$x^2 - 5x + 6 = (x - 2)(x - 3),$$
then, we can use what we learn in Example 1.46 to dissolve a quadratic equation into two first-order equations first:
$$\text{either } x - 2 = 0 \quad \text{or} \quad x - 3 = 0.$$
Then, we obtain
$$\text{either } x = 2 \quad \text{or} \quad x = 3.$$

So, the key step in solving the equation
$$x^2 + ax + b = 0,$$
is to factorize $x^2 + ax + b$:
$$x^2 + ax + b = (x - x_1)(x - x_2)$$
for some numbers x_1 and x_2.

Recall in "Introductory Algebra", we already did some basic factorization exercises.

Exercise 1.26 Factorize:

1).
$$x^2 - 3x;$$

2).
$$x^3 - 3x^2 - x + 3.$$

For given two numbers a and b, we compute
$$(x + a)(x + b) = x(x + b) + a(x + b)$$
$$= x^2 + bx + ax + ab$$
$$= x^2 + (b + a)x + ab.$$

Conversely, we can do factorization:
$$x^2 + (b+a)x + ab = (x+a)(x+b).$$

We usually use a cross-product form to represent the above identity.

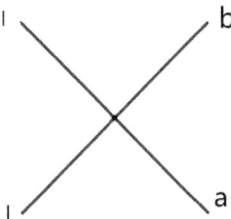

Figure 1.4: Cross product

For a given second degree polynomial
$$x^2 + px + q,$$
we need to find a and b so that $q = a \cdot b$ and $a + b = p$, then
$$x^2 + px + q = (x+a)(x+b).$$

✎ **Exercise 1.27** Factorize

1).
$$x^2 + 7x + 12;$$

2).
$$x^2 - 7x + 12;$$

3).
$$2x^2 + 7x + 5;$$

4).
$$2x^2 - 7x + 5.$$

(Note, the cross-product form for 3) and 4) needs to be modified since the leading coefficients are not 1.)

Later, after we introduce radical expressions and irrational numbers, we will give a general solution formula for quadratic equations and the method to solve higher-order equations.

After we learn other elementary functions, we will list the corresponding algebraic equations in Chapter 3.

1.3.5 Exercises after class

1. **Basic skills**. Solve equations (x is variable):
   (1) $2x - 3 = 7$;   (2) $\frac{2}{x} - 3 = 7$;
   (3) $3x - 4 = 9$;   (4) $\frac{3}{x} - 4 = 9$;
   (5) $3x + 4 = 9$;   (6) $\frac{3}{x} + 4 = 9$;
   (7) $3x + 2 = 1 - 2(x + 2)$;   (8) $3x - 2 = 3 - 5(x - 2)$;
   (9) $\frac{x}{3} - \frac{1}{2} = 1 - \frac{2x}{3}$;   (10) $\frac{x}{3} - 2 = 3 - \frac{x-2}{5}$.
   (11) $x - a = 1 - \frac{x}{2}$;   (12) $x - 2 = 3(2 - a) - x$;
   (13) $\frac{x-a}{2} = 1 - \frac{x}{2}$;   (14) $\frac{x-a}{2} = 1 + \frac{x}{2}$;
   (15) $x - a = ax$;   (16) $x - b = ax$.

2. **Basic skills**. Solve 2×2 systems (x, y are variables)
   (1)
   $$\begin{cases} 5x + y = 17 \\ 2x - y = 4 \end{cases}$$

   (2)
   $$\begin{cases} 3x + 2y = 13 \\ 2x + 3y = 12 \end{cases}$$

   (3)
   $$\begin{cases} 3x + 5y = 4 \\ 5x - 3y = 1 \end{cases}$$

   (4)
   $$\begin{cases} 4x - 9y = -2 \\ 12x + 3y = 34 \end{cases}$$

   (5)
   $$\begin{cases} \frac{1}{2}x + \frac{3}{2}y = 7 \\ \frac{x}{5} + \frac{y}{3} = 2 \end{cases}$$

   (6)

$$\begin{cases} \frac{3x}{5} + \frac{2y}{3} = \frac{1}{2} \\ \frac{x}{4} - \frac{3y}{4} = -\frac{5}{4} \end{cases}$$

3. **Basic skills and substitution**. Solve 2×2 systems (x, y are variables)

    (1)
    $$\begin{cases} 3x + y = 9 \\ 2x - y = 3 \end{cases}$$

    (2)
    $$\begin{cases} \frac{3}{x} + \frac{2}{y} = 13 \\ \frac{2}{x} + \frac{3}{y} = 12 \end{cases}$$

    (3)
    $$\begin{cases} \frac{3}{x+1} + \frac{2}{y-2} = 13 \\ \frac{2}{x+1} + \frac{3}{y-2} = 12 \end{cases}$$

4. **Factorize polynomials in the rational number set**[12]

    (1) $x^2 + 3x + 2$;   (2) $x^4 + 3x^2 + 2$;

    (3) $x^2 - 3x + 2$;   (4) $x^4 - 3x^2 + 2$;

    (5) $x^2 + 2x - 24$; (6) $x^4 + 2x^2 - 24$;

    (7) $2x^2 + 7x + 3$; (8) $2x^4 + 7x^2 + 3$;

    (9) $4x^2 + 11x - 3$; (10) $4x^4 + 11x^2 - 3$;

    (11) $x^3 + 8$; (12) $x^6 + 8$;

    (13) $x^2 - 64$; (14) $x^4 - 64$;

    (15) $x^3 - 64$; (16) $x^6 - 64$;

    (17) $2(x-1)^2 + 3(x-1) - 2$; (18) $x^4 + x^2 + 1$.

5. **More on solving equations**. Solve equations in the rational number set.

    (1) $x^2 - 4x + 3 = 0$;   (2) $x^2 - 5x + 6 = 0$;

    (3) $x^2 - 13x + 36 = 0$;   (4) $x^4 - 13x^2 + 36 = 0$;

    (5) $2x^2 - 7x - 4 = 0$;   (6) $2x^4 - 7x^2 - 4 = 0$;

    (7) $2x^2 - 7x - 4 = 0$;   (8) $2x^4 - 7x^2 - 4 = 0$;

---

[12] Rational numbers refer to all fraction numbers.

(9) $2(x+1)^2 - 3(x+1) - 2 = 0$;   (10) $4(x+1)^4 + 5(x+1)^2 - 2 = 0$.

6. **Word problems**

    (1) Alex's mom had some pocket money that day. She gave Alex some money and told him that was $\frac{1}{3}$ of her money. Alex used half of it to buy a two-dollar ice cream roll. Can you find out how much pocket money Alex's mom had that day?

    (2) Today Anna's age is $\frac{1}{6}$ of her mom's age. Ten years later, her age is $\frac{3}{8}$ of her mom's age. Can you find out what is Anna's age now?

    (3) **Chicken and rabbit problem** In a farm, only chickens and rabbits are raised. The farmer counts: there is a total of 45 heads and 144 legs. Can you help the farmer to figure out: how many chickens and how many rabbits are there in that farm?

7. **Challenging problems** For which values of a can you find an x value that satisfies the equation?

    (1) $ax = 1$,

    (2) $ax = x + 1$,

    (3) $\frac{a}{x} = \frac{2}{3}$.

8. **Challenging problems**

    (1) For which values of a does x only have one solution?

    $$(a+1)x = 0$$

    (2) Assume a, b are given parameters. Find x:

    $$ax - 1 = b - x$$

9. **Challenging problems**

    If $A = B$ and $C = D$, prove:

    (1)

    $$A + C = B + D.$$

    (2)

    $$AC = BD.$$

1.4 Properties of inequality

If we replace "equal" with "less than""or "greater than" in equations, we get inequalities.

1.4.1 Definition of inequality

Natural numbers have a "natural" property: they are ordered. 1 is smaller than 9, and a two-digit number is smaller than any three-digit number. We introduce the following notation:

$$1 < 9, \quad 101 > 98.$$

In general, we use

$$a \geq b$$

to indicate that number a is larger than or equal to number b.

We use

$$a > b$$

to indicate number a is larger than number b. Sometimes, we emphasize that ">" is a strict inequality. Similarly, we introduce the notation "$\leq$" and "$<$" for the opposite meaning.

1.4.2 Properties of inequality

The following rules hold for inequality.

Proposition 1.9. Addition invariance

For any number C,

$$A < B \iff A + C < B + C.$$

Example 1.48 Since $2 < 5$, by Proposition 1.9 (Choose $C = -5$), we have

$$2 - 5 < 5 - 5,$$

that is

$$-3 < 0.$$

□

1.4 Properties of inequality

> **Proposition 1.10. Invariance under positive number multiplication**
>
> For any positive number C (that is C > 0),
> $$A < B \iff AC < BC.$$

Example 1.49 Since $2 < 5$,
$$2 \times \frac{1}{10} < 5 \times \frac{1}{10},$$
that is
$$\frac{1}{5} < \frac{1}{2}.$$
Here we used Proposition 1.10 (choosing $C = \frac{1}{10}$).

> **Proposition 1.11. Reverse inequality under negative number multiplication**
>
> For any negative number C (that is C < 0),
> $$A < B \iff AC > BC.$$

Example 1.50 Since $1 < 10$,
$$1 \times (-1) > 10 \times (-1),$$
that is:
$$-10 < -1.$$
Here we used proposition 1.11 (choosing $C = -1$).

> **Proposition 1.12. Symmetric property**
>
> $$A > B \iff B < A.$$

Similar to equality, the transitive property of inequality is also very important.

> **Proposition 1.13. Transitive property**
>
> If $A < B$ and $B < C$, then
> $$A < C.$$

Remark **Observation and thinking**: Invariance under positive natural number multiplication can be derived from Addition invariance and the transitive property. For example, from
$$2 < 5$$

we can derive
$$2 \times 2 < 5 + 2 < 2 \times 5.$$

In the above four propositions, if we replace the strict inequality "<" with "≤", we obtain corresponding propositions.

1.4.3 Solve inequality

Just as we utilize properties of equality to solve equations, the properties of inequality allow us to solve inequalities using similar techniques.

Example 1.51 Find all numbers that satisfy
$$\frac{2}{3}x - 1 \leq \frac{1}{3}.$$

Solution:

$$\frac{2}{3}x - 1 \leq \frac{1}{3} \Leftrightarrow \frac{2}{3}x \leq \frac{1}{3} + 1 \quad \text{(Proposition 1.9)}$$
$$\Leftrightarrow \frac{2}{3}x \leq \frac{4}{3} \quad \text{(Simplify)}$$
$$\Leftrightarrow x \leq \frac{4}{3} \cdot \frac{3}{2} \quad \text{(Proposition 1.10)}$$
$$\Leftrightarrow x \leq 2. \quad \text{(Simplify)}$$

□

Similarly, in order to solve general inequalities, we need to classify them first.

1.4.4 Classification of algebraic inequalities

As we did with the classification of equations, we first consider polynomial inequalities. If the inequality only involves polynomials and the highest degree is n, we call the inequality a one-variable n-th order inequality. As we discussed in the previous subsection, we can solve first-order inequalities via the properties of inequality. To solve a 2nd-order inequality, we use the principles such as: If $AB < 0$, then either $A < 0$ and $B > 0$ or $A > 0$ and $B < 0$. (**Can you give the reason for such a principle?**)

Example 1.52 Solve 2nd-order inequality
$$2x^2 - 3x + 2 \leq 0.$$

Solution:

$$2x^2 - 3x - 2 \leq 0 \iff (2x+1)(x-2) \leq 0. \quad \text{(Factorization)}$$

Case 1: $2x+1 \leq 0$ and $x-2 \geq 0$. Then $x \leq -\frac{1}{2}$ and $x \geq 2$. But no number satisfies both inequalities.

Case 2: $2x+1 \geq 0$ and $x-2 \leq 0$. Then $x \geq -\frac{1}{2}$ and $x \leq 2$. Combining two inequalities together, we have $-\frac{1}{2} \leq x \leq 2$.

Conclusion: the solution to the inequality is $-\frac{1}{2} \leq x \leq 2$.

□

Exercise 1.28 Solve inequality.

(1) $x^2 - 4x + 3 < 0$;

(2) $2x^2 - 3x - 2 > 0$.

Later in Section 3.1.2.2 we will show that using function graphs can help us easily solve polynomial inequalities.

Besides polynomial inequalities, we will learn other types of inequalities, such as absolute value inequalities, which will be discussed below. More complicated inequalities we will learn later include exponential inequalities, logarithmic inequalities, and trigonometric inequalities. In particular, we will learn an advanced inequality related to convex (concave) functions: Jensen's inequality.

1.4.5 Absolute value

Once we know numbers are ordered, we can introduce the number line.

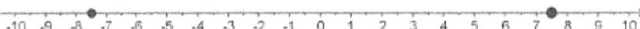

Figure 1.5: The number line contains three elements: The origin, the direction, and the unit length

On the number line, the distance from the number x to the origin is called the absolute value of x. We denote $|x|$ as the absolute value of x. There are two cases: Case 1, $x \geq 0$ (on the number line, x is at the origin or on its right side), then $|x| = x$; Case 2, $x < 0$ (on the number line, x is on the left side of the origin), then $|x| = -x$. We can write

$$|x| = \begin{cases} x, & \text{if } x \geq 0; \\ -x, & \text{if } x < 0. \end{cases}$$

Example 1.53 Assume $|x| = 1$, then $x = 1$ (if $x \geq 0$, or x is on the right of the origin on the number line); or $x = -1$ (if $x < 0$, or x is on the left of the origin on the number line).

So, to get rid of the absolute value, we need to consider two cases.

Example 1.54 Solve equation
$$|x - 1| = 2.$$

Solution: We consider two cases.

(1) If $x - 1 \geq 0$, then
$$x - 1 = 2,$$
so $x = 3$.

(2) If $x - 1 < 0$, then
$$-(x - 1) = 2,$$
so $x = -1$.

The final answer is: $x = 3$ or $x = -1$.

Example 1.55 Solve inequality
$$|x - 1| > 2.$$

Solution: We consider two cases.

(1) If $x - 1 > 0$, that is $x > 1$, then

$\qquad x - 1 > 2$ \hfill (Get rid of the absolute value)

$\qquad \Leftrightarrow x > 3.$ \hfill (Solve inequality)

The common solution to the above two inequalities ($x > 1$ and $x > 3$) is $x > 3$.

(2) If $x - 1 < 0$, that $x < 1$, then

$\qquad -(x - 1) > 2$ \hfill (Get rid of the absolute value)

$\qquad \Leftrightarrow x < -1.$ \hfill (Solve inequality)

The common solution to the above two inequalities is $x < -1$.

The final answer is $x > 3$ or $x < -1$.

Example 1.56 Solve inequality
$$|x - 1| \leq 2.$$

Solution: We consider two cases.

(1) If $x - 1 \geq 0$, that is $x \geq 1$, then

$$x - 1 \leq 2 \qquad \text{(Get rid of the absolute value)}$$
$$\Leftrightarrow x \leq 3. \qquad \text{(Solve inequality)}$$

Combining the above two inequalities, we obtain $1 \leq x \leq 3$.

(2) If $x - 1 \leq 0$, that is $x \leq 1$, then

$$-(x - 1) \leq 2 \qquad \text{(Get rid of the absolute value)}$$
$$\Leftrightarrow x \geq 3. \qquad \text{(Solve inequality)}$$

The common solution to the above two inequalities is the empty set, $\emptyset$.

The final answer is $-1 \leq x \leq 3$.

□

In Section 3.1.2.2, we will use function graphs to help us solve absolute value inequalities. We can also convert an absolute value inequality to a polynomial inequality.

1.4.6 Exercises after class

1. **Basic skills**. Solve inequalities for x:

    (1) $x + 5 \geq -2$; (2) $3 - 2x < 7$;

    (3) $\frac{1}{3}x + 5 \geq -2x$; (4) $3 - \frac{2}{3}x < 2x$;

    (5) $\frac{2x+2}{7} + 1 \leq \frac{3x-4}{6}$; (6) $\frac{x-1}{3} \geq \frac{2}{a}$;

    (7) $ax - 2 \geq 2$; (8) $ax - 2 < 2a$;

    (9) $ax - 2 \geq 2b$; (10) $ax - 2 < 2bx$.

2. **More basic skills**. Solve inequalities for x:

    (1) $x^2 - 4 \geq 0$; (2) $x^2 + 6x - 7 < 0$;

    (3) $|x - 4| \leq 1$; (4) $|x + 5| > 2$;

    (5) $\frac{x-2}{x+3} \leq 0$; (6) $\frac{x-1}{x-4} > 0$.

3. **Proof**.

    (1) Assume a, b are two positive numbers. Prove: if $a < b$, then $a^2 < b^2$.

    (2) Suppose $a^2 < b^2$ and $a + b > 0$. Prove: $a < b$.

    (3) Suppose $a < b$, prove: $a^3 < b^3$.

    (4) If $p > 0$, use the proof by contradiction to prove $\frac{1}{p} > 0$.

    (5) For any positive number C, prove
    $$|x| > C \Leftrightarrow x^2 > C^2.$$

    (6) Prove: for any numbers x, y, z,
    $$x^2 + y^2 + z^2 - xy - xz - yz \geq 0.$$

    (7) Prove: for any positive numbers x, y, z,
    $$x^3 + y^3 + z^3 \geq 3xyz.$$

1.5 Real numbers

1.5.1 n-th root

Whenever we consider the inverse exponential operators, we will deal with radical expressions. For example, let us consider the square and ask the following question:

If the square of a number is 4, what is the number?

Yes, 2 is one answer. But do not forget about -2.

Now we ask: if the square of a number is 2, what is this number? We can set up an equation to find this number. Assume this number is x, then

$$x^2 = 2. \qquad (1.13)$$

If there is a number 2^p that solves the above equation[13], then

$$(2^p)^2 = 2.$$

Suppose we assume the Same Base Exponent rule 1.4 holds for other exponents (**caution: we need the base to be a positive number**): then $(2^p)^2 = 2^{2p}$. So we have $p = \frac{1}{2}$, and $2^{\frac{1}{2}}$ is a solution to equation (1.13).

Thus, we introduce the n-th root of a positive number as the following.

> **Definition 1.20. n-th root**
>
> Suppose n is a positive natural number. For any given positive number b, we define $b^{\frac{1}{n}}$ as a positive number that satisfies
>
> $$\left(b^{\frac{1}{n}}\right)^n = b.$$
>
> We call $b^{\frac{1}{n}}$ the "b's n-th root". We also notate it as $\sqrt[n]{b}$.
>
> If m is another natural number, we define
>
> $$b^{\frac{m}{n}} = \left(b^{\frac{1}{n}}\right)^m, \quad \text{and} \quad b^{-\frac{m}{n}} = \left(b^{\frac{m}{n}}\right)^{-1} = \frac{1}{b^{\frac{m}{n}}}.$$

Note In the above definition, If b is negative, we will have trouble. It is not yet time to explain what $(-1)^{\frac{1}{2}}$ is. We shall cover this topic when we are learning complex numbers. Due to this, we suggest that avoiding expressions like $(-8)^{\frac{1}{3}}$ (or $\sqrt[3]{-8}$).

1.5.2 Exponent rules

Once we introduce fractional exponents, we have the following more general exponential operation rules.

---

[13] You may want to ask: is there a solution to the above equation? This will be answered in a college math course later.

1.5 Real numbers

> **Proposition 1.14. Exponent rules(Exponential operation rules)**
>
> Suppose a, b are two positive numbers, r and s are two fractions (note that integers are also considered fractions), then
> $$a^r \cdot a^s = a^{r+s},$$
> $$(a^r)^s = a^{rs},$$
>
> and
> $$(ab)^r = a^r b^r.$$

Proof: Using a common multiplier, we can assume
$$r = \frac{m}{n}, \quad s = \frac{l}{n},$$
where n is a positive integer, m, l are integers. So $r + s = \frac{m+l}{n}$, $rs = \frac{ml}{n^2}$.

We have:

$$\begin{aligned}
a^r \cdot a^s &= a^{\frac{m}{n}} \cdot a^{\frac{l}{n}} & \text{(Assumption)} \\
&= (a^{\frac{1}{n}})^m \cdot (a^{\frac{1}{n}})^l & \text{(Definition 1.20)} \\
&= (a^{\frac{1}{n}})^{m+l} & \text{(Proposition 1.4)} \\
&= a^{\frac{m+l}{n}} & \text{(Definition 1.20)} \\
&= a^{r+s}.
\end{aligned}$$

The first rule is proved.

Again,

$$\begin{aligned}
[(a^r)^s]^n &= (\{[(a^{\frac{1}{n}})^m]^{\frac{1}{n}}\}^l)^n & \text{(Assumption)} \\
&= (\{[(a^{\frac{1}{n}})^m]^{\frac{1}{n}}\}^n)^l & \text{(Proposition 1.4)} \\
&= (a^{\frac{1}{n}})^{ml}.
\end{aligned}$$

So,

$$\begin{aligned}
[(a^r)^s]^{n^2} &= \{[(a^r)^s]^n\}^n & \text{(Proposition 1.4)} \\
&= [(a^{\frac{1}{n}})^{ml}]^n & \text{(Proposition 1.4)} \\
&= [(a^{\frac{1}{n}})^n]^{ml} & \text{(Using the above result)} \\
&= a^{ml} & \text{(Definition 1.20)} \\
&= \{[a^{ml}]^{\frac{1}{n^2}}\}^{n^2} & \text{(Definition 1.20)} \\
&= \{a^{\frac{ml}{n^2}}\}^{n^2} & \text{(Definition 1.20)} \\
&= [a^{rs}]^{n^2}.
\end{aligned}$$

The second rule is proved.

Finally

$$[(ab)^r]^n = [(ab)^{\frac{m}{n}}]^n \qquad \text{Assumption}$$
$$= (ab)^m \qquad \text{(The second rule)}$$
$$= a^m b^m. \qquad \text{(Proposition 1.4)}$$

On the other hand

$$[a^r b^r]^n = (a^r)^n (a^s)^n \qquad \text{(Proposition 1.4)}$$
$$= a^m b^m \qquad \text{(The second rule)}$$
$$= [(ab)^r]^n \qquad \text{(Using the above result)}.$$

The third rule is proved. $\square$

Example 1.57 Simplify

(1)
$$\frac{3^{\frac{3}{2}} 5^{-2} 2^4}{5^{-3} 3^{\frac{1}{2}} 2^3};$$

(2)
$$\frac{y^{\frac{4}{3}} x^{\frac{2}{3}} z^2}{z^{\frac{1}{2}} y^{\frac{4}{3}} x^{-\frac{1}{3}}}.$$

Solution

(1)
$$\frac{3^{\frac{3}{2}} 5^{-2} 2^4}{5^{-3} 3^{\frac{1}{2}} 2^3} = 3^{\frac{3}{2}} 5^{-2} 2^4 5^3 3^{-\frac{1}{2}} 2^{-3}$$
$$= 3^{\frac{3}{2} - \frac{1}{2}} 5^{-2+3} 2^{4-3}$$
$$= 3 \times 5 \times 2$$
$$= 30.$$

(2)
$$\frac{y^{\frac{4}{3}} x^{\frac{2}{3}} z^2}{z^{\frac{1}{2}} y^{\frac{4}{3}} x^{-\frac{1}{3}}} = x^{\frac{2}{3}} y^{\frac{4}{3}} z^2 x^{\frac{1}{3}} y^{-\frac{4}{3}} z^{-\frac{1}{2}}$$
$$= x^{\frac{2}{3} + \frac{1}{3}} y^{\frac{4}{3} - \frac{4}{3}} z^{2 - \frac{1}{2}}$$
$$= xz^{\frac{3}{2}}.$$

1.5.3 Rational and irrational numbers

A pretty difficult question is: is there a number, whose square is 2?

In the future (we have to wait for the proof. Usually this will be discussed in an "analysis" course offered for math major students in college), with certain axioms, we will define the real number system, in which there are "rational numbers" as well as "irrational numbers" [14]. The system sure contains the number whose square is 2.

> **Definition 1.21. Fraction and rational number**
> Assume n, m are two integers. If m is not 0, we call $\frac{n}{m}$ a fraction. The rational number refers to all fraction numbers.

Apparently, all integers n are rational since $n = \frac{n}{1}$. All decimal numbers are rational. For example, $0.12 = \frac{12}{100}$.

We will use an indirect way to define "irrational numbers" [15].

> **Definition 1.22. Irrational number**
> Any number that can not be written as a fraction is called an irrational number. Rational numbers and irrational numbers together are called real numbers. We use the notation $\mathbb{Q}$ for the rational number set, $\mathbb{R}$ for the real number set.

With the above definition, we can show that a certain number is not rational. Here is a famous example.

Example 1.58 Prove: $\sqrt{2}$ can not be represented by a fraction (this indicates that $\sqrt{2}$ is an irrational number).

Euclid's proof: We shall prove it via the contradiction argument.

Assume that $\sqrt{2}$ can be represented as a fraction:

$$\sqrt{2} = \frac{p}{q}, \tag{1.14}$$

where p, q are two natural numbers. After simplification (canceling the common factors), we assume that the greatest common divisor is 1. That is $(p, q) = 1$.

We square both sides of equation (1.14), and have (we use the result in Section 1.3.5 question 9 part (b))

$$2 = \frac{p^2}{q^2}. \tag{1.15}$$

---

[14] We will NOT use the confusing definition used by standard textbooks: rational number are integers, decimal numbers and repeating decimal numbers. As we pointed out in Section 1.14: Logically, using long division to derive a repeating decimal is not correct. Understanding a repeating decimal requires students to know the concept of limit. we shall discuss this in a later Chapter.

[15] The main concern that we have is: are there such numbers? We will confirm this in some advanced math courses in college

So p^2 is an even number (since $p^2 = 2q^2$). From this, we know that p must be an even number (since an odd number multiplying an odd number is still an odd number). We can write $p = 2n$ for another natural number n, and bring it into equation (1.15) and simplify to get
$$q^2 = 2n^2.$$

This implies that number q is an even number, that is, $q = 2m$ for another natural number m. Thus, we observe that 2 is a common factor. This contradicts the condition that the greatest common divisor of p, q is 1. We hereby complete the proof.

$\square$

As we discussed for simplifying polynomials, to simplify an algebraic expression, we need to combine "like terms". It takes a lot of practice to learn how to identify "like terms" involving fractional exponents.

Example 1.59 Find the "like terms":
$$-\sqrt{2},\ 2^{\frac{1}{3}}, 2^{\frac{1}{2}},\ 6\sqrt[3]{2},\ 2\sqrt{12},\ 2^{-\frac{1}{2}},\ 5\sqrt{3}.$$

Solution:

First group: $2^{\frac{1}{3}}$, $6\sqrt[3]{2}$;

Second group: $2\sqrt{12}$, $5\sqrt{3}$;

Third group: $-\sqrt{2}$, $2^{\frac{1}{2}}$, $2^{-\frac{1}{2}}$.

$\square$

Group 1 is easy to recognize: once we can recognize $2^{\frac{1}{3}} = \sqrt[3]{2}$ are different writing. For Group 2: we notice that
$$2\sqrt{12} = 2(12)^{\frac{1}{2}} = 2(2^2 \cdot 3)^{\frac{1}{2}} = 2 \cdot 2 \cdot 3^{\frac{1}{2}} = 4\sqrt{3}.$$

It is a bit hard to understand why $2^{-\frac{1}{2}}$ is in Group 3. According to the definition, we know
$$2^{-\frac{1}{2}} = \frac{1}{2^{\frac{1}{2}}} = \frac{1}{\sqrt{2}}.$$

This is a fraction involving radical expression, and even worse: the radical expression is in the denominator. We need to get rid of the radical expression in the denominator.

The process to get rid of the radical expressions in the denominators is called "rationalizing denominators". We need to find a conjugate of a radical number.

Definition 1.23. Conjugate-1

Suppose $m < n$ are two positive integers. For a given positive number b, The

> *conjugate of the exponential number $b^{\frac{m}{n}}$ is*
> $$b^{\frac{n-m}{n}}.$$

For example, the conjugate of $\sqrt{2}$ is $\sqrt{2}$. If we need to rationalize the denominator
$$\frac{1}{\sqrt{2}},$$
we multiply it by
$$\frac{\sqrt{2}}{\sqrt{2}}.$$
We have
$$\frac{1}{\sqrt{2}} = \frac{1}{\sqrt{2}} \cdot \frac{\sqrt{2}}{\sqrt{2}} = \frac{\sqrt{2}}{2} = \frac{1}{2}\sqrt{2}.$$

So $2^{-\frac{1}{2}}$ is a like term to other terms in Group 3.

The difference of squares formula can also be used in rationalizing a fraction with a radical denominator.

> **Definition 1.24. Conjugate-2**
>
> *For two positive rational numbers a and b, we call $\sqrt{a} + \sqrt{b}$ and $\sqrt{a} - \sqrt{b}$ are conjugate to each other.*

Example 1.60 Simplify
$$\frac{1}{\sqrt{3}-\sqrt{2}} + \frac{1}{2-\sqrt{3}}.$$

Solution:
$$\frac{1}{\sqrt{3}-\sqrt{2}} + \frac{1}{2-\sqrt{3}} = \frac{\sqrt{3}+\sqrt{2}}{3-2} + \frac{2+\sqrt{3}}{4-3} \quad \text{(Rationalizing denominator)}$$
$$= \sqrt{3} + \sqrt{2} + 2 + \sqrt{3} \quad \text{(Simplify)}$$
$$= 2\sqrt{3} + \sqrt{2} + 2. \quad \text{(Combining like terms)}$$

□

Exercise 1.29 Prove: for any positive number x,
$$\frac{1}{\sqrt{x+1}+\sqrt{x}} = \sqrt{x+1} - \sqrt{x}.$$

1.5.4 The Fundamental Assumption of Elementary Math

In this chapter, we learned that the exponential operation rule holds for fractional exponents (Proposition 1.14). Now we ask: (1) is there a number like $3^{\sqrt{2}}$? (2) If there is such a number, does the following operation rule

$$3^{\sqrt{2}} \cdot 3^{2-\sqrt{2}} = 3^{\sqrt{2}+2-\sqrt{2}} = 3^2$$

still hold? More generally, we want to know whether the exponential operation rule holds for all real-number exponents (Proposition 1.14 holds for all real-number exponents).

The answers to both questions are affirmative. From the following assumption, we can easily see that the exponential operation rule holds for all real-number exponents.

> **Proposition 1.15. The Fundamental Assumption of Elementary Mathematics**
>
> *The operation rules that hold for all rational numbers are true for all real numbers[a].*
>
> ---
> [a]The introduction of this assumption is inspired by H. Wu's book: *Teaching School Mathematics: Pre-Algebra*, AMS, 2016.

In the book "Introductory Geometry and Proofs", we use this assumption to explain the reason that we have the area formula for a rectangle (Chapter 1, Theorem 1.3). In Chapter 7 of this book, we will explain: why we have such a basic assumption.

1.5.5 General exponential operation rules

From Proposition 1.15 and 1.14, we can easily obtain the general exponential operation rules for all real-number exponents.

> **Theorem 1.3. General exponential operation rules**
>
> *Suppose a, b are two positive numbers, and r and s are two real numbers. Then*
>
> $$a^r \cdot a^s = a^{r+s},$$
> $$(a^r)^s = a^{rs}$$
>
> *and*
>
> $$(ab)^r = a^r b^r.$$

Example 1.61 Calculate and simplify

(a) $9^{\frac{1}{6}} \cdot 9^{\frac{1}{3}}$;

(b) $2^{1-\sqrt{2}} \cdot 2^{\sqrt{2}}$;

(c) $(2^{2-\sqrt{2}})^{2+\sqrt{2}}$.

Solution:

(a)
$$9^{\frac{1}{6}} \cdot 9^{\frac{1}{3}} = 9^{\frac{1}{6}+\frac{1}{3}}$$
$$= 9^{\frac{1}{2}}$$
$$= 3.$$

(b)
$$2^{1-\sqrt{2}} \cdot 2^{\sqrt{2}} = 2^{1-\sqrt{2}+\sqrt{2}}$$
$$= 2.$$

(c)
$$(2^{2-\sqrt{2}})^{2+\sqrt{2}} = 2^{(2-\sqrt{2})\cdot(2+\sqrt{2})}$$
$$= 2^{4-2}$$
$$= 4.$$

□

From Theorem 1.3 we know that for any given positive number $a \neq 1$, we can assign a number y for any real number x. We call it an exponential function:

$$y = a^x.$$

(If $a = 1$, this is a constant function). In Chapter 3, we will show: this is a one-to-one function.

Example 1.62 Solve equation (this is an exponential equation)

$$2^{x-2} = 16.$$

Solution: First, we observe

$$16 = 4^2 = 2^4.$$

Since the exponential function is a one-to-one function, we know from $2^{x-2} = 2^4$, that the exponents on both sides must be the same:

$$x - 2 = 4.$$

We thus obtain $x = 6$.

□

More discussion on exponential functions will be given in Chapter 3.

1.5.6 logarithm

Once we define exponentiation, we can introduce its inverse operator and the corresponding numbers.

> **Definition 1.25. Logarithm**
>
> *For given two positive numbers B and P (here we also require $B \neq 1$), we define a new number L, which satisfies*
> $$B^L = P.$$
> *We call such a number "logarithm", and use the following notation:*
> $$L = \log_B P.$$

♣

From the definition of the logarithm, we know that for any two positive numbers $b \neq 1$ and y,

$$x = \log_b y \quad \text{if and only if} \quad y = b^x.$$

Traditionally, we prefer to use x for the variable and y for the function value. So, if we switch x with y in the above, we then obtain the inverse function of an exponential function: the logarithmic function

$$y = \log_b x.$$

Example 1.63 Solve equation

(1)
$$\log_2 x = 3$$

(2)
$$\log_x 16 = 2$$

Solution:

(1) From the definition of the logarithm, we have
$$x = 2^3 = 8.$$

(2) From the definition of the logarithm, we have
$$x^2 = 16.$$

So $x = 4$ or $x = -4$. However, the base for a logarithmic function must be a positive number. So we have to throw away the negative solution. That is $x = 4$ is the only solution.

□

The following rules hold for logarithmic operation

> **Theorem 1.4. Logarithmic operation rules**
>
> *Suppose a, b, c, d are positive numbers, and $a \neq 1, b \neq 1$, r is any real number. Then*
>
> (a)
> $$\log_b(cd) = \log_b c + \log_b d,$$
>
> (b)
> $$\log_b c^r = r \log_b c,$$
>
> *and*
>
> (c)
> $$\log_b d = \frac{\log_a d}{\log_a b}.$$

Proof: (a) From the definition of the logarithm, we have, for any two positive numbers $a \neq 1$ and x, that
$$a^{\log_a x} = x.$$

So
$$b^{\log_b(cd)} = cd, \quad b^{\log_b c + \log_b d} = b^{\log_b c} \cdot b^{\log_b d} = cd.$$

Since exponential function is one to one, we conclude
$$\log_b(cd) = \log_b c + \log_b d.$$

(b) First, we have
$$b^{r \log_b c} = (b^{\log_b c})^r = c^r.$$

From the definition of the logarithm, we have,
$$\log_b c^r = r \log_b c.$$

(c) Define
$$x = \frac{\log_a d}{\log_a b}.$$

Then
$$\log_a d = x \log_a b$$
$$= \log_a b^x. \quad \text{(Rule (b))}$$

Thus $d = b^x$, that is $x = \log_b d$. We conclude
$$\log_b d = \frac{\log_a d}{\log_a b}.$$

□

In Chapter 7 we will reveal the definition of the natural constant e (we also call it the Euler constant). We call the logarithm with base e "natural logarithm" and denote it as $\ln x = \log_e x$.

1.5.7 Exercises after class

1. **Basic skills.** Simplify

   (1) $2^{\frac{1}{2}} \cdot 3^2 \cdot 2^{-2} \cdot \sqrt{2}$; (2) $\dfrac{3^2 5^{\frac{1}{2}} 5^{\frac{1}{4}}}{3^2 5^{\frac{5}{4}}}$;

   (3) $\dfrac{x^2 y^{\frac{1}{2}} x^3 y^{\frac{1}{4}}}{x^2 y^{\frac{1}{4}}}$; (4) $\dfrac{x^2 y^{\frac{1}{2}} x^{-1} y^{\frac{1}{4}}}{x^{-2} y^{-\frac{1}{4}}}$;

   (5) $\sqrt{12} + \sqrt{3} - \sqrt{2}$; (6) $\sqrt{32} + \sqrt{3} - \sqrt{2}$;

   (7) $\sqrt{5} - \sqrt{3} + \dfrac{1}{\sqrt{5}} + \dfrac{1}{\sqrt{3}}$; (8) $\sqrt{\dfrac{1}{2}} + \sqrt{\dfrac{1}{3}} - \sqrt{3} - \sqrt{2}$;

   (9) $\sqrt{18} + \sqrt{8} - \dfrac{1}{\sqrt{2}}$; (10) $\sqrt{\dfrac{1}{8}} + \dfrac{1}{\sqrt{8}} - \dfrac{1}{\sqrt{2}}$;

   (11) $\log_3 54 - \log_3 2$; (12) $\log_4 30 - \log_4 15$;

   (13) $\dfrac{1}{\sqrt{3}-1} + \dfrac{1}{1-\sqrt{3}}$; (14) $\dfrac{\sqrt{2}}{\sqrt{5}-\sqrt{2}} - \dfrac{\sqrt{5}}{\sqrt{5}-\sqrt{2}}$.

2. **More calculations.** Simplify

   (1) $\dfrac{1}{\sqrt{3}-\sqrt{2}} + \sqrt{2}$; (2) $\dfrac{2}{\sqrt{7}+\sqrt{5}} - \dfrac{6}{\sqrt{7}-\sqrt{5}}$;

   (3) $\sqrt{5 + 2\sqrt{6}} - \sqrt{3}$; (4) $\sqrt{5 + 2\sqrt{6}} + \sqrt{2}$;

   (5) $\sqrt{10 + 4\sqrt{6}} - \sqrt{6}$; (6) $\dfrac{\log_3 32}{\log_3 4}$;

3. **Logical argument.**

   (1) Suppose a is an irrational number, and b is a nonzero rational number. Prove: ab is an irrational number.

   (2) Prove: If $p > 1$ is a prime number, the $\sqrt{p}$ must be an irrational number.

   (3) Using the result from (2) to show that there are infinitely many irrational numbers.

   (4) Prove:
   $$\log_2 3 \cdot \log_3 2 = 1.$$

4. **Solve equation for** x.

   (1) $\log_x 16 = 2$.

   (2) $\log_{x^2} 16 = 2$.

1.6 Complex numbers

In the book "Introductory Geometry and Proofs", we discussed, from the viewpoint of Algebra (how to do operations) and from the viewpoint of Geometry (Phase angles and vector addition), how to do the operations on complex numbers and their applications.

Using the imaginary number i as a square root of -1 was first proposed in the first century by Greek mathematician Hero of Alexandria (see "Introductory Geometry and Proofs" exercise 6 in Chapter 4). It was later used by Girolamo Cardano in solving third-order equations. With the discovery of de Moivre formula and the Euler formula, complex numbers were finally widely used in many mathematical fields, and have broad applications in the electrical engineering field.

1.6.1 Definition and operation

The original motivation could be: to find two solutions to
$$x^2 = -1 \qquad (1.16)$$
so we will have a complete theory: any quadratic equation has two solutions (could be repeated solutions).

We first recall the definition of the imaginary number i from the book "Introductory Algebra".

> **Definition 1.26. Imaginary symbol i**
>
> The imaginary symbol i is defined as
> $$i = \sqrt{-1}.$$
> The main property for i is that $i^2 = -1$, or i is one solution to the following equation
> $$x^2 = -1. \qquad (1.17)$$

We can also verify $(-i)^2 = (-1) \times (-1) \times i \times i = -1$. That is $x_2 = -i$ is the second solution to the equation (1.17).

A general complex number is introduced in the following.

> **Definition 1.27. Complex number**
>
> For two given real numbers a, b, number $z = a + bi$ is called a complex number. a is called the real part (we use notation: $a = ReZ$), b is called the imaginary part (we use notation: $b = ImZ$).

With the expansion of numbers to complex numbers, we now have a so-called complete number system with respect to all algebraic operations. For instance, we will soon see that every number has two square roots. In particular, we will introduce (the proof will be given in a "Complex Analysis" course in college) the **Fundamental Theorem of Algebra** in Chapter 3.

Since complex numbers are numbers, all rules for addition and multiplication still hold for them.

Example 1.64 Calculate:

(1) $3 + i + (4 - 2i)$,

(2) $(1 + \sqrt{3}i)^2$.

Solution: (1)

$$\begin{aligned} 3 + i + (4 - 2i) &= 3 + i + 4 - 2i & \text{(Associative rule)} \\ &= 3 + 4 + i - 2i & \text{(Commutative rule)} \\ &= 7 + (1 - 2)i & \text{(Associative and distributive rules)} \\ &= 7 - i; \end{aligned}$$

(2)

$$\begin{aligned} (1 + \sqrt{3}i)^2 &= (1 + \sqrt{3}i) \cdot (1 + \sqrt{3}i) & \text{(Definition of square)} \\ &= 1 + 2\sqrt{3}i + (\sqrt{3})^2 i^2 & \text{(Operation rules)} \\ &= 1 + 2\sqrt{3}i + 3 \cdot (-1) & \text{(Definition of i)} \\ &= -2 + 2\sqrt{3}i. \end{aligned}$$

□

We now can solve all quadratic equations in the complex number set.

Example 1.65 Solve quadratic equations in the complex number set:

(1) $x^2 = -3$,

(2) $x^2 - 2x + 4 = 0$.

Solution:

(1) We first observe: $(\sqrt{3}i)^2 = 3i^2 = -3$; Similarly $(-\sqrt{3}i)^2 = 3i^2 = -3$. So the equation has two solutions[16]: $x_1 = \sqrt{3}i$ and $x_2 = -\sqrt{3}i$.

---

[16]One common error in standard textbooks is this: from $x^2 = -3$, we take a square root on both sides of the equation to get $x = \pm\sqrt{-3}$. Then $\sqrt{-3} = \sqrt{3} \cdot \sqrt{-1} = \sqrt{3}i$, and get the right answer $x = \pm\sqrt{3}i$.

Yes, the answer is correct, but there is an error in arguing $\sqrt{-3} = \sqrt{3} \cdot \sqrt{-1} = \sqrt{3}i$. Using the same argument, we would have:

$$1 = \sqrt{(-1) \cdot (-1)} = \sqrt{-1} \cdot \sqrt{-1} = i^2 = -1!$$

Besides the above wild guess, we can also solve the equation using the following logical argument. The equation is equivalent to $(\frac{x}{\sqrt{3}})^2 = -1$. From the solutions to equation (1.17) we know $\frac{x}{\sqrt{3}} = i$ or $\frac{x}{\sqrt{3}} = -i$. We thus obtain two solutions: $x_1 = \sqrt{3}i$ and $x_2 = -\sqrt{3}i$.

(2)

$$x^2 - 2x + 4 = 0 = 0 \iff x^2 - 2x + 1 + 3 = 0$$
$$\iff (x-1)^2 = -3$$
$$\iff x - 1 = \sqrt{3}i \text{ or } x - 1 = -\sqrt{3}i.$$

So two solutions to the equation are $x_1 = 1 + \sqrt{3}i$ and $x_2 = 1 - \sqrt{3}i$.

□

Exercise 1.30 Verify $x = 1 + \sqrt{3}i$ is a solution to $x^2 - 2x + 4 = 0$.

Once we learn the square root of -1, we may further ask: what is the cubic root of -1? How about the n-th root of -1?, How about the square root for general complex numbers?

1.6.2 Modulus and argument

In order to answer the above-mentioned questions, we need to learn another expression for a complex number. Here we need some basic knowledge on trigonometric functions[17].

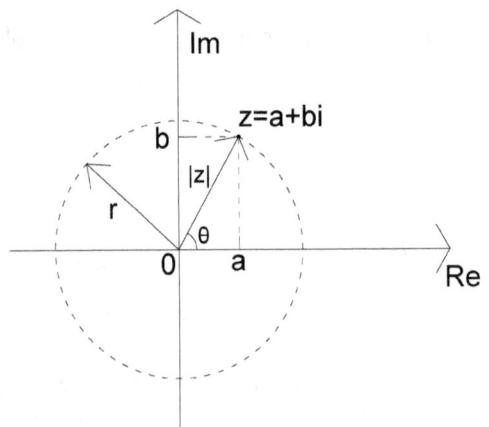

Figure 1.6: Complex plane

---

[17]One can come back to learn this section after the study of Chapter 4

Let $z = a + bi$ be a nonzero complex number, where a, b are two real numbers. Then
$$z = a + bi = \sqrt{a^2 + b^2} \cdot \left(\frac{a}{\sqrt{a^2 + b^2}} + \frac{bi}{\sqrt{a^2 + b^2}}\right).$$

We call $\sqrt{a^2 + b^2}$ the modulus of complex number z (or just call it the absolute value), denote as $|z|$. We write
$$r = |z|, \quad \cos\theta = \frac{a}{\sqrt{a^2 + b^2}} \quad \text{and} \quad \sin\theta = \frac{b}{\sqrt{a^2 + b^2}}.$$
The θ angle here is called the argument of the complex number.

On the complex plane, we use point (a, b) to represent complex number $z = a + bi$. Moreover, for nonzero complex number z, if we restrict $\theta \in [0, 2\pi)$, then there is a one-to-one correspondence between the point (a, b) in the complex plane and the pair (r, θ). We call such a θ the principal argument (or the principal phase angle).

Example 1.66 Find the modulus, the principal argument and the argument of the complex number $z = 1 + i$.

Solution: Its modulus is
$$|z| = \sqrt{1^2 + 1^2} = \sqrt{2}.$$

Its principal argument θ satisfies
$$\cos\theta = \frac{1}{\sqrt{2}} \quad \text{and} \quad \sin\theta = \frac{1}{\sqrt{2}}.$$
So the principal argument $\theta = \frac{\pi}{4}$. The argument is $\frac{\pi}{4} + 2n\pi$ for any integer n.

Later, we can use an infinite series to define the number $e^{i\theta}$ (see Chapter 8 for more details). Based on this expression, we can prove the following important, yet, quite astonishing Euler formula:

> **Proposition 1.16. Euler formula**
> *For any given radian θ,*
> $$e^{i\theta} = \cos\theta + i\sin\theta.$$

So, we can write the complex number z with modulus r and principal argument θ as
$$z = re^{i\theta}.$$

We call this the polar form of the complex number, and call (r, θ) the polar coordinates of the complex number.

On the other hand, Abraham de Moivre discovered (1722) the following formula[18]:

---

[18] Historically, when $\alpha = \beta$, this formula was first derived by de Moivre, thus is called de Moivre formula. When $\alpha \neq \beta$, it can be derived in a similar way. We thus call it de Moivre formula as well.

1.6 Complex numbers

> **Proposition 1.17. de Moivre formula**
>
> For any two radians α and β,
> $$(\cos\alpha + i\sin\alpha)(\cos\beta + i\sin\beta) = \cos(\alpha+\beta) + i\sin(\alpha+\beta).$$

One can derive the following corollary. For any radian α and positive integer n,
$$(\cos\alpha + i\sin\alpha)^n = \cos n\alpha + i\sin n\alpha. \tag{1.18}$$

Exercise 1.31 Use de Moivre formula (Proposition 1.17) and mathematical induction to prove formula (1.18).

The proof of the Euler formula usually is given in a Calculus course (see Chapter 8 in this book); The proof of de Moivre formula will be given in Chapter 4 later.

For any complex number $z = a + bi$, we define
$$e^{a+bi} = e^a \cdot e^{ib} = e^a \cos b + ie^a \sin b.$$

With the aid of the Euler formula as well as de Moivre formula, we now can establish the exponential operation rules for complex exponents.

> **Proposition 1.18. Complex exponential operation rules**
>
> For any two complex numbers z and w, we have
> $$e^z \cdot e^w = e^{z+w}$$

Proof: Let $z = a + bi$, $w = c + di$, where a, b, c, d are real numbers. Then
$$e^{z+w} = e^{(a+c)+(b+d)i}.$$

On the other hand

$$\begin{aligned}
e^z \cdot e^w &= e^a \cdot e^{bi} \cdot e^c \cdot e^{di} && \text{(From the definition)}\\
&= e^a \cdot e^c \cdot e^{bi} \cdot e^{di} && \text{(Commutative rule)}\\
&= e^{a+c} \cdot e^{(b+d)i} && \text{(De Moivre formula)}\\
&= e^{(a+c)+(b+d)i} && \text{(Definition)}
\end{aligned}$$

We hereby complete the proof. □

Apparently, for any integer n and any complex number z, we have
$$(e^z)^n = e^{nz}.$$

But the above formula may not hold if n is not an integer. In fact, in order to avoid confusion, we will not use $i^{\frac{1}{2}}$ or $\sqrt{i}$ to represent the square root of i (since we do know by now that any complex number has TWO square roots). See more discussions in Section 4.2.6.

1.6.3 Solving polynomial equations in the complex number set

Not too hard to verify
$$i = \cos\frac{\pi}{2} + i\sin\frac{\pi}{2} = e^{\frac{\pi}{2}i}.$$

So
$$x_1 = e^{\frac{\pi i}{4}} = \cos\frac{\pi}{4} + i\sin\frac{\pi}{4} = \frac{\sqrt{2}}{2} + \frac{\sqrt{2}}{2}i$$

is one of i's square roots (since we use the principal argument of i, we call this square root the principal square root of i). Another square root is
$$x_2 = e^{\frac{1}{2}\cdot(2\pi i + \frac{\pi i}{2})} = \cos\frac{5\pi}{4} + i\sin\frac{5\pi}{4} = -\frac{\sqrt{2}}{2} - \frac{\sqrt{2}}{2}i.$$

Example 1.67 Solve polynomial equations in the complex number set:

(1)
$$x^2 = 1 + \sqrt{3}i;$$

(2)
$$x^4 = -1.$$

Solution:

(1) Observe:
$$1 + \sqrt{3}i = 2(\frac{1}{2} + \frac{\sqrt{3}}{2}i) = 2e^{2\pi ki + \frac{\pi}{3}i}, \text{ for } k = 0, 1.$$

So
$$x_1 = \sqrt{2}e^{\frac{\pi}{6}i} = \frac{\sqrt{6}}{2} + \frac{\sqrt{2}}{2}i,$$
$$x_2 = \sqrt{2}e^{\pi i + \frac{\pi}{6}i} = -\frac{\sqrt{6}}{2} - \frac{\sqrt{2}}{2}i$$

are two solutions.

(2) Observe:
$$-1 = e^{2\pi ki + \pi i}, \text{ for } k = 0, 1, 2, 3.$$

So
$$x_k = e^{\frac{2\pi k}{4}i + \frac{\pi}{4}i}, \text{ for } k = 0, 1, 2, 3$$

are all solutions to $x^4 = -1$. That is
$$x_1 = e^{\frac{\pi}{4}i} = \frac{\sqrt{2}}{2} + \frac{\sqrt{2}}{2}i, \quad \text{(The principal 4-th root of -1)}$$
$$x_2 = e^{\frac{2\pi}{4}i + \frac{\pi}{4}i} = -\frac{\sqrt{2}}{2} + \frac{\sqrt{2}}{2}i,$$
$$x_3 = e^{\frac{4\pi}{4}i + \frac{\pi}{4}i} = -\frac{\sqrt{2}}{2} - \frac{\sqrt{2}}{2}i,$$

1.6 Complex numbers

$$x_4 = e^{\frac{6\pi}{4}i + \frac{\pi}{4}i} = \frac{\sqrt{2}}{2} - \frac{\sqrt{2}}{2}i.$$

□

Note *Generally speaking, we use $z^{\frac{1}{n}}$ to represent the n-th principal root of z (corresponding to the principal argument). Unfortunately, many textbooks misuse the notation. For example, they use $(-8)^{\frac{1}{3}}(=-2)$ to represent the real cubic root for -8. We suggest eliminating the notations such as $(-8)^{\frac{1}{3}}$.*

Exercise 1.32 Find the principal cubic root of -8.

1.6.4 Exercises after class

1. **Basic skills.** Operation on complex numbers
   (1) $-1+2i-2-3i$;   (2) $(1-2i)(2-i)$;
   (3) $(2-i)^2$;   (4) $(5-4i)(5+4i)$;
   (5) $(1+i)^3$;   (6) $1+2i+\frac{10}{2-i}$;
   (7) $2e^{\frac{\pi}{3}i} \cdot 2e^{\frac{\pi}{6}i}$;   (8) $(\frac{\sqrt{3}}{2}+\frac{1}{2}i) \cdot 2e^{\frac{\pi}{3}i}$.

2. **More computations.** Find the modulus and argument of a given complex number.
   (1) $\frac{1}{2}+\frac{\sqrt{3}}{2}i$;   (2) $\frac{1}{2}-\frac{\sqrt{3}}{2}i$;
   (3) $\frac{\sqrt{3}}{2}+\frac{1}{2}i$;   (4) $\frac{\sqrt{3}}{2}-\frac{1}{2}i$;
   (5) $\frac{\sqrt{3}}{2}+\frac{\sqrt{3}}{2}i$;   (6) $\frac{\sqrt{2}}{2}-\frac{\sqrt{2}}{2}i$;
   (7) Find the principal cubic root of -8;
   (8) Find the principal 4-th root of -16.

3. **Solving equations in complex number set.**
   (1) $x^2=-2$;   (2) $x^2+5=0$;
   (3) $x^2-x+2=0$;   (4) $x^2+3x+6=0$;
   (5) $x^2+i=0$;   (6) $x^3-i=0$;
   (7) $x^3+x-2=0$.

4. **Comprehensive questions.**
   (1) If $x^2+2x+4=0$, find x^3.
   (2) If $x^4+2x^3+4x^2+8x+16=0$, find x^5.

1.7 Chapter review and exercises

1.7.1 Chapter 1 review

We summarize the main contents covered in this Chapter.

1. Two operations and three operation rules

Proposition 1.1-1.3 For any three numbers A, B, C, we have

$$A + B = B + A \qquad \text{(Commutative rule for addition)}$$
$$AB = BA \qquad \text{(Commutative rule for multiplication)}$$
$$A + B + C = (A + B) + C = A + (B + C) \qquad \text{(Associative rule for addition)}$$
$$ABC = (AB)C = A(BC) \qquad \text{(Associative rule for multiplication)}$$
$$A(B + C) = AB + AC \qquad \text{(Distributive rule)}$$

2. Exponent, exponent rules

   2.1. **Exponent rules for integer exponents**.

For any two integers n, m, any two nonzero complex numbers a, b,

$$a^n \cdot a^m = a^{n+m}$$
$$(a^n)^m = a^{nm}$$
$$(ab)^n = a^n b^n$$

   2.2. **Exponent rules for real number exponents** (caution: we have restrictions on the bases)

For any two real numbers r, s, any two positive numbers a, b,

$$a^r \cdot a^s = a^{r+s}$$
$$(a^r)^s = a^{rs}$$
$$(ab)^r = a^r b^r$$

   2.3. **Exponent rules for complex number exponents**

For any two complex numbers z, w,

$$e^z \cdot e^w = e^{z+w}$$

3. Properties for equality

   3.1. (**Addition invariance**): For any number C,

$$A = B \iff A + C = B + C.$$

3.2. **(Multiplication invariance)**: For any $C \neq 0$,
$$A = B \iff AC = BC.$$

3.3. **(Symmetric property)**:
$$A = B \iff B = A.$$

3.4. **(Transitive property)**: If $A = B$ and $B = C$, then
$$A = C.$$

4. **Properties for inequality**

    4.1. **(Addition invariance)**: For any number C,
    $$A < B \iff A + C < B + C.$$

    4.2.1. **(Invariance under positive multiplication)**: For any positive number C (that is $C > 0$),
    $$A < B \iff AC < BC.$$

    4.2.2. **(Reverse inequality under negative multiple)**: For any negative number C (that is $C < 0$),
    $$A < B \iff AC > BC.$$

    4.3. **(Symmetric property)**:
    $$A > B \iff B < A.$$

    4.4. **(Transitive property)**: If $A < B$ and $B < C$, then
    $$A < C.$$

5. **Concepts on Complex numbers**

    5.1. **Imaginary number** i:
    $$i^2 = 1, \quad i = \sqrt{-1}.$$

    5.2. **Modulus and argument of complex numbers**:

    $$z = a + bi = re^{i\theta}$$
    $$r = \sqrt{a^2 + b^2}$$
    $$\cos\theta = \frac{a}{\sqrt{a^2 + b^2}} \quad \text{and} \quad \sin\theta = \frac{b}{\sqrt{a^2 + b^2}}.$$

    5.3. **Euler formula**: For a given radian θ,
    $$e^{i\theta} = \cos\theta + i\sin\theta.$$

5.4. **de Moivre formula**: For two radians α and β,
$$(\cos\alpha + i\sin\alpha)(\cos\beta + i\sin\beta) = \cos(\alpha+\beta) + i\sin(\alpha+\beta).$$

1.7.2 Chapter 1 test

1. Simplify. The answer shall not involve negative exponents, no radical expressions and imaginary numbers in the denominators.

   (1).
   $$\frac{1}{2} - \frac{1}{3} \cdot 1\frac{1}{2} - 3\frac{1}{3} + \frac{1}{6} + \frac{2}{9}$$

   (2).
   $$\frac{1}{\sqrt{2}} + \frac{\sqrt{2}}{4} - \frac{1}{1+\sqrt{2}}$$

   (3).
   $$\frac{1}{1 \times 2} + \frac{1}{2 \times 3} + \cdots + \frac{1}{39 \times 40}$$

   (4).
   $$\frac{1}{\sqrt{2}+1} + \frac{1}{\sqrt{3}+\sqrt{2}} + \cdots + \frac{1}{\sqrt{1024}+\sqrt{1023}}$$

   (5).
   $$\log_4 \sqrt{2} + \log_{\sqrt{2}} 4$$

   (6).
   $$(a^3)^{-2}(b^3)^2(a^3 b^{-2})^2$$

   (7).
   $$\left(\frac{ab^{-2}c^3}{a^3 b^{-2}}\right)^{-2}$$

   (8).
   $$(2-i)(3+2i)$$

   (9).
   $$\frac{4-8i}{1+i}$$

   (10).
   $$(2+i)(3-2i)$$

   (11).
   $$\frac{4+8i}{1-i}$$

   (12).
   $$(1+x)(3-x) - (2-x)(2+x)$$

(13).
$$\frac{x^2 + 2x - 3}{2x^2 - x - 3} \cdot \frac{2x - 3}{x - 1}$$

(14).
$$(x^2 + 2x + 1)^4$$

2. Solve equations/inequalities (x, y are variables)

   (1).

   (1a).
   $$\frac{x}{2} - \frac{1}{3} = \frac{1}{2}$$

   (1b).
   $$\frac{1}{2(x + 1)} - \frac{1}{3} = \frac{1}{2}$$

   (2).

   (2a).
   $$\frac{x}{a} = \frac{1}{2}$$

   (2b).
   $$\frac{a + 1}{x} = \frac{1}{2}$$

   (3).

   (3a).
   $$x^2 + 11x + 28 = 0$$

   (3b).
   $$6x^2 - 7x - 3 = 0$$

   (4).

   (4a).
   $$x^2 - 11x + 28 = 0$$

   (4b).
   $$6x^2 + 7x - 3 = 0$$

   (5).

   (5a).
   $$x^2 - 4x + 9 = 0$$

   (5b).
   $$x^3 + x + 2 = 0$$

(6).

(6a).
$$x + 2 = \sqrt{x^2 + 16}$$

(6b).
$$\sqrt{x+3} - \sqrt{x-2} = 1$$

(7).

(7a).
$$-x + 1 \leq 0$$

(7b).
$$\frac{1}{x} > 2$$

(8).

(8a).
$$x(x+1) \leq 0$$

(8b).
$$\frac{x+1}{x} \leq \frac{1}{2}$$

(9).

(9a).
$$|x - 2| - 5 \geq 0$$

(9b).
$$ax \leq 3.$$

3. Proof

(1). For any positive number a, prove
$$\frac{a+1}{a} > \frac{a+2}{a+1}.$$

(2). Prove: if $A > B, x > y$, then
$$A + x > B + y.$$

(3). Suppose that A is an irrational number, B is a nonzero rational number. Prove $A + B$, AB and $\frac{A}{B}$ are all irrational numbers.

(4). Assume $x, y > 0$, prove
$$\ln(x^2 + y^2) - \ln 2 \geq \ln x + \ln y.$$

(5). Prove: for any natural number n,
$$2^n \geq n+1.$$

Chapter 2 Function

> **Introduction**
>
> - Definition and expression for set
> - Operation on sets
> - Number sets
> - Cardinality of set
> - Arrangement/Permutation
> - Addition principle
> - Multiplication principle
> - Combinatorial number
> - Mapping
> - Graph of function
> - Symmetry of function
> - Monotonicity of function
> - Composite and inverse functions

In the book "Introductory Algebra" we introduced some basic concepts of functions. Here we recall and study: what is a set, operation on sets, basic counting principles; What is a function, the domain, the range, and some basic properties of a function, etc. These preparations will enable us to thoroughly understand elementary functions in the next chapter.

2.1 Sets

We first recall: what is a set.

2.1.1 Set and the representation of a set

Set is a basic concept in mathematics. It usually refers to a collection of certain objects (we usually call them elements). If these elements are numbers, the set is called a "number set"; if these elements are lines, we call the set a "line set". We sometimes use a bracket to include elements, and sometimes use a circle to include all elements in one set.

Recall: in the book "Introductory Algebra" (Chapter 13, Section 13.4), we introduce the following notation to represent a set:

$$S = \{x \text{ has some property} \mid x \text{ has more special property}\}.$$

For example, set A:

$$A = \{x \text{ is integer} \mid 1 < x < 10\}$$

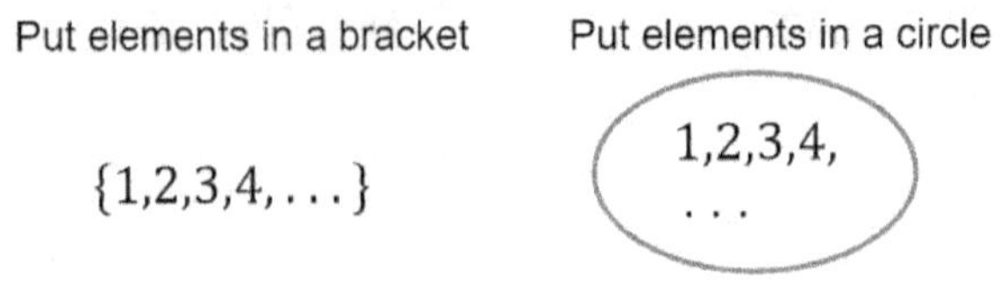

2.1 Sets

Put elements in a bracket

$\{1, 2, 3, 4, \ldots\}$

Put elements in a circle

1, 2, 3, 4, ...

Figure 2.1: Graphic expression of set

represents set
$$\{2, 3, 4, 5, 6, 7, 8, 9\}.$$

In the book "Introductory Algebra" we also introduced some well-known sets:

$\mathbb{N} = \{\text{All natural functions}\}$,

$\mathbb{Z} = \{\text{All integers}\}$, $\mathbb{Q} = \{\text{All rational numbers}\}$,

$\mathbb{R} = \{\text{All real numbers}\} = \{\text{all rational and irrational numbers}\}$,

and

$\mathbb{C} = \{\text{All complex numbers}\}$.

We recall that *zero* plays an essential role for us to expand natural numbers into all integers. Here we introduce the empty set, which will play an important role in set operations.

A set that does not have any element is called an empty set. We use the notation $\emptyset$ for an empty set.

If x is an element in set A, we write $x \in A$ (read: x belongs to set A). If x is not an element in set A, we write $x \notin A$ (read: x does not belong to set A).

> **Definition 2.1. Subset and complement**
>
> *For two given sets A and M, if all elements in set A are elements in set M, we say that A is a subset of set M. We use the notation: $A \subseteq M$.*
>
> *If $A \subseteq M$, we define the complement of A in M, using the notation A^c, as*
>
> $$A^c = \{x \in M \mid x \notin A\}.$$

Example 2.1 Define
$$M = \{\text{All teachers in Norman public schools}\},$$
$$A = \{\text{All male teachers in Norman public schools}\}.$$

Then $A \subseteq M$.

Example 2.2 From the definition, we know $\mathbb{Q} \subseteq \mathbb{R}$; The complement of $\mathbb{Q}$ in $\mathbb{R}$ is the set of all irrational numbers.

Sometimes, A^c could be an empty set $\emptyset$.

> **Definition 2.2. Proper subset**
>
> *For two given sets A and M, if $A \subseteq M$, and there is at least one element in M that is not an element in A, we then call A a proper subset of set M. We denote this as $A \subset M$.*

For example: $\mathbb{Z} \subset \mathbb{Q}$; $\mathbb{Q} \subset \mathbb{R}$.

A bit abstract concept is: we define that the empty set $\emptyset$ is a subset of any set.

2.1.2 Operation on sets

With the definitions and notations for sets, we introduce the following operations on sets.

> **Definition 2.3. Operation on sets**
>
> *For two given sets A and B, we define the intersection of A and B, using the notation $A \cap B$, as*
> $$A \cap B = \{x \mid x \in A \text{ and } x \in B\}.$$
>
> *We define the union of A and B,*
> $$A \cup B = \{x \mid x \in A \text{ or } x \in B\}.$$

From the definition, easy to see
$$A \cap B = B \cap A, \quad A \cup B = B \cup A.$$

The operation on sets can be illustrated by the following Venn diagram.

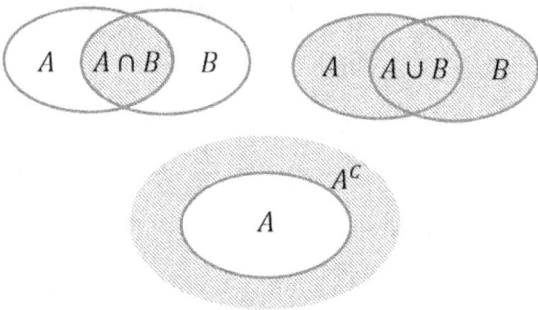

Figure 2.2: Venn diagram

Example 2.3 Prove
$$A \cap B \subseteq A, \quad \text{and} \quad B \cap A \subseteq B.$$

Proof: Suppose $A \cap B \neq \emptyset$, say $x \in A \cap B$. Then $x \in A$ and $x \in B$. Thus $A \cap B \subseteq A$

and $A \cap B \subseteq B$.

Suppose $A \cap B = \emptyset$, then, by definition, $\emptyset \subseteq A$ and $\emptyset \subseteq B$. Thus $A \cap B \subseteq A$ and $A \cap B \subseteq B$.

□

Example 2.4 Suppose $A \subseteq B$, and $B \subseteq D$. Prove: $A \subseteq D$.

Proof: If $A \neq \emptyset$, then for any element $x \in A$, we know $x \in B$ (since A is a subset of B), thus $x \in D$ (since B is a subset of D). From the definition, we know $A \subseteq D$.

If $A = \emptyset$, then by definition we know $\emptyset \subseteq D$, thus $A \subseteq D$.

□

Exercise 2.1 Suppose $A = \{x \in \mathbb{N} : x \text{ is divisible by } 2\}$, $B = \{x \in \mathbb{N} : x \text{ is divisible by } 3\}$.

(1) Find $A \cap B$ and $A \cup B$. Is $A \cup B = \mathbb{N}$? Why?

(2) Find the complement of A and complement of B in $\mathbb{N}$.

(3) Find $A^c \cap B^c$ and $A^c \cup B^c$.

We now derive the following rules on the operations of sets.

Proposition 2.1. Set operation rules

Considering three sets A, B and C, we have

(1) (Commutative rule): $A \cap B = B \cap A$; $A \cup B = B \cup A$.

(2) (Associative rule): $(A \cap B) \cap C = A \cap (B \cap C)$; $(A \cup B) \cup C = A \cup (B \cup C)$.

(3) (Distributive rule): $A \cap (B \cup C) = (A \cap B) \cup (A \cap C)$; $A \cup (B \cap C) = (A \cup B) \cap (A \cup C)$.

Proof: We only give the proof for $A \cap (B \cup C) = (A \cap B) \cup (A \cap C)$. The proofs for other rules are left for exercise.

We use the following standard strategy to prove two sets are the same: we show these two sets are a subset of each other.

Step 1. Prove: $A \cap (B \cup C) \subseteq (A \cap B) \cup (A \cap C)$.

Choose an element $x \in A \cap (B \cup C)$. Then $x \in A$ and $x \in B$ or $x \in C$, thus $x \in A$ and $x \in B$, or $x \in A$ and $x \in C$. That is $x \in (A \cap B) \cup (A \cap C)$. This shows $A \cap (B \cup C) \subseteq (A \cap B) \cup (A \cap C)$.

Step 2. Prove: $(A \cap B) \cup (A \cap C) \subseteq A \cap (B \cup C)$.

Choose an element $x \in (A \cap B) \cup (A \cap C)$. Then $x \in A$ and $x \in B$, or $x \in A$ and $x \in C$. Thus $x \in A$, and $x \in B$ or $x \in C$. That is $x \in A \cap (B \cup C)$. This shows $(A \cap B) \cup (A \cap C) \subseteq A \cap (B \cup C)$.

The proof is completed.

□

Exercise 2.2 Prove parts (2) and (3) in Proposition 2.1.

For complement sets, we have the following rules.

Proposition 2.2. Operation rules on complement sets

(1) $(A \cup B)^c = A^c \cap B^c$.

(2) $(A \cap B)^c = A^c \cup B^c$.

Proof: We only give the proof for part (1) and leave the proof of (2) as an exercise.

Choose an element $x \in (A \cup B)^c$. Then $x \notin A \cup B$. So $x \notin A$ and $x \notin B$. That is $x \in A^c \cap B^c$. This shows $(A \cup B)^c \subseteq A^c \cap B^c$.

On the other hand, choose an element $x \in A^c \cap B^c$. Then x does not belong to A, x does not belong to B either. That is $x \notin (A \cup B)$. This shows $A^c \cap B^c \subseteq (A \cup B)^c$.

$\square$

Exercise 2.3 Prove part (2) in Proposition 2.2.

2.1.3 Number sets and operation

If the elements in a set are all numbers, we call this set a number set. Let us recall the number sets we learned before.

In the book "Introductory Algebra", we first learned natural number set $\mathbb{N}$

$$\mathbb{N} = \{0, 1, 2, 3, \cdots\}.$$

Counting natural numbers leads us naturally to discover two operations on numbers: Addition and multiplication, and three rules satisfied by these two operations:

Proposition 2.3. Three operation rules

The addition and multiplication of numbers satisfy the following rules:

  1. (Commutative rule)

  $$a + b = b + a, \quad a \times b = b \times a.$$

  2. (Associative rule)

  $$a + b + c = (a + b) + c = a + (b + c),$$

  and

  $$a \times b \times c = (a \times b) \times c = a \times (b \times c).$$

  3. (Distributive rule)

  $$a \times (b + c) = a \times b + a \times c.$$

Note *Though, we summarize the above three rules from the operations on natural*

numbers. In the following study, we define that all three rules hold for all complex numbers.

We then define the opposite numbers, then expand natural numbers to integer numbers $\mathbb{Z}$:
$$\mathbb{Z} = \{\cdots, -3, -2, -1, 0, 1, 2, 3, \cdots\}.$$

And proposition 2.3 holds for $\mathbb{Z}$.

For nonzero integers, we introduce their reciprocals, thus extending the number set further to the rational number set $\mathbb{Q}$:
$$\mathbb{Q} = \{\frac{p}{q} \mid p, q \in \mathbb{Z}, q \neq 0\}.$$

For nonzero rational number a and integer m, we also introduce another operation: exponential operator: a^m. And there are three operation rules held for the exponential operator:

> **Proposition 2.4. Exponent rules for integer exponents**
>
> *For given two integers m, n and two nonzero numbers a, b, we have*
>
> 1.
> $$a^m \cdot a^n = a^{m+n},$$
>
> 2.
> $$(a^m)^n = a^{mn}$$
>
> 3.
> $$(ab)^m = a^m b^m.$$

For positive bases and fractional exponents, we introduce n-th root and radical expressions and further expand numbers to all real numbers $\mathbb{R}$. Again we define that all operation rules in Proposition 2.3 hold for real numbers. Need to point out, for positive bases a, b, Proposition 2.4 holds for all real numbers (see, Theorem 1.3). We repeat Theorem 1.3 here again:

> **Theorem 2.1. General exponential operation rules**
>
> *Suppose a, b are two positive numbers, and r and s are two real numbers. Then*
> $$a^r \cdot a^s = a^{r+s},$$
> $$(a^r)^s = a^{rs}$$
>
> *and*
> $$(ab)^r = a^r b^r.$$

Once we study an exponent with a negative base, for example, the square root of -2, we naturally introduce imaginary numbers, thus expand numbers to the complex number set $\mathbb{C}$.

$$\mathbb{C} = \{a + bi \ : \ a \in \mathbb{R}, \ b \in \mathbb{R}\}.$$

From the definition of complex numbers, we can prove that operation rules (Proposition 2.3) hold for complex numbers.

Example 2.5 Suppose $z_1 = a_1 + b_1 i$ and $z_2 = a_2 + b_2 i$ are two complex number, where a_1, a_2, b_1, b_2 are all real numbers. Prove:

$$z_1 \cdot z_2 = z_2 \cdot z_1.$$

Proof:

$$\begin{aligned}
z_1 \cdot z_2 &= (a_1 + b_1 i)(a_2 + b_2 i) \\
&= a_1 a_2 - b_1 b_2 + (a_1 b_2 + b_1 a_2)i \\
&= a_2 a_1 - b_2 b_1 + (a_2 b_1 + b_2 a_1)i \\
&= (a_2 + b_2 i)(a_1 + b_1 i) \\
&= z_2 \cdot z_1.
\end{aligned}$$

□

Warning: Theorem 2.1 does not hold for complex numbers or those exponents with non-positive bases. See Proposition 4.8 later for more details.

Example 2.6 (a).
$$1 = 1^i = (e^{2\pi i})^i \neq e^{-2\pi};$$

(b).
$$1 = 1^{\frac{1}{2}} = (e^{\pi i} \cdot e^{\pi i})^{\frac{1}{2}} \neq e^{\frac{\pi i}{2}} \cdot e^{\frac{\pi i}{2}} = -1.$$

□

Note *In the book "Introductory Geometry and Proofs", we showed that the operation rules for numbers may not hold for operations on vectors or matrices. For example, the commutative rule does not hold for matrix multiplication.*

Other number sets include interval sets: (a, b), $[a, b]$, $[a, b)$ and $(a, b]$.

$$(a, b) = \{x \mid a < x < b\}, \quad [a, b] = \{x \mid a \leq x \leq b\},$$
$$[a, b) = \{x \mid a \leq x < b\}, \quad (a, b] = \{x \mid a < x \leq b\}.$$

2.1.4 Exercises after class

1. **Basic skills**. Consider the following subset of the natural number set $\mathbb{N}$:
$$A = \{2n \mid n \in \mathbb{N}\}; \quad B = \{3n \mid n \in \mathbb{N}\}.$$

   (1) Find $A \cap B$; (2) Find A^c;

   (3) Prove:
   $$A \cup B \subset \{2n + 3m \mid n, m \in \mathbb{N}\}.$$

2. **Basic skills and computations**. Consider the following subset of $\mathbb{R}$:
$$A = \{x \in \mathbb{R} \mid x > 2\}; \quad B = \{x \in \mathbb{R} \mid |x - 2| < 10\}.$$

   (1) Find $A \cap B$; (2) Find $A \cup B$;

   (3) Find $A^c \cup B^c$ (4) Prove $(A \cap B)^c = A^c \cup B^c$.

3. **More practice**. Consider the following 4th order equation:
$$x^4 = 1.$$

   (1) Find all solutions in the complex number set.

   (2) Let $\mathbb{S} = \{x_1, x_2, x_3, x_4\}$ be the solution set. Show that $x_1 x_2, x_1 x_3, x_1 x_4$ are all in $\mathbb{S}$.

2.2 Counting

A simple question in counting is to count how many elements there are in a given set. This is quite similar to the introduction of natural numbers in arithmetic: how many sheep are there in a herd?

2.2.1 Cardinality and Addition Principle

Definition 2.4. Cardinality of a finite set

If set S has finite elements, we call the number of elements in set S the Cardinality of S. We denote it as $card(S)$.

Example 2.7 Consider the set of all fifth graders in school A. Then $card(S)$ is the number of all fifth graders.

□

If we consider the set S_1 of those fifth graders in school A who are younger than or equal to 11 years old, it is a subset of S: $S_1 \subset S$; Similarly, the set S_2 of those fifth graders in school A who are older than or equal to 11 years old is another subset of S: $S_2 \subset S$. We can see that

$$S = S_1 \cup S_2.$$

Exercise 2.4 Prove the above equality.

We also observe that

$$S_1 \cap S_2 = \{\text{fifth graders in school A whose age is 11}\}.$$

We conclude that

$$card(S) = card(S_1) + card(S_2) - card(S_1 \cap S_2).$$

Generally, for finite sets, we have the following Addition Principle.

Theorem 2.2. Addition Principle

Suppose that sets A and B are two finite sets. Then

$$card(A \cup B) = card(A) + card(B) - card(A \cap B).$$

A quite interesting and challenging question is: For a given finite set, how many subsets does it have? We need to learn more counting principles and will give the answer in Section 2.2.4.

Note *If a set has infinite elements, we call this set an infinite set. The way to define the cardinality for an infinite set is different. This will be taught in a college analysis course, in which, we will learn that there are countable infinite sets, as well as uncountable sets.*

2.2.2 Multiplication Principle

Let us start with the following example.

Example 2.8 Consider direct flights from three cities in China to five cities in the USA. Three cities in China are Beijing (using symbol C_1 to represent it), Shanghai (using symbol C_2), and Hongkong (using symbol C_3); Five cities in the USA are NYC (using U_1), Boston (using U_2), Chicago (using symbol U_3), San Francisco (using symbol U_4) and Los Angles (using symbol U_5).

We first check the flights with NYC as the destination: there are three flights: Beijing to NYC (using symbol C_1U_1), Shanghai to NYC (using symbol C_2U_1), and Hongkong to NYC (using symbol C_3U_1). In summary, since there are three different starting locations, there are three different choices.

On the other hand, there are five different destinations for choosing from, we thus have the following arrangement set:

$$\{C_jU_k \mid j = 1, 2, 3; k = 1, 2, 3, 4, 5\}$$

Easy to see: there are $3 \times 5 = 15$ elements.

□

> **Theorem 2.3. Multiplication Principle**
>
> If $A = \{A_j \mid j = 1, \cdots J\}$ and $B = \{B_k \mid k = 1, \cdots, K\}$ are two finite sets, where $J = card(A)$, $K = card(B)$, then the following arrangement set
>
> $$R_{AB} = \{A_jB_k \mid j = 1, \cdots, J; k = 1, \cdots, K\}$$
>
> has total $JK = card(A) \cdot card(B)$ elements. That is:
>
> $$card(R_{AB}) = card(A) \cdot card(B).$$

Example 2.9 The set of one-digit numbers $\{0, 1, 2, 3, 4, 5, 6, 7, 8, 9\}$ has ten elements. So there are 10 different one-digit numbers.

□

Example 2.10 Similarly, we can list all two-digit numbers:

$$\{10, 11, \cdots, 98, 99\}$$

There are total of 90 elements. So there are 90 different two-digit numbers.

Example 2.11 We can also use Multiplication Principle to count how many there are two-digit numbers. First, we can choose any number from 1 to 9 at the tens place. We have 9 choices; then we can choose any number from 0 to 9 at the ones place. There are 10 choices. Using the Multiplication Principle, we totally have

$$9 \times 10 = 90$$

different two-digit numbers.

2.2.3 Permutation

We consider the following practical problem.

Example 2.12 Suppose: there are 10 athletes in a game. How many possible ways are there to choose 1st place, 2nd place, and 3rd place winners?

For the first place, we have ten athletes to choose from: 10 choices.

For the second place, we have nine athletes left to choose from (since one is chosen for the first place): 9 choices.

For the third place, we have eight athletes left to choose from (since two are chosen for the first and the second places): 8 choices.

Totally, we have

$$10 \times 9 \times 8 = 720$$

different arrangements.

> **Theorem 2.4. Permutation**
> *If we choose ordered k elements from a finite set $A = \{A_j \mid j = 1, \cdots n\}$, there are*
> $$A_n^k = n \times (n-1) \times \cdots \times (n-k+2) \times (n-k+1)$$
> *different arrangement. We call A_n^k the "k permutation of n".*

For $k = n$, we introduce a new notation $n!$ (read: n factorial):

$$n! = n \times (n-1) \times \cdots \times 3 \times 2 \times 1.$$

For consistency, we use the convention: $0! = 1$.

Example 2.13 Prove

$$A_n^k = nA_{n-1}^{k-1}.$$

Proof: We show two ways to verify the above identity.

Method one: Using the definition of the permutation (for A_n^k). First, we choose one element from n elements. There are n choices. Then, in the rest $n-1$ elements, we can do "$n-1$ choosing $k-1$" ordered arrangement, and we have total A_{n-1}^{k-1} choices. Using multiplication principle, we thus have nA_{n-1}^{k-1} ordered arrangements for "n choosing k", thus obtain the desired equality.

Method two: We use algebraic calculation (roughly speaking: we can use a machine to verify the identity).

$$\begin{aligned} A_n^k &= n \times (n-1) \times \cdots \times (n-k+2) \times (n-k+1) \\ &= n \times [(n-1) \times \cdots \times (n-k+2) \times (n-k+1)] \\ &= nA_{n-1}^{k-1}. \end{aligned}$$

□

Example 2.14 Assume we arrange three girls and four boys sitting in a row. We ask:

(1) How many different arrangements do we have?

(2) If we want to sit girls together, and boys sitting together, then how many different arrangements do we have?

Solution:

(1) Permutation for 7 students is:

$$A_7^7 = 5040.$$

So, we have total 5040 different arrangements to sit 7 students.

(2) There are two ordered arrangements for group girl and group boy (which group sits on the left). Then, for group boy, there are permutations of 4 students, and for group girl there are permutations of 3 students. So, by multiplication principle, totally we have

$$2 \cdot (A_4^4 \cdot A_3^3) = 288.$$

different arrangements.

□

2.2.4 Combination

We consider the following example (very similar to Example 2.12).

Example 2.15 If we choose 3 athletes from 10 athletes to compete in a game. How many choices do we have?

Discussion and Solution: It looks like that we are choosing the 1st place, the 2nd place and the 3rd place athletes again. But the main difference between this example and Example 2.12 is that: here we really do not care who is the first, the second and the third! Once we choose the top three persons, this will be counted as one choice. Using Theorem 2.4, we know $3! = 6$ permutations among the top three athletes. Thus, we have total

$$\frac{A_{10}^3}{3!} = \frac{10 \times 9 \times 8}{3 \times 2 \times 1} = 120$$

arrangement to send the top three athletes for a game!

More generally, we have

Theorem 2.5. Combination
If we choose k elements (without considering the order) from a finite set $A = \{A_j \mid j = 1, \cdots n\}$, there are

$$C_n^k = \frac{A_n^k}{k!} = \frac{n \times (n-1) \times \cdots \times (n-k+2) \times (n-k+1)}{k \times (k-1) \times \cdots \times 2 \times 1}$$

different arrangements. We call C_n^k the "k of n combination".

From $0! = 1$, we know

$$C_n^n = C_n^0 = 1.$$

We can also observe that

$$C_n^k = \frac{n!}{(n-k)!k!} = C_n^{n-k}.$$

Example 2.16 There are ten elements in one-digit number set $M = \{0, 1, 2, 3, 4, 5, 6, 7, 8, 9\}$. So there are ten different one-digit numbers.

Example 2.17 There is one subset of $M = \{0, 1, 2, 3, 4, 5, 6, 7, 8, 9\}$ that contains no element: the empty set. And there is only one subset that is the same as set M.

Example 2.18 For any natural number k from 0 to 10, there are C_{10}^k subsets of set $M = \{0, 1, 2, 3, 4, 5, 6, 7, 8, 9\}$, which contains exactly k elements.

So

Example 2.19 Set $M = \{0, 1, 2, 3, 4, 5, 6, 7, 8, 9\}$ has total

$$C_{10}^0 + C_{10}^1 + \cdots + C_{10}^9 + C_{10}^{10}$$

subsets.

2.2 Counting

Exercise 2.5 Verify

$$C_{10}^0 + C_{10}^1 + \cdots + C_{10}^9 + C_{10}^{10} = 2^{10} = 1024.$$

We will see the above equality again once we learn the binomial formula.

The above discussion indicates the following theorem.

> **Theorem 2.6. Number of subsets to a finite set**
> For a given finite set S, there are total $2^{Card(S)}$ subsets.

Proof: We can also use the multiplication principle to prove this theorem.

Let $k = card(S)$. We first list all elements in set S:

$$S = \{s_1, s_2, \cdots, s_k\}.$$

To make any subset of S, we can do the following:

For the first element s_1, we either choose it into a subset, or not choose it into a subset. So there are 2 choices. Similarly, for any element $s_i \in S$, there are always two options: either we choose it into a subset, or we do not choose it. Thus, by the multiplication principle, there are total

$$\underbrace{2 \times 2 \times \cdots, \times 2}_{k \text{ times}} = 2^k$$

ways to make a subset. That is: there are total 2^k subsets.

□

2.2.5 Exercises after class

1. **Basic counting**

   (1) There are 55 students in A class. Among them, there are 5 students with ages older than or equal to 12; there are 52 students with ages younger than or equal to 12.

   (a) How many students with ages older than 12 are there in the class?

   (b) How many students with ages younger than 12 are there in the class?

   (2) Among the natural numbers 0 to 100 (including 0 and 100),

   (a) How many of them divisible by 2 are there?

   (b) How many of them divisible by 3 are there?

   (c) How many of them divisible by 6 are there?

   (d) How many of them divisible by 2 but not divisible by 3 are there?

   (e) How many of them divisible by 3 but not divisible by 2 are there?

   (3) Number of factors.

   (a) How many different factors does the natural number $2^4 5^3 7^5$ have?

   (b) How many different factors does the natural number 10000 have?

   (c) How many different factors that are not divisible by 20 does the natural number 10000 have?

2. **Counting subsets**. Consider a given set $S = \{A, B, C, D, E\}$.

   (1) Count the numbers of the two-element subsets. Then list them all.

   (2) How many pairs of two-elements subsets of S which do not intersect each other? Please list these subsets.

2.3 Concepts of functions

2.3.1 Algebraic expression

We usually call any expressions with letters or variables and algebraic operations "Algebraic expressions".

Recall, when we count natural numbers, we use the decimal system. For example

$$1042 = 1 \times 10^3 + 4 \times 10^1 + 2 \times 10^0.$$

In computer language (we can use 0 to represent the circuit is on, use 1 to represent it is off), we usually use the binary system. Here is the expression of the number 1024 in the binary system:

$$1042 = 1024 + 16 + 2 = 1 \times 2^{10} + 1 \times 2^4 + 1 \times 2^1 = 10000010010_2.$$

Motivated by the different base number systems, we introduced the polynomial expression in the book "Introductory Algebra":

$$c_n x^n + c_{n-1} x^{n-1} + \cdots + c_1 x + c_0,$$

where $c_0, c_1, \cdots, c_n$ are called the coefficients, n is called the order or the degree of the polynomial. For example, $3x^4 - 5x + 9$ is a 4-th order polynomial. And we call the ratio of two polynomials (unless the numerator is divisible by the denominator) a rational polynomial:

$$\frac{a_n x^n + a_{n-1} x^{n-1} + \cdots + a_1 x + a_0}{b_m x^m + b_{m-1} x^{m-1} + \cdots + b_1 x + b_0}.$$

For example,

$$\frac{x^4 + 3x^2 - 4}{x^3 + x + 3}$$

is a rational polynomial.

Using the inverse power operation, we also introduced the n-th root for any positive number x: $x^{\frac{1}{n}}$. Then, using the Fundamental Assumption of Elementary Math, we introduced the general power operation for any real number exponent α:

$$x^\alpha.$$

For example: $x^{\sqrt{2}}$. If α is not an integer, we usually call x^α an irrational expression.

Example 2.20 We define[1] $4\hat{2}1 = 4 \times 100 - 2 \times 10 + 1$, $2\hat{1} = 2 \times 10 - 1$. Compute
(1) $4\hat{2}1 \times 21$; (2) $421 \times 2\hat{1}$.

Solution: Instead of doing addition and subtraction first, then doing the multiplication,

---
[1] Hat notation "$\hat{a}$" ©Meijun Zhu and Helen Wu

here, we use the vertical form to carry out the multiplication directly. Students who are familiar with multiplication formulas can try to calculate these mentally.

(1) We did the same exercise in Exercise 1.6:

$$
\begin{array}{rrrr}
 & 4 & -2 & 1 \\
\times & & 2 & 1 \\
\hline
 & 4 & -2 & 1 \\
+ \quad 8 & -4 & 2 & \\
\hline
8 & 0 & 0 & 1.
\end{array}
$$

So $4\hat{2}1 \times 21 = 8001$.

(2) Similarly

$$
\begin{array}{rrrr}
 & 4 & 2 & 1 \\
\times & & 2 & -1 \\
\hline
 & -4 & -2 & -1 \\
+ \quad 8 & 4 & 2 & \\
\hline
8 & 0 & 0 & -1.
\end{array}
$$

So $4\hat{2}1 \times 21 = 8000 - 1 = 7999$.

$\square$

For exponents, if the powers are letters or variables, we have the exponential expression (again, we require that the base a is positive and is not 1):

$$a^x.$$

For example, 2^x, π^x, etc.

Considering the inverse exponential operation, we introduced the logarithmic operations and expressions:

$$\log_b x, \quad \ln x = \log_e x,$$

where b, x are positive, and e is the Euler constant. We call $\ln x = \log_e x$ natural logarithmic. For example, $\log_{10} x$, $\ln(2x)$, etc. are logarithmic expressions.

We also introduced algebraic expressions involving more than one variable. For example, we learned first-order, two-variable polynomial:

$$ax + by + c,$$

and second-order, two-variable polynomial (sometimes, we call this "quadratic form")

$$ax^2 + bxy + cy^2$$

The relation between a quadratic form and general inner products on the plane was discussed in Chapter 8 of the book "Introductory Geometry and Proofs".

In the book "Introductory Geometry and Proofs", we also introduced trigonometric expressions:

$$\sin x, \quad \cos x, \quad \tan x, \quad ctan x, \quad sec x, \quad csc x.$$

We will learn more about trigonometric functions in next the three chapters.

2.3.2 Mapping

For two given sets A and B, the assignment that assigns elements in set A to elements in set B is called a mapping from set A to set B. For example, consider a set A of a family

$$A = \{\text{Father, Mother, Daughter, Son}\}$$

and a set B of names

$$B = \{\text{John, Nancy, Anna, Jacob, Lillian}\}.$$

We can use a mapping to identify each family member's name: $Father \to John$, $Mother \to Nancy$, $Daughter \to Lillian$, and $Son \to Jacob$. We can also use the following Figure 2.3 to represent this mapping between two sets:

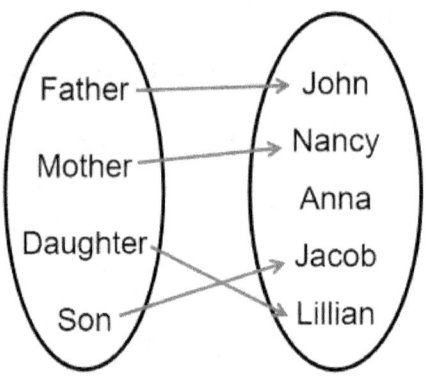

Figure 2.3: Family member mapping

We conveniently call the elements in the left side set the "inputs" of the mapping, and the ones in the right the "outputs".

It is possible that the daughter later changes her name to Anna, we then use $Father \to John$, $Mother \to Nancy$, $Daughter \to Lillian$ and $Anna$, $Son \to Jacob$ to represent this new mapping.

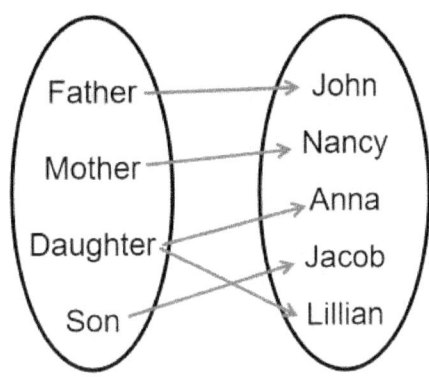

Figure 2.4: Two outputs for one input in mapping

2.3.3 Function

A function is a mapping between two number sets A and B, which assigns a number in set A (we call such a number: a "pre-image" of the function) to at most one number in set B (we call this number: an "image" of the function). The set of all pre-images is called the domain of the function, and the set of all images is called the range of the function.

The requirement for functions that we can only assign at most one number in the target set B to a number in set A guarantees that we can draw a graph for a given function in the Cartesian coordinate system.

We use the notation $s \in S$ to represent an element s in set S. Generally we use $y = f(x): x \in A \to y = f(x) \in B$ to represent a function from A to B. Set

$$R = \{f(a) : a \in A\}$$

represents the range for the function; Set

$$D = \{a : a \in A \text{ and } f(a) \in R\}$$

represents the domain for the function.

For example, if we consider a mapping from set $\mathbb{N}$ to $\mathbb{N}$ itself by assigning any natural number to its double: $x \in \mathbb{N} \to 2x \in \mathbb{N}$. We usually write the original number as x, and call it "variable"; Use y to represent the target number and call it "function value", and write the map as:

$$y = 2x, \quad \text{for all } x \in \mathbb{N}.$$

Its domain is the natural number set $\mathbb{N}$, its range is the set of all even natural numbers: $\{2n \mid n \in \mathbb{N}\}$.

Example 2.21 Find the domain and the range (in real number system) for function $y = \sqrt{x+2}$.

Solution: Under the square root, the number must be non-negative. So we know the restriction on x is
$$x + 2 \geq 0.$$
That is: $x \geq -2$. So the domain is: $\{x \in \mathbb{R} : x \geq -2\}$.

For $x = -2$, the function value is the smallest one: 0. So the range for the function is: $\{y \in \mathbb{R} : y \geq 0\}$.

□

Generally, it is relatively hard to find the range for a given function. To verify a number y_0 is in the range of function $y = f(x)$, we need to check whether there is a number x_0 in the domain so that $y_0 = f(x_0)$. Recall, if y_0 is in the range, we call y_0 an image of the function $f(x)$, and x_0 is the pre-image of y_0.

Example 2.22 Find the domain and the range (in the real number system) for the function $y = \frac{x+1}{x-1}$.

Solution: The denominator of a fraction can not be zero. So we know the restriction on x is
$$x - 1 \neq 0.$$
That is: $x \neq 1$. So the domain is: $\{x \in \mathbb{R} : x \neq 1\}$.

To find the range, we need to see: for what value of y_0, we can find its pre-image x_0. That is: for a given number y_0, solve the following equation for x_0:
$$\frac{x_0 + 1}{x_0 - 1} = y_0.$$
This equation is equivalent to
$$(y_0 - 1)x_0 = y_0 + 1.$$
Easy to see: if $y_0 = 1$, there is no solution for x_0; On the other hand, if $y_0 \neq 1$, we can always find $x_0 = \frac{y_0+1}{y_0-1}$, that is, for any number $y_0 \neq 1$, it is in the range of the function. So the range for the function is: $\{y \in \mathbb{R} : y \neq 1\}$.

□

Exercise 2.6 Find the domain and the range (in the real number system) for the function
$$y = \frac{1}{x^2 - 3x + 2}.$$

2.3.4 Graphs of functions

A picture is worth a thousand words. To "see" a function, it is the best way to "watch" its graph.

Usually, we assign a value for x, find the corresponding y value, and create a box with these values (see figure 2.5). These pairs of numbers will be the points in the coordinate system. We plot them in the system and link the dots to obtain the graph. indexGraph of function

Apparently, for each given x value, there is only one function value y corresponding to the variable by the definition of a function, thus we have no confusion in linking the dots (say, from left to right direction).

Example 2.23 Graph function $y = x + 2$.

Solution: We choose $x = 0, 1, 2, 3, 4, 6$; We obtain the corresponding y value: $y = 2, 3, 4, 5, 6, 8$.

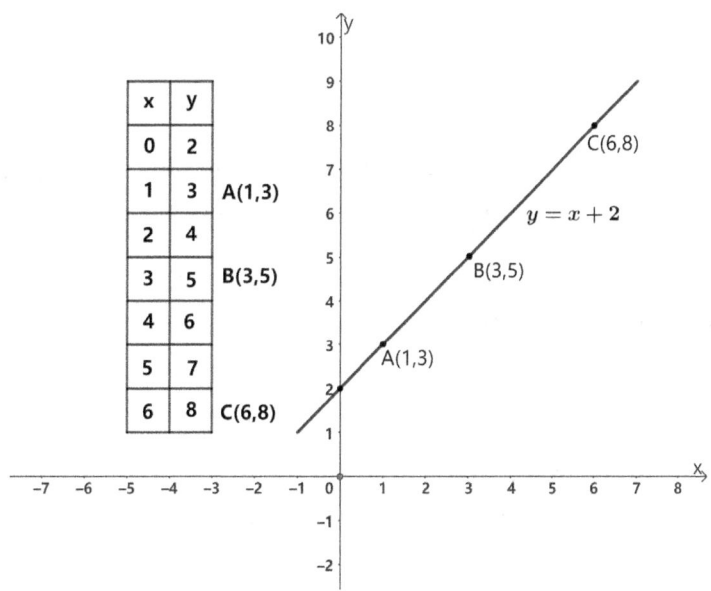

Figure 2.5: Graph of $y = x + 2$: a straight line

We connect any two points (Say, for example, $A(1, 3)$ and $B(3, 5)$; or $A(1, 3)$ and $C(6, 8)$), we obtain a straight line.

□

Example 2.24 Draw the graph for function $y = x^3$.

Solution: We choose $x = 0, 1, -1, 2, -2$; We obtain the corresponding y value: $y = 0, 1, -1, 8, -8$.

Connecting these points, we obtain the graph.

□

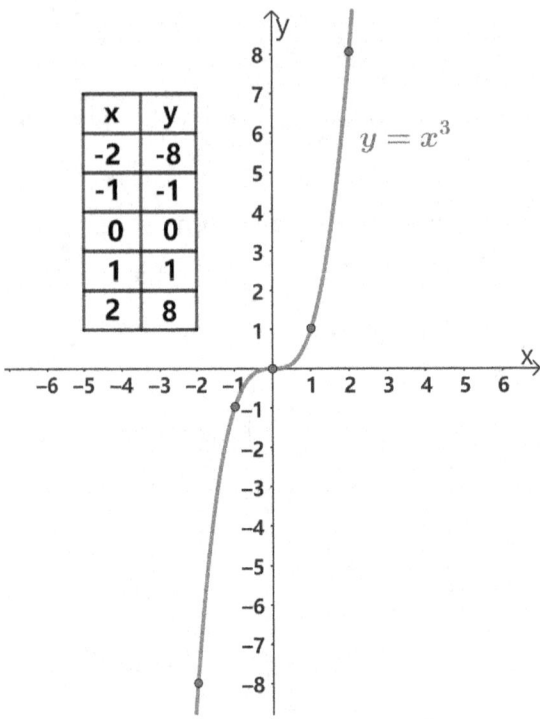

Figure 2.6: Graph of $y = x^3$

2.3.5 Translation of graph

Horizontal translation of graph. We compare graph T_1 for function $y = f(x)$ with graph T_2 for function $y = f(x + c)$. We observe that the point $(0, f(0))$ is shifted to the point $(-c, f(0))$ on graph T_2. So we can obtain graph T_2 by shifting T_1 by c unit to the left (here, we assume $c > 0$; if $c < 0$, we shift T_1 by $|c|$ unit to the right to obtain T_2).See figure 2.7.

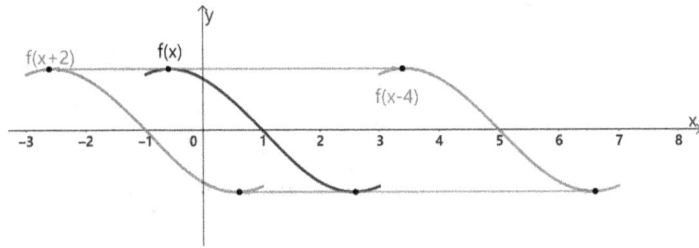

Figure 2.7: Graph of the function shifts to the right by 4 units, and Graph of the function shifts to the left by 2 units

Vertical translation of graph. We compare graph T_1 for function $y = f(x)$ with graph T_3 for function $y = f(x) + c$. We observe that the point $(0, f(0))$ is shifted to the point $(0, f(0) + c)$ on graph T_3. So we can obtain graph T_3 by shifting T_1 by c unit upward (here, we assume $c > 0$; if $c < 0$, we shift T_1 by $|c|$ unit downward to obtain T_3). See figure 2.8.

We put horizontal and vertical translations into one coordinate system to show the

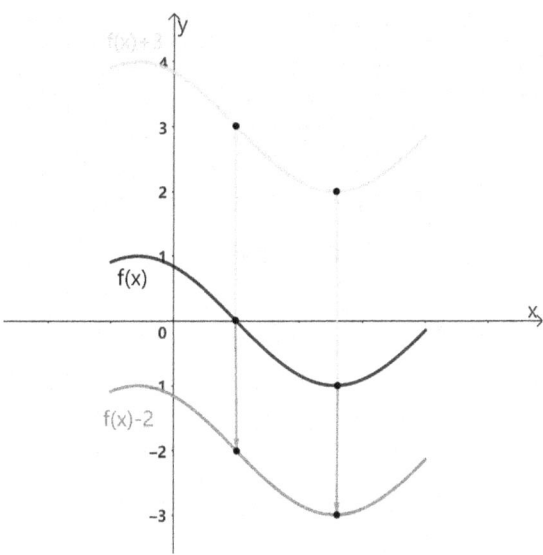

Figure 2.8: Graph of the function shifts upward by 3 units, and Graph of the function shifts downward by 2 units

relations among these graphs. See figure 2.9.

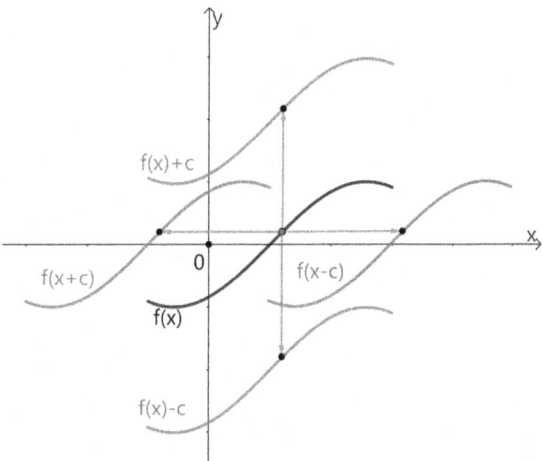

Figure 2.9: Translation of graphs

2.3.6 Concepts and properties of functions

In this section, we introduce some basic concepts and properties of functions.

Definition 2.5. Monotone functions

Let $y = f(x)$ be a function defined on an interval I.

(1). If $f(x_1) \leq f(x_2)$ for all $x_1 < x_2$ in I, we call $y = f(x)$ an increasing function; If $f(x_1) < f(x_2)$ for all $x_1 < x_2$ in I, we call $y = f(x)$ a strictly increasing function.

(2). If $f(x_1) \geq f(x_2)$ for all $x_1 < x_2$ in I, we call $y = f(x)$ a decreasing function; If $f(x_1) > f(x_2)$ for all $x_1 < x_2$ in I, we call $y = f(x)$ a strictly decreasing

function.

Example 2.25 Prove

(1) Function $f(x) = x^2 + 1$ is strictly increasing on interval $I = (0, \infty)$; (2) Function $f(x) = \frac{x-1}{x+1}$ is strictly increasing on interval $I = (0, \infty)$.

Proof: (1) For any two numbers $x_1, x_2 \in I = (0, \infty)$ and $x_1 < x_2$, by Theorem 2.1 in the book "Introductory Geometry and Proofs", we know that

$$f(x_1) = 1 + x_1^2 < f(x_2) = 1 + x_2^2.$$

So function $f(x) = 1 + x^2$ is strictly increasing on interval $I = (0, \infty)$.

(2) For any two numbers $x_1, x_2 \in I = (0, \infty)$ and $x_1 < x_2$, we have $0 < 1 + x_1 < 1 + x_2$. So, $0 < \frac{1}{1+x_2} < \frac{1}{1+x_1}$. We have

$$-\frac{2}{1+x_1} < -\frac{2}{1+x_2}.$$

Adding 1 on both sides of the above inequality, we obtain

$$1 - \frac{2}{1+x_1} < 1 - \frac{2}{1+x_2},$$

thus

$$f(x_1) = \frac{x_1 - 1}{x_1 + 1} = 1 - \frac{2}{1+x_1} < 1 - \frac{2}{1+x_2} = \frac{x_2 - 1}{x_2 + 1} = f(x_2).$$

So function $f(x) = \frac{x-1}{x+1}$ is strictly increasing on interval $I = (0, \infty)$.

□

Exercise 2.7 Prove:

(1) $f(x) = \sqrt{x}$ is a strictly increasing function on interval $I = (0, \infty)$.

(2) $f(x) = 2^x$ is a strictly increasing function on interval $I = (-\infty, \infty)$[2].

Definition 2.6. Symmetric property

Let $I = (-a, a)$ be a symmetric interval on the number line $\mathbb{R}$, and $y = f(x)$ is a function defined on I.

(1) If for all $x \in I$, $f(x) = f(-x)$, we call this function an even function. The graph of this function is mirror symmetric with respect to the y-axis[a].

(2) If for all $x \in I$, $f(x) = -f(-x)$, we call this function an odd function. The graph of this function is symmetric with respect to the origin $(0, 0)$[b].

---

[a]The rigorous definition for the graph to be mirror-symmetric with respect to the y-axis is: $(x, f(x))$ is a point on the graph if and only if $(-x, f(x))$ is a point on the graph for all $x \in I$.

---

[2]This is a hard exercise. Please check back in Section 3.3.

> [b] The rigorous definition for the graph to be symmetric with respect to the origin $(0,0)$ is: $(x, f(x))$ is a point on the graph if and only if $(-x, -f(x))$ is a point on the graph for all $x \in I$.

Example 2.26 Show (1) $y = x^2$ is an even function on $\mathbb{R}$; (2) $y = x^3$ is an odd function on $\mathbb{R}$; (3) $y = 2^x$ is not an even function, or an odd function on $\mathbb{R}$.

Proof: (1) For any $x \in \mathbb{R}$,
$$f(-x) = (-x)^2 = x^2$$
$$= f(x).$$
So $f(x) = x^2$ is an even function.

(2) For any $x \in \mathbb{R}$,
$$f(-x) = (-x)^3 = -x^3$$
$$= -f(x).$$
So $f(x) = x^3$ is an odd function.

(3) We observe: $2^{-1} \neq 2^1$, so $f(x) = 2^x$ is not an even function; Also, we observe: $2^{-1} \neq -2^1$, so $f(x) = 2^x$ is not an odd function.

$\square$

Example 2.27 Prove: if $y = f(x)$ is an odd function on interval $I = (-a, a)$, then $f(0) = 0$.

Proof: Since $f(x)$ is an odd function on $I = (-a, a)$, so for any $x \in (-a, a)$, $f(-x) = -f(x)$. In particular, for $x = 0$, $f(0) = f(-0) = -f(0)$. This yields $2f(0) = 0$. Dividing both sides by 2, we have $f(0) = 0$,

$\square$

Example 2.28 For any given function $y = f(x)$ defined on the interval $I = (-a, a)$, prove: $y = f(x) + f(-x)$ is an even function.

Proof: We write $F(x) = f(x) + f(-x)$. Then, for any $x \in (-a, a)$, $F(-x) = f(-x) + f(x) = f(x) + f(-x) = F(x)$. So $F(x)$ is an even function.

$\square$

Exercise 2.8 For any given function $y = f(x)$ defined on the interval $I = (-a, a)$, prove: there are an even function $F(x)$ and an odd function $G(x)$, both defined on the same interval I, such that, for any $x \in I$,
$$f(x) = F(x) + G(x).$$

We also introduce periodic functions.

2.3 Concepts of functions

> **Definition 2.7. Periodic function**
> Let $y = f(x)$ be a function defined on the whole number line $\mathbb{R}$. If there is a smallest positive number T such that for all $x \in \mathbb{R}$, $f(x+T) = f(x)$, we call $f(x)$ periodic function with the smallest period T.

Example 2.29 In the book "Introductory Geometry and Proofs", we studied the sine function $f(x) = \sin x$ and learned from its definition that it is a periodic function with period 2π.

□

Example 2.30 For any real number $x \in \mathbb{R}$, we define function $f(x) = [x]$— the decimal part of x (that is: $x =$ the largest integers that is less than or equal to x + $[x]$. For example, $[1.2] = 0.2$, $[\sqrt{2}] = \sqrt{2} - 1$, and $[-1.2] = 0.8$. Then $f(x) = [x]$ is a periodic function (whose period is 1).

□

Next, we introduce "one-to-one" corresponding functions.

> **Definition 2.8. One-to-one function**
> Assume that $y = f(x)$ is a function defined on interval I. If for any two different numbers $x_1 \neq x_2$ in the interval, we have $f(x_1) \neq f(x_2)$, then we call such a function a "one-to-one" function on the interval I.

From the definition, we easily understand that a strictly monotone function is a "one-to-one" function.

Exercise 2.9 Prove: A strictly increasing function on interval I must be a "one-to-one" function.

Conversely, a "one-to-one" function may not be monotone. See the following example.

Example 2.31 On interval $[0, 2]$ we define the following function:

$$f(x) = \begin{cases} x, & x \in [0, 1] \\ 4 - x, & x \in (1, 2]. \end{cases}$$

From Figure 2.10, we can see that $f(x)$ is a "one-to-one" function defined on interval $[0, 2]$, but it is not a monotonic function.

Finally, we introduce composite functions and inverse functions.

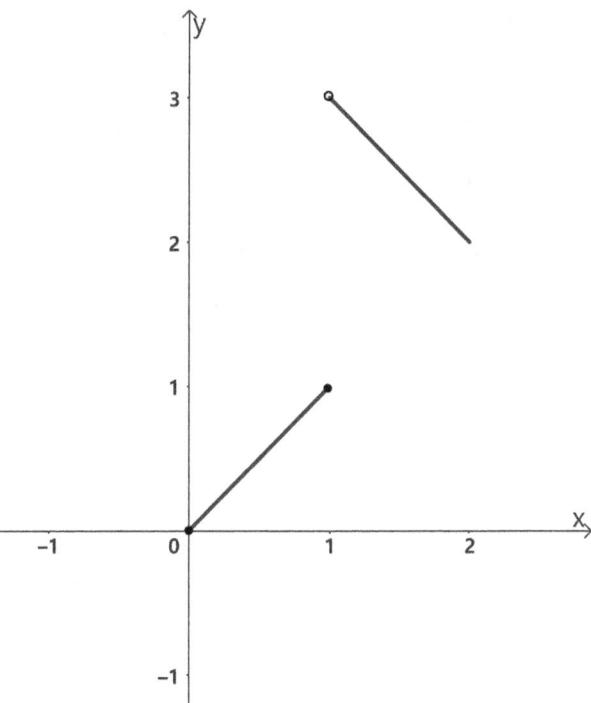

Figure 2.10: Graph of the function in Example 2.32

> **Definition 2.9. Composite function**
>
> *Suppose $f(x)$ is a function defined on interval I with range in interval J, $g(x)$ is a function defined on K which contains interval J. The composite of $g(x)$ with $f(x)$ is defined as $g[f(x)]$. We often use notation $g \circ f(x)$.*

Example 2.32 $f(x) = e^x$ is defined on $\mathbb{R}$ with range $\mathbb{R}_+ = \{a \in \mathbb{R} : a > 0\}$. $g(x) = x^2$ is defined on $\mathbb{R}$ with range $\mathbb{R}_+ = \{a \in \mathbb{R} : a > 0\}$.

So the composition of $f(x)$ with $g(x)$ is

$$g \circ f(x) = g[f(x)] = (f(x))^2 = e^{2x}.$$

The composition of $g(x)$ with $f(x)$ is

$$f \circ g(x) = f[g(x)] = e^{(g(x))} = e^{x^2}.$$

□

From the above example we see that for certain functions $f(x)$ and $g(x)$, $g \circ f(x) \neq f \circ g(x)$[3].

Using composition functions, we can introduce the inverse functions for one-to-one functions

---

[3] If we view the composition as an operation, then the commutative rule does not hold for such an operation.

2.3 Concepts of functions

> **Definition 2.10. Inverse function**
>
> Assume $f(x)$ is a one-to-one function from interval I to interval J. If $g(x)$ is a one-to-one function from interval J to interval I, satisfying:
> $$f \circ g(x) = x, \quad g \circ f(x) = x,$$
> we call $g(x)$ is the inverse function of function $f(x)$. We denote it as $g(x) = f^{-1}(x)$. Apparently, $f(x)$ is also the inverse function of $g(x)$. We use the notation $f(x) = g^{-1}(x)$.

Example 2.33 Consider $y = f(x) = x^2$ on $\mathbb{R}_+$. It is a one-to-one function from $\mathbb{R}_+$ to itself. For any given $y \in \mathbb{R}_+$, we can solve x:
$$x = \sqrt{y}, \quad \text{for any } y \in \mathbb{R}_+.$$

Traditionally, we use x for variable and y for function value. We thus define $g(x) = \sqrt{x}$ on $\mathbb{R}_+$. We verify: for $x \geq 0$,
$$\begin{aligned} g \circ f(x) &= g(x^2) \\ &= \sqrt{x^2} \\ &= x; \end{aligned}$$

and
$$\begin{aligned} f \circ g(x) &= f(\sqrt{x}) \\ &= (\sqrt{x})^2 \\ &= x. \end{aligned}$$

So $f(x)$ and $g(x)$ are inverse functions to each other.

□

> **Theorem 2.7. Inverse function of a monotonic function**
>
> Suppose $f(x)$ (from interval I to interval J) is the inverse function to $g(x)$ (from interval J to interval I). Then $f(x)$ is a strictly increasing function on I if and only if $g(x)$ is a strictly increasing function on interval J.

Proof: We only need to show: if $f(x)$ is strictly increasing, then $g(x)$ is strictly increasing.

For $x_1 < x_2$ in interval I, we need to prove
$$g(x_1) < g(x_2).$$

We prove the above by contradiction. If $g(x_1) \geq g(x_2)$, then using the monotonic

property of $f(x)$, we have
$$f[g(x_1)] \geq f[g(x_2)].$$
thus $x_1 = f[g(x_1)] \geq f[g(x_2)] = x_2$. Contradiction!

$\square$

In the next chapter we will derive radical expression functions as the inverse of power functions, and logarithmic functions as the inverse of exponential functions thus will cover all elementary functions except trigonometric functions.

2.3.7 Exercises after class

1. **Basic concept**. Graph functions
   (1) $y = 2x$;   (2) $y - 1 = 2(x + 2)$;
   (3) $y = 4x^2$;   (4) $y - 1 = -4(x + 1)^2$;
   (5) $y = \frac{1}{x}$;   (6) $y = \frac{1}{x+1}$.

2. **Concepts and properties for functions**.
   (1) Find the domains and the ranges for the following functions
   (i) $y = \sqrt{2 - x}$;   (ii) $y = \frac{x+1}{x-1}$;
   (iii) $y = x^2 - 6x + 7$;   (iv) $y = \sqrt{x^2 - 3}$;

   (2) Let $f(x) = e^x$; $g(x) = x^2 - 1$. Find
   (i) $f[g(x)]$;   (ii) $g[f(x)]$.

   (3) Consider both function $f(x) = x \ln x$ and $g(x) = e^x$ on the set $\mathbb{R}_+$. Find
   (i) $f[g(x)]$;   (ii) $g[f(x)]$.

3. **Comprehensive questions**.
   (1) Suppose that $y = f(x)$ and $y = g(x)$ are even function defined on $\mathbb{R}$. Prove: $y = f(x)g(x)$ is also an even function.

   (2) Assume the domain of $y = f(x)$ is $\mathbb{R}$, and $y = g(x)$ is an even function on $\mathbb{R}$. Prove: $y = f[g(x)]$ is an even function.

   (3) Assume the domain of $y = f(x)$ is $\mathbb{R}$. Prove: there are an even function $g(x)$ and an odd function $h(x)$ both defined on $\mathbb{R}$, such that $f(x) = g(x) + h(x)$.

   (4) Suppose both $y = f(x)$ and $y = g(x)$ are increasing function on $\mathbb{R}$. Prove: $y = f(x) + g(x)$ is an increasing function. Is $y = f(x)g(x)$ an increasing function? Why?

   (5) Suppose both $y = f(x)$ and $y = g(x)$ are increasing function on $\mathbb{R}$. Prove: $y = f[g(x)]$ is an increasing function.

   (6) Assume the domain of $y = f(x)$ is $\mathbb{R}$, and $y = g(x)$ defined on $\mathbb{R}$ is a periodic function. Prove: $y = f[g(x)]$ is a periodic function.

   (7) Suppose both $y = f(x)$ and $y = g(x)$ are periodic function on $\mathbb{R}$ Discuss: Under what condition, $y = f(x) + g(x)$ must be a periodic function? Under what condition $y = f(x) + g(x)$ can not be a periodic function?

2.4 Chapter review and exercises

2.4.1 Chapter 2 review

We summarize the main contents covered in this chapter.

We use the following set notations: $\mathbb{N}$={All natural numbers }, $\mathbb{Z}$={All integers }, $\mathbb{Q}$={All rational numbers}, $\mathbb{R}$={All real numbers }, $\mathbb{C}$={All complex numbers }.

1. Concepts and operations for sets.

Operations on sets (Proposition 2.1-2.2): For given three sets A, B and C, we have

(1) (Commutative rule):
$$A \cap B = B \cap A; \quad A \cup B = B \cup A.$$

(2) (Associative rule):
$$(A \cap B) \cap C = A \cap (B \cap C); \quad (A \cup B) \cup C = A \cup (B \cup C).$$

(3) (Distributive rule):
$$A \cap (B \cup C) = (A \cap B) \cup (A \cap C); \quad A \cup (B \cap C) = (A \cup B) \cap (A \cup C).$$

(4) (Rules on complements)
$$(A \cup B)^c = A^c \cap B^c; \quad (A \cap B)^c = A^c \cup B^c.$$

2. Counting

2.1. Addition Principle (Theorem 2.2). Suppose that sets A and B are two finite sets. Then
$$card(A \cup B) = card(A) + card(B) - card(A \cap B).$$

2.2. Multiplication Principle (Theorem 2.3). If $A = \{A_j \mid j = 1, \cdots J\}$ and $B = \{B_k \mid k = 1, \cdots, K\}$ are two finite sets, where $J = card(A)$, $K = card(B)$, then the following arrangement set
$$R_{AB} = \{A_j B_k \mid j = 1, \cdots, J; k = 1, \cdots, K\}$$
has total $JK = card(A) \cdot card(B)$ elements. That is:
$$card(R_{AB}) = card(A) \cdot card(B).$$

2.3. Permutation (Theorem 2.4). If we choose ordered k elements from a finite set $A = \{A_j \mid j = 1, \cdots n\}$, there are
$$A_n^k = n \times (n-1) \times \cdots \times (n-k+2) \times (n-k+1)$$
different arrangement. We call A_n^k the "k permutation of n".

For $k = n$, we introduce a new notation $n!$ (read: n factorial):
$$n! = n \times (n-1) \times \cdots \times 3 \times 2 \times 1.$$
And use the convention: $0! = 1$.

2.4. Combination (Theorem 2.5). If we choose k elements (without considering the order) from a finite set $A = \{A_j \mid j = 1, \cdots n\}$, there are
$$C_n^k = \frac{A_n^k}{k!} = \frac{n \times (n-1) \times \cdots \times (n-k+2) \times (n-k+1)}{k \times (k-1) \times \cdots \times 2 \times 1}.$$
different arrangements. We call C_n^k the "k of n combination".

3. Concepts and properties on functions

Concepts: Domain, Range, Monotone, Symmetric properties, periodic property.

Properties: Translations of functions; Odd and Even functions and their symmetric properties, Smallest period, Composition of two functions, inverse functions.

2.4.2 Chapter 2 test

1. Questions on set.

   (1). For given sets $A = \{2n \mid n \in \mathbb{N}\}$, $B = \{6n \mid n \in \mathbb{N}\}$, calculate $A \cap B$, $A \cup B$.

   (2). For given sets $A = \{2n \mid n \in \mathbb{N}\}$, $B = \{6n \mid n \in \mathbb{N}\}$ and $C = \{3n \mid n \in \mathbb{N}\}$, compute $A \cap B$, $A \cup B$ and $A \cap (B \cup C)$.

   (3). Suppose $card(A) = 8$, $card(B) = 7$ and $card(A \cup B) = 14$, compute $card(A \cap B)$.

   (4). If a set has a total 16 subset, what is the cardinality of this set?

   (5). If a set has more than 65 subsets but less than 130 subsets, what is the cardinality of this set? How many subsets does it have?

2. Questions on functions.

   (1). Find the domain for function $y = \sqrt{x-2} - \sqrt{3-x}$.

   (2). Find the domain for function $y = \sqrt{-x^2 + 4x - 3}$.

   (3). Find the domain and range for function $y = \frac{1}{x-1}$.

   (4). Find the domain and range for function $y = \frac{x-1}{x+3}$.

   (5). Find the domain and range for function $y = \frac{1}{x^2+4x-5}$.

   (6). Find the domain and range for function $y = \frac{1}{-x^2+4x-3}$.

   (7). Let $f(x) = x^2 + 4x + 1$ and $g(x) = \frac{1}{2}x + 2$. Find
   (i) $f(x) + g(x)$;   (ii) $f(g(x))$;
   (iii) $g \circ f(x)$;   (iv) $g \circ g(x)$.

   (8). Let $f(x) = x^2 + 4x + 1$ and $g(x) = \frac{1}{2}x + 2$. Find
   (i) $g \circ g(x)$;   (ii) $g^{-1}(x)$;
   (iii) Find the range of $f(x)$;   (iv) Find $f^{-1}(-3)$.

3. Comprehensive questions

   (1). Let $f(x) = x^2 - 5$. Find the set (it is called the 4-sublevel set of $f(x)$)
   $$\{x \in \mathbb{R} \mid 4 > f(x)\}.$$

   (2). Assume $y = f(x)$ is a strictly increasing function defined on the real number line. Show it can not be an even function.

   (3). Let $y = f(x)$ be a function defined on the real number line with period T. For any positive integer k, find the period for function $f(kx)$ and $f(\frac{x}{k})$.

(4). Let $y = f(x)$ be a function defined on the real number line. If $f(ab) = f(a)f(b)$ for all natural numbers a and b, and $f(0) \neq 0$. Find $f(2022)$.

Chapter 3 Elementary Functions

> **Introduction**
>
> - Monomial
> - Polynomial
> - Operations on polynomials
> - First-degree polynomials and the line equations
> - Second-degree polynomials and parabolas
> - Power functions
> - Coefficients and exponents
> - Exponential operations
> - Exponential functions
> - Euler constant
> - Natural exponential functions
> - Logarithmic and the operation
> - Logarithmic functions
> - Natural logarithmic functions

In this chapter, we will learn all elementary functions, except trigonometric functions (which will be covered in the next two chapters in detail). We usually refer to polynomial functions, power functions, exponential functions, logarithmic functions, trigonometric functions, and their combinations and compositions, as elementary functions.

3.1 Polynomials

Recall how we represent a number in the decimal system. For example:
$$3125_{10} = 3 \times 10^3 + 1 \times 10^2 + 2 \times 10 + 5 \times 10^0;$$
We usually omit the subscribe 10, and use 3125 to represent the above decimal number.

Look at a number in the base-5 system. For example:
$$241_5 = 2 \times 5^2 + 4 \times 5 + 1 \times 5^0 = 521;$$
where we convert it into a decimal number. Or look at a binary number
$$10101_2 = 1 \times 2^4 + 1 \times 2^2 + 1 \times 2^0 = 21;$$
where it is also converted into a decimal number. The representation of numbers in different number systems may motivate people to introduce polynomials.

3.1.1 Monomials

> **Definition 3.1. Monomial**
>
> For a given positive integer n, we define a monomial
> $$f(x) = cx^n,$$
> where $c \neq 0$ is called the coefficient, n is called the degree or the order. The domain for this monomial is $\mathbb{R}$. We call
> $$f(x) = c$$
> a constant function. We also call it a zero-degree monomial.

Example 3.1 Graph functions $y = x^3$, $y = x^2$, $y = x$ in one coordinate system, and compare them with function $y = x^{\frac{1}{2}}$ and function $y = \frac{1}{x}$.

Solution: See function graphs in Figure 3.1. Function $y = x^{\frac{1}{2}}$ is a convex function(see the definition in Section 6.1), function $y = \frac{1}{x}$ is a decreasing function.

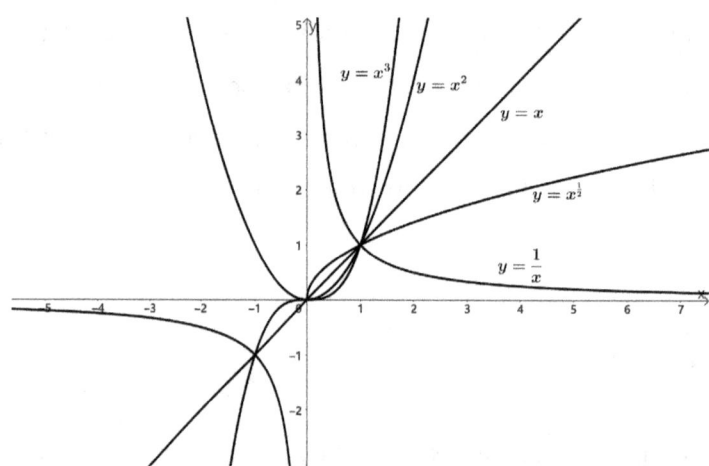

Figure 3.1: Graphs of $y = x$, $y = x^2$, $y = x^3$

Monomial functions have the following properties.

> **Proposition 3.1. Properties of monomial functions**
>
> (1) If the power n is an even number, function $f(x) = x^n$ is an even function; If the power n is an odd number, function $f(x) = x^n$ is an odd function.
>
> (2) For any positive integer n, function $y = x^n$ is a strictly increasing function on interval $I = (0, \infty)$.

Proof: (1) If $n = 2m$, where $m \in \mathbb{N}$, then
$$f(-x) = (-x)^{2m} = x^{2m} = f(x).$$
That is: $f(x) = x^{2m}$ is an even function.

If $n = 2m + 1$, where $m \in \mathbb{N}$, then
$$f(-x) = (-x)^{2m+1} = -x^{2m+1} = -f(x).$$
That is: $f(x) = x^{2m+1}$ is an odd function.

(2) We first prove $f(x) = x$ is increasing on interval $I = (0, \infty)$: For any two positive numbers $0 < x_1 < x_2$, we have
$$f(x_1) = x_1 < x_2 = f(x_2).$$
Proved.

Next, we show $f(x) = x^2$ is increasing on interval $I = (0, \infty)$: For any two positive numbers $0 < x_1 < x_2$, we have
$$x_1^2 < x_1 \cdot x_2 < x_2^2.$$
Thus
$$f(x_1) = x_1^2 < x_2^2 = f(x_2).$$
Proved.

Motivated by the above argument, we shall prove, via mathematical induction, for all $n \in \mathbb{N}$, and any two positive numbers $0 < x_1 < x_2$, we have
$$x_1^n < x_2^n.$$

For $n = 1$, it is proved.

For $n = k$, we assume
$$x_1^k < x_2^k. \tag{3.1}$$

Then, for $n = k + 1$,
$$\begin{aligned} x_1^{k+1} &= x_1^k \cdot x_1 \\ &< x_1^k \cdot x_2 \quad &\text{(Multiplication invariant for inequality)} \\ &< x_2^{k+1}. \quad &\text{(Induction assumption (3.1))} \end{aligned}$$

Proved!

$\square$

Further, we can show that for an odd natural number n, $f(x) = x^n$ is a strictly increasing function on whole number line. For an even positive natural number n, $f(x) = x^n$ is a strictly decreasing function on interval $I = (-\infty, 0)$.

3.1.2 Polynomial function

> **Definition 3.2. Polynomial function**
>
> For any positive integer n, we define polynomial function
> $$f(x) = c_n x^n + c_{n-1} x^{n-1} + \cdots + c_1 x + c_0,$$
> where $c_n \neq 0$, $c_0, c_1, \cdots, c_n$ are coefficients, the highest order/degree n is called the order/degree of the polynomial function. The domain of the polynomial function is $\mathbb{R}$.

Example 3.2 (1). $f(x) = x^2 - x + 5$ is a second degree polynomial function; (2). $f(x) = x^3 - 4x + 3$ is a 3rd degree polynomial function.

□

3.1.2.1 First degree polynomial $y = ax + b$

In the Cartesian coordinate system, the graph of function $y = ax + b$ is a straight line (we can use similar triangles to prove this fact).

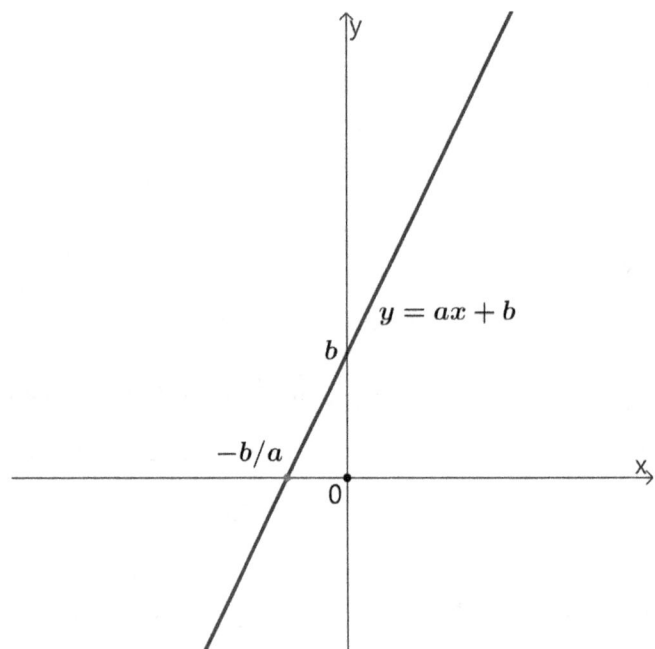

Figure 3.2: Graph of $y = ax + b$ and intercepts

If $x = 0$, $y = b$. We call the y-intercept is b. Here a is the slope of the line. In fact, we can verify: for two points on the line (x_1, y_1) and (x_2, y_2) (here $x_1 \neq x_2$), we have:
$$y_2 = ax_2 + b, \quad y_1 = ax_1 + b.$$

The first equation subtracts the second equation, we have
$$a = \frac{y_2 - y_1}{x_2 - x_1}.$$
This is the definition of the slope for the straight line.

To find *x*-**intercept**, we need to solve a first-order one-variable equation: $ax+b = 0$. For $a \neq 0$, we obtain: $x = -\frac{b}{a}$.

For $a = 0$, $y = b$ is a line parallel to *x*-axis. If $b \neq 0$, then there is no intersection between the line and *x*-axis; If $b = 0$, $y = 0$ is the *x*-axis, so any number can be the *x*-intercept.

Example 3.3 For the line function $y = 2x - 3$, find: (1) its slope; (2) *y*-intercept; (3) *x*-intercept.

Solution: The slope of the line is $a = 2$, Its *y*-intercept is -3, its *x*-intercept is $3/2$.

□

3.1.2.2 Second order polynomial function $y = ax^2 + bx + c$

In the Cartesian coordinate system, the graph of function $y = ax^2 + bx + c$ is a parabola. It opens upward for $a > 0$, and downward for $a < 0$.

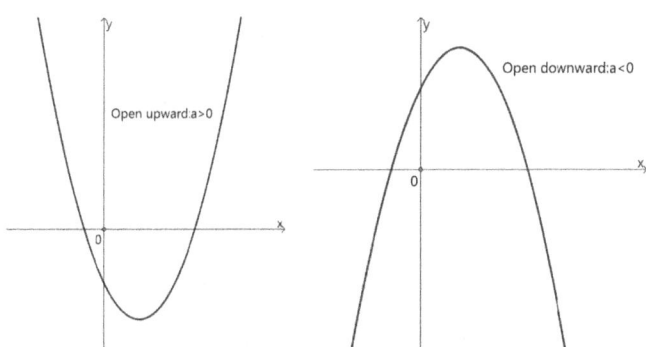

Figure 3.3: Graph of $y = ax^2 + bx + c$

By completing the square, we have
$$\begin{aligned} y &= ax^2 + bx + c \\ &= a(x^2 + \frac{b}{a}x) + c \\ &= a(x + \frac{b}{2a})^2 + c - \frac{b^2}{4a}, \end{aligned}$$
thus, the vertex of the parabola is $(-\frac{b}{2a}, c - \frac{b^2}{4a})$. At $x = -\frac{b}{2a}$, function y achieves its largest value (for $a < 0$, the parabola opens downward), or the smallest value (for $a > 0$, the parabola opens upward): $c - \frac{b^2}{4a}$.

3.1 Polynomials

We can also obtain the graph of the function $y = ax^2+bx+c$ by shifting the parabola $y = ax^2$ with the vertex $(0,0)$ shifting to $(-\frac{b}{2a}, c - \frac{b^2}{4a})$.

To find the intersection between $y = ax^2 + bx + c$ and x-axis, we need to solve the following quadratic equation:

$$ax^2 + bx + c = 0.$$

According to the discriminant $\Delta = b^2 - 4ac$, we have three cases: (1) $\Delta > 0$, there are two intersections with x-axis; (2) $\Delta < 0$, there is no intersection with x-axis; (3) $\Delta = 0$, there is one intersection with x-axis.

Example 3.4 Find the smallest value for the function $y = x^2-4x-5$, and its intersections with x-axis.

Solution: By completing the square, we have

$$\begin{aligned} y &= x^2 - 4x - 5 \\ &= x^2 - 4x + 4 - 9 \\ &= (x-2)^2 - 9, \end{aligned}$$

So the vertex of the parabola is $(2, -9)$. At $x = 2$, function y achieves its smallest value -9.

If $y = 0$,

$$(x-2)^2 - 9 = 0 \implies x_1 = 5, \ x_2 = -1.$$

So the intersection points between the parabola and x-axis are $(5,0)$ and $(-1,0)$

□

In the book "Introductory Geometry and Proofs", we studied the tangent line for a circle. Now we learn how to find a tangent line for a parabola.

Example 3.5 Find the tangent line for the parabola $y = x^2$ at $(1,1)$.

Solution: The line equation that passes through point $(1,1)$ and does not parallel to y-axis is:

$$y = m(x-1) + 1.$$

We need to find the right slope m so that there is only one intersection point between the line and the parabola[1]. That is: there is only one solution to

$$x^2 = m(x-1) + 1$$

---

[1] It appears that using "the line that interests only one point with the parabola is called the tangent line" is not accurate: In fact, any line that parallels to y-axis intersects the parabola at only one point, but they are not the tangent lines. Here, we use "any line that does not parallel to y-axis intersects the parabola at only one point" to describe the tangent line.

The above equation is equivalent to a quadratic equation

$$x^2 - mx + m - 1 = 0.$$

This equation has only one solution (a repeated root) if and only if its discriminant $\Delta = (-m)^2 - 4 \times (m - 1) = 0$. From this, we obtain $m = 2$. So the tangent line equation is

$$y = 2(x - 1) + 1. \quad \text{Or after simplification: } y = 2x - 1.$$

□

Exercise 3.1 Find the tangent line for the parabola $y = x^2 + 2$ at point $(1, 3)$.

Note *After we learn the derivative in Chapter 8, we will use a simpler method to find the tangent line for general curves.*

Using the graph of a parabola, we can easily solve quadratic inequalities and those inequalities that can be reduced into quadratic inequalities.

Example 3.6 Solve inequality

$$-x^2 - 2x + 3 > 0.$$

Solution: We first observe that there are two roots to $-x^2 - 2x + 3 = 0$: $x_1 = 1$ and $x = -3$. Graphing the parabola (see figure 3.4), we then easily see the solution to the inequality is:

$$-3 < x < 1.$$

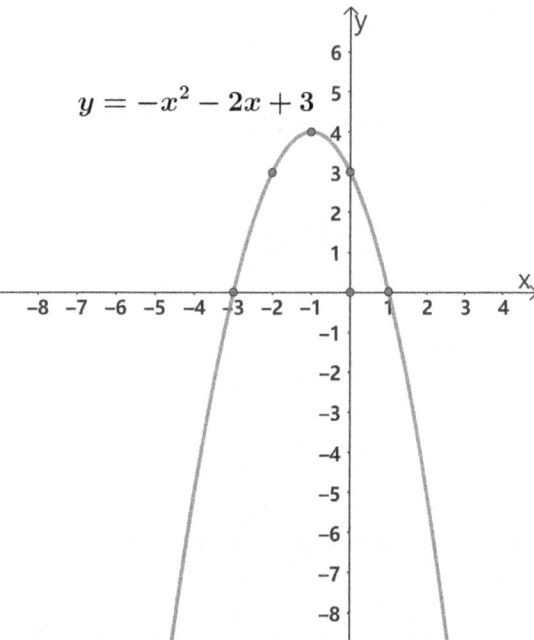

Figure 3.4: Graph of $y = -x^2 - 2x + 3$

□

Example 3.7

Solve absolute inequality

(a)
$$|x - 2| > 3.$$

(b)
$$|x + 3| \leq \frac{1}{2}.$$

Solution: Using definition of absolute value, we can show (see, for example, Question 3(4) in Section 1.4.6): for any positive number $C > 0$,

$$|x| > C \Leftrightarrow x^2 > C^2.$$

(a)
$$|x - 2| > 3 \Leftrightarrow (x - 2)^2 > 3^2$$
$$\Leftrightarrow (x - 5)(x + 1) > 0$$
$$\Leftrightarrow x > 5 \text{ or } x < -1.$$

The last step can be seen from the graph of the function $y = (x - 5)(x + 1)$.

(b).

$$|x + 3| \leq \frac{1}{2} \Leftrightarrow (x + 3)^2 \leq (\frac{1}{2})^2$$
$$\Leftrightarrow (x + \frac{5}{2})(x + \frac{7}{2}) \leq 0$$
$$\Leftrightarrow -\frac{7}{2} \leq x \leq -\frac{5}{2}.$$

The last step can be seen from the graph of the function $y = (x + \frac{5}{2})(x + \frac{7}{2})$.

□

Example 3.8 Solve the rational inequalities.

(a)
$$\frac{x + 3}{x - 1} > 0.$$

(b)
$$\frac{x + 3}{x - 1} \leq 0.$$

Solution: By multiplying a positive polynomial, we can reduce rational inequalities into polynomial inequalities.

(a) Multiplying both sides of the inequality by $(x - 1)^2$, we have the equivalent inequality:

$$(x - 1)(x + 3) > 0.$$

From figure 3.4, we know $x < -3$ or $x > 1$.

(b) Multiplying both sides of the inequality by $(x-1)^2$, for $x \neq 1$, we have the equivalent inequality:
$$(x-1)(x+3) \leq 0 \quad \text{and} \quad x \neq 1.$$

From the graph of function $y = (x-1)(x+3)$ we know $-3 \leq x < 1$.

□

3.1.3 Polynomial equations

In this section, we study the zero points to real number coefficient polynomial functions $f(x) = c_n x^n + c_{n-1} x^{n-1} + \cdots c_1 x + c_0$. That is: how to find solutions to the following polynomial equation:
$$c_n x^n + c_{n-1} x^{n-1} + \cdots c_1 x + c_0 = 0.$$

We first recall how to do long division for polynomials if the divisor is $x - a$, where a is a real number.

Example 3.9 Compute $(x^2 + 5x + 7) \div (x+1)$, and find the remainder polynomial.

Solution:

$$
\begin{array}{r}
x + 4 \\
x+1 \overline{) x^2 + 5x + 7} \\
\underline{-(x^2 + x)} \\
4x + 7 \\
\underline{-(4x + 4)} \\
3
\end{array}
$$

So the remainder for the division $(x^2 + 5x + 7) \div (x+1)$ is a constant polynomial 3.

□

It is not hard to see that the degree of the remainder polynomial is less than the degree of the divisor polynomial. So, if we consider $c_n x^n + c_{n-1} x^{n-1} + \cdots c_1 x + c_0 = 0$ divided by $x - a$, we know the degree of the remainder polynomial can only be zero. That is: it is a constant R. Therefore
$$c_n x^n + c_{n-1} x^{n-1} + \cdots c_1 x + c_0 = 0 = (x-a)(d_{n-1} x^{n-1} + d_{n-2} x^{n-2} + \cdots + d_1 x + d_0) + R.$$

Choosing $x = a$ in the above equation, we obtain

3.1 Polynomials

> **Theorem 3.1. Remainder Theorem**
>
> Let R be the remainder for polynomial function $f(x) = c_n x^n + c_{n-1} x^{n-1} + \cdots c_1 x + c_0$ divided by the first degree polynomial $x - a$, then
> $$R = f(a).$$

Example 3.10 Calculate the remainder for $(x^2 + 5x + 7) \div (x + 1)$.

Solution: Let $f(x) = x^2 + 5x + 7$. By the Remainder Theorem, we have
$$R = f(-1) = (-1)^2 + 5 \times (-1) + 7 = 3.$$

□

From Theorem 3.1 we obtain the following result.

> **Corollary 3.1. Divisibility**
>
> Consider real number coefficient polynomial function $f(x) = c_n x^n + c_{n-1} x^{n-1} + \cdots c_1 x + c_0$. For any real number a, $f(a) = 0$ if and only if $f(x)$ is divisible by $x - a$.

If $f(x)$ is divisible by $x - a$, we say that $x - a$ is a factor to polynomial $f(x)$.

Example 3.11 Verify $x^3 - 3x^2 + 2$ is divisible by $x - 1$.

Solution: Let $f(x) = x^3 - 3x^2 + 2$. Then $f(1) = 1 - 3 + 2 = 0$. By Corollary 3.1 we know that $x^3 - 3x^2 + 2$ is divisible by $x - 1$.

□

We sure can verify the above statement via the long division:

$$
\begin{array}{r}
x^2 \quad -2x \quad -2 \\
x-1 \overline{\smash{)}\; x^3 \quad -3x^2 \quad +0x \quad +2} \\
\underline{-(x^3 \quad -x^2)} \\
-2x^2 \quad +0x \\
\underline{-(-2x^2 \quad +2x)} \\
-2x \quad +2 \\
\underline{-(-2x \quad +2)} \\
0.
\end{array}
$$

It is clear that using long division is relatively tedious to verify the divisibility.

We already saw before: some equations have no solutions in the real number set (for example $x^2 + 1 = 0$); some polynomial equations have solutions, but the numbers of solutions are different to the degree of the polynomials (for example $x^3 + 1 = 0$). It seems that it is hard to have some general statements concerning the solution sets to polynomial equations. However, Gauss was able to prove the following great theorem[2].

> **Theorem 3.2. Fundamental Theorem of Algebra**
>
> Consider n-th order ($n \geq 1$) polynomial equation
> $$f(x) = c_n x^n + c_{n-1} x^{n-1} + \cdots c_1 x + c_0 = 0 \tag{3.2}$$
> where the leading coefficient is not zero (that is: $c_n \neq 0$), and all coefficients are complex numbers. There is a complex number r, such that $f(r) = 0$. That is: there is always a complex solution to equation (3.2).

From Corollary 3.1 and the Fundamental Theorem of Algebra (Theorem 3.2) we have

> **Theorem 3.3. Factorization of polynomial**
>
> Consider polynomial equation (3.2). If the leading coefficient is not zero (that is: $c_n \neq 0$), and all coefficients are complex numbers, then there are n complex numbers $x_1, x_2, \cdots, x_n$, such that
> $$f(x) = c_n x^n + c_{n-1} x^{n-1} + \cdots c_1 x + c_0 = c_n (x - x_1)(x - x_2) \cdots (x - x_n).$$

Proof: Without loss of generality, we assume $c_n = 1$.

By the Fundamental Theorem of Algebra (Theorem 3.2), there is a complex number x_1, such that $f(x_1) = 0$. By Corollary 3.1 we know: there is a $n - 1$ degree polynomial $f_{n-1}(x)$, such that
$$f(x) = (x - x_1) f_{n-1}(x).$$

Comparing the leading coefficients on both sides of the above equation, we know that the leading coefficient for f_{n-1} is also 1.

Again, by the Fundamental Theorem of Algebra, there is a complex number x_2, such that $f_{n-1}(x_2) = 0$. By Corollary 3.1 we know: there is a $n - 2$ degree polynomial $f_{n-2}(x)$ with leading coefficient 1, such that
$$f_{n-1}(x) = (x - x_2) f_{n-2}(x).$$

That is
$$f(x) = (x - x_1)(x - x_2) f_{n-2}(x).$$

---

[2]Historically, d'Alembert first gave an incomplete proof in 1746. Later, many great mathematicians tried to give proof (for example, Euler (1749), Lagrange (1772), and Laplace (1795), gave partial proof). The proof given by Gauss in 1799 is also not complete. It is Alexander Ostrowski who eventually gave a complete proof in 1920. After Chapter 4, readers shall be able to understand the proof of the Fundamental Theorem of Algebra for the second order equations.

We iterate the above procedure and finally get

$$f(x) = (x - x_1)(x - x_2) \cdots (x - x_n).$$

$\square$

Consider the following polynomial equation, whose leading coefficient is 1 and all other coefficients are complex numbers:

$$f(x) = x^n + p_{n-1}x^{n-1} + \cdots + p_1 x + p_0 = 0.$$

It has n complex numbers: $x_1, x_2, \cdots, x_{n-1}, x_n$. Thus

$$x^n + p_{n-1}x^{n-1} + \cdots + p_1 x + p_0 = (x - x_1)(x_2) \cdots (x - x_{n-1})(x - x_n).$$

Multiplying out the right side and comparing the coefficients on both sides, we obtain the following Vieta's formula:

Theorem 3.4. Vieta's formula

Suppose the following complex coefficient polynomial equation

$$f(x) = x^n + p_{n-1}x^{n-1} + \cdots + p_1 x + p_0 = 0$$

has n complex solutions: $x_1, x_2, \cdots, x_{n-1}, x_n$. Then

$$x_1 + x_2 + \cdots + x_n = -p_{n-1}$$

$$x_1 x_2 + \cdots + x_1 x_n + x_2 x_3 + \cdots x_2 x_n + \cdots + x_{n-1} x_n = (-1)^2 p_{n-2}$$

$$\cdots \cdots$$

$$x_1 x_2 \cdots x_{n-1} x_n = (-1)^n p_0$$

Example 3.12 (1) Assume $x^2 - 2x + i = 0$ has two solutions x_1 and x_2. Find $\frac{1}{x_1} + \frac{1}{x_2}$.

(2) Assume $x^3 - 4x^2 + 2ix + 4 = 0$ has three solutions x_1, x_2 and x_3. Find $\frac{1}{x_1} + \frac{1}{x_2} + \frac{1}{x_3}$.

(3) Suppose x_1, x_2, x_3, x_4 and x_5 are the five solutions to $5x^5 - 4x^4 + 3x^3 - 2x^2 + x - 1 = 0$. Find $\frac{1}{x_1} + \frac{1}{x_2} + \frac{1}{x_3} + \frac{1}{x_4} + \frac{1}{x_5}$.

Solution:

(1).
$$\frac{1}{x_1} + \frac{1}{x_2} = \frac{x_1 + x_2}{x_1 x_2}$$
$$= \frac{-(-2)}{i} \qquad \text{(Vieta's formula)}$$
$$= -2i. \qquad \text{(Simplify)}$$

(2).
$$\frac{1}{x_1}+\frac{1}{x_2}+\frac{1}{x_3}=\frac{x_2x_3+x_1x_3+x_1x_2}{x_1x_2x_3}$$
$$=\frac{2i}{-4} \qquad \text{(Vieta's formula)}$$
$$=-\frac{i}{2}. \qquad \text{(Simplify)}$$

(3). We could mimic the calculations in (1) and (2), but the computation is tedious. Here, we use the substitution to simplify the computation. Let $y=\frac{1}{x}$, then y satisfies

$$5-4y+3y^2-2y^3+y^4-y^5=0.$$

It is equivalent to the following equation with the leading coefficient 1:

$$y^5-y^4+2y^3-3y^2+4y-5=0.$$

By Vieta's formula, we have $y_1+y_2+y_3+y_4+y_5=1$. Thus

$$\frac{1}{x_1}+\frac{1}{x_2}+\frac{1}{x_3}+\frac{1}{x_4}+\frac{1}{x_5}=y_1+y_2+y_3+y_4+y_5=1.$$

□

We now consider a polynomial equation whose leading coefficient is 1 and whose other coefficients are integers:

$$x^n+c_{n-1}x^{n-1}+\cdots c_1x+c_0=0.$$

If the equation has an integer solution $x=r$, then

$$r(r^{n-1}+c_{n-1}r^{n-2}+\cdots+c_1)=-c_0.$$

So, $|r|$ must be a factor of integer $|c_0|$.

Example 3.13 Solve equation: $x^3-5x^2+4x+6=0$.

Solution: We first observe $x=3$ is a solution (3 is a factor of 6). Using long division, we obtain

$$x^3-5x^2+4x+6=(x-3)(x^2-2x-2) \qquad \text{(By long division)}$$
$$=(x-3)(x-1+\sqrt{3})(x-1+\sqrt{3}). \qquad \text{(Factorization)}$$

The original equation is reduced to

$$(x-3)(x-1+\sqrt{3})(x-1+\sqrt{3})=0.$$

We thus obtain all solutions:

$$x_1=3, \quad x_2=1-\sqrt{3}, \quad x_3=1+\sqrt{3}.$$

□

Exercise 3.2 Solve equation:

$$x^3-2x^2-x+2=0.$$

Example 3.14 Assume $x^2 + 5x + 25 = 0$, evaluate x^3.

Solution: If we solve the equation, we will obtain two complex solutions. Then it will be tedious to compute x^3. Here we use a small trick, based on the multiplication formula we learned in the book "Introductory Algebra" (see, for example, Proposition 1.6 in Chapter 1).

We first observe $x \neq 5$. We then multiply both sides of the equation by $(x - 5)$ and obtain:
$$x^3 - 5^3 = 0.$$
Thus $x^3 = 5^3 = 125$.

$\square$

3.1.4 Binomial formula

At the end of this section, we learn some special polynomials: the polynomials generated by taking the power of a first-degree binomial.

For any natural number n, we consider the function
$$f(x) = (x+1)^n = \underbrace{(x+1)(x+1)\cdots(x+1)}_{\text{n copies}}$$
$$= x^n + C_{n-1}x^{n-1} + \cdots + C_1 x + C_0.$$

It is easy to see that the constant term $C_0 = 1$ (we can also choose $x = 0$ to obtain this fact).

We use combination numbers to figure out the coefficient C_k. The coefficient for x^k is the numbers of the terms that contain x^k when we multiply the n-th power out — it actually is "taking k (with x) from n" same multiplier $(x+1)$ to have all the terms containing x^k. The number is the combination
$$C_n^k = \frac{n!}{(n-k)!k!}.$$

So
$$(x+1)^n = x^n + C_n^{n-1}x^{n-1} + \cdots + C_n^k x^k + \cdots + C_n^1 x + 1. \tag{3.3}$$

> **Proposition 3.2. Relations Among the Coefficients in Binomial Formula**
> *There are the following relations among the coefficients in Binomial Formula:*
> $$C_{n-1}^{k-1} + C_{n-1}^k = C_n^k.$$

Proof:

$$C_{n-1}^{k-1} + C_{n-1}^k = \frac{(n-1)!}{(n-k)!(k-1)!} + \frac{(n-1)!}{(n-k-1)!k!}$$
$$= \frac{(n-1)!k}{(n-k)!k!} + \frac{(n-1)!(n-k)}{(n-k)!k!}$$
$$= \frac{n!}{(n-k)!k!}$$
$$= C_n^k.$$

□

The above proposition indicates that the coefficients in the Binomial Formula form an elegant triangle shape shown below[3].

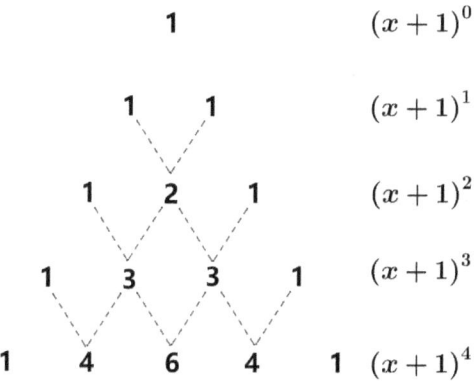

Figure 3.5: Pascal/Jian Xian's triangle

In formula (3.3), if we choose $x = 1$, we obtain

$$C_n^n + C_n^{n-1} + \cdots + C_n^1 + C_n^0 = 2^n. \tag{3.4}$$

One of the applications of the formula (3.4) in counting is to count the number of total subsets of a given finite set. Consider a finite set with n elements

$$S_n = \{E_1, E_2, \cdots, E_n\}.$$

Let k be a natural number less than or equal to n. Then, set S_n has C_n^k different subsets with k elements. So set S_n has total

$$C_n^n + C_n^1 + \cdots + C_n^1 + C_n^0 = 2^n$$

subsets. This gives another proof of Theorem 2.6 in the previous chapter.

---

[3]With a written note, we now know that this triangle was discovered by an ancient Chinese scholar: Xian Jia in the eleven century, though it is often called Pascal triangle in Western countries. Pascal discovered and used the triangle around 1653.

3.1.5 Exercises after class

1. **Basic skill**.

   (1) Find the slope, the x-intercept and y-intercept for $y - 2 = -2x - 3$.

   (2) Find the intersection points between $y = 4x^2 - 6x + 1$ and x-axis.

   (3) If parabola $y = 2x^2 - 6x + m$ does not intersect with x-axis, what is the range for m?

   (4) If parabola $y = -2x^2 - 6x + m$ does not intersect with x-axis, what is the range for m?

   (5) Solve inequality: $\frac{x}{2} - \frac{1}{3} > \frac{1}{2}$.

   (6) Solve inequality: $x^2 - 9 > 0$.

   (7) Solve inequality: $\frac{1}{x} < \frac{2}{3}$.

   (8) Solve inequality: $\frac{x-1}{x-3} < \frac{1}{2}$.

2. **Comprehensive questions**.

   (1) Suppose
   $$ax^2 + bx + c \equiv x^2 + 2x + 3$$
   Prove: $a = 1$, $b = 2$ and $c = 3$.

   (2) Solve equation:
   $$x^3 - 5x^2 - 16x + 80 = 0.$$

   (3) Consider equation:
   $$x^6 - 5x^4 - 16x^2 + 80 = 0.$$

   (i) Solve the equation in the rational number system.

   (ii) Solve the equation in the real number system.

   (iii) Find
   $$\frac{1}{x_1} + \frac{1}{x_2} + \frac{1}{x_3} + \frac{1}{x_4} + \frac{1}{x_5} + \frac{1}{x_6}.$$

   (4) Consider the polynomial $(x^2 - 1)^{10}$. Find the coefficient of x^4.

   (5) Consider polynomial $(x^2 + 4x + 4)^5$. Find the coefficient of x^4.

3.2 Power functions

3.2.1 Integer exponent power functions

For a given nonzero number k, the first-order power function is defined as (yes, we also covered this function while we were learning polynomial functions):

$$f(x) = kx$$

It is often called a proportional function. Its domain and range are $\mathbb{R}$.

For a given nonzero number k, the function

$$f(x) = \frac{k}{x}$$

is called an inversely proportional function. Its domain is $\mathbb{R} \setminus \{0\} = \{x \in \mathbb{R} \mid x \neq 0\}$, its range is $\{y \in \mathbb{R} \mid y \neq 0\}$.

For positive integer n, the power function

$$f(x) = x^n$$

has $\mathbb{R}$ as its domain. It is an strictly increasing function on interval $\mathbb{R}_+ = \{x \in \mathbb{R} \mid x > 0\}$.

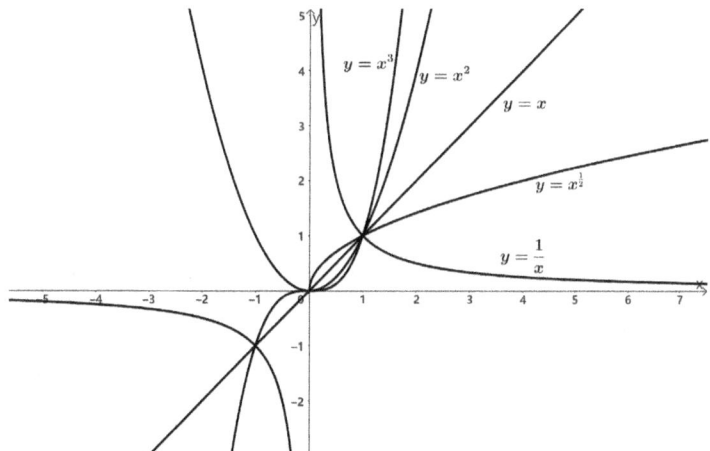

Figure 3.6: Graphs of $y = x$, $y = x^2$, $y = x^3$, $y = \frac{1}{x}$ and $y = x^{\frac{1}{2}}$.

3.2.2 Fraction exponent power functions

Let n be a positive integer. On the positive number interval $\mathbb{R}_+$, we consider the inverse function of $y = f(x) = x^n$.

First, we solve x in terms of y:

$$x = y^{\frac{1}{n}}.$$

Thus, we define its inverse function: $y = g(x) = x^{\frac{1}{n}}$.

Let us check: for any $x \in \mathbb{R}_+$,
$$g(f(x)) = g(x^n) = (x^n)^{\frac{1}{n}} = x, \quad f(g(x)) = f(x^{\frac{1}{n}}) = (x^{\frac{1}{n}})^n = x.$$

Thus, by the definition, we know that $g(x) = f^{-1}(x)$ on interval $[0, +\infty)$.

If the exponents are fractions, we call the power expressions "radical expressions". Usually, when we plug a rational number into a radical expression we obtain an irrational number. So often, we call radial expressions "irrational expressions".

When we deal with radical expressions, we often use substitution or take power to get rid of the radical expressions.

Example 3.15 Solve equations

(1)
$$\frac{\sqrt{x}+2}{5-\sqrt{x}} = \frac{4}{3}.$$

(2)
$$\sqrt{x+5} - \sqrt{x} = 1.$$

Solution: (1) We use substitution to get rid of the radical expressions. Let $a = \sqrt{x}$, then
$$\frac{\sqrt{x}+2}{5-\sqrt{x}} = \frac{4}{3} \Leftrightarrow \frac{a+2}{5-a} = \frac{4}{3}$$
$$\Leftrightarrow 3a + 6 = 20 - 4a \quad \text{and} \quad a \neq 5$$
$$\Leftrightarrow a = 2.$$

So $x = a^2 = 4$.

(2) We square the equation to get rid of the radical expressions. Caution: after we square the equation, the new equation may not be equivalent to the old equation. At the end of solving equations, we need to check the answers to avoid "extraneous solutions".

First,
$$\sqrt{x+5} - \sqrt{x} = 1 \Leftrightarrow \sqrt{x+5} = \sqrt{x} + 1.$$

We square both sides of the right equation to get
$$x + 5 = x + 2\sqrt{x} + 1.$$

Caution: this new equation is not equivalent to the original equation! Simplifying it, we obtain: $\sqrt{x} = 2$, thus $x = 4$. Bringing it to the original equation, we verify that $x = 4$ is the solution.

□

Example 3.16 Solve equation
$$\sqrt{2x^2 - 31} = x + 3.$$

Solution: From $\sqrt{2x^2 - 31} = x + 3 \implies 2x^2 - 31 = x^2 + 6x + 9 \implies x^2 - 6x - 40 = 0 \implies x = 10$ or $x = -4$.

After squaring, the above equations may not be equivalent anymore. So we need to check our answers. Bringing solutions to the original equation we verify: $x = 10$ is a solution to the original equation, but $x = -4$ is not (bringing $x = -4$ into the original equation, the right side is a negative number. Impossible!).

Answer: $x = 10$ is the only solution.

□

3.2.3 Exercises after class

1. **Basic skill**.

   (1) Graph function

   (i) $y = x^2$; (ii) $y - 1 = (x+2)^2$;

   (iii) $y = -4x^2$; (iv) $y - 1 = -4(x+1)^2$.

   (2) Domain, range, and inverse function

   (i) Find the domain, range, and the inverse function of the function $y = \sqrt{x+4}$.

   (ii) Find the domain, range, and the inverse function of the function $y = x^2 - 2x + 4$ on the interval $I = (1, \infty)$.

2. **Radical equations**.

   (1) Solve equation
   $$\sqrt{x-2} - 1 = \frac{1}{2}.$$

   (2) Solve equation
   $$1 - \sqrt[3]{5-x} = -3.$$

   (3) Solve equation
   $$x - \sqrt{x-3} = 3.$$

   (4) Solve equation
   $$x - \sqrt{x-3} = 5.$$

   (5) Solve equation
   $$\sqrt{2x-4} = \sqrt{x+5} + 1.$$

3.3 Exponential functions

If x is an integer or a fraction, we have no difficulty understanding the number 2^x. For example, we know

$$2^4 = 2 \times 2 \times 2 \times 2; \quad 2^{-4} = \frac{1}{2^4}.$$

We also know that $2^{\frac{1}{3}}$ is a positive number whose cube is 2; and $2^{\frac{2}{3}} = (2^{\frac{1}{3}})^2$. But what is $2^{\sqrt{2}}$?

The good news is: every irrational number can be "approximated" by a sequence of rational numbers. So, using the Fundamental Assumption of Elementary Math, we can expand the exponential operation rules for irrational powers. We repeat Theorem 1.3 here.

Proposition 3.3. General exponential operation rules

Assume a, b are two positive numbers, r and s are two real numbers. We have

$$a^r \cdot a^s = a^{r+s},$$

$$(a^r)^s = a^{rs}$$

and

$$(ab)^r = a^r b^r.$$

For a given positive number $a > 0$ (we usually do not consider $a = 1$, since this is a trivial case), the domain for the exponential function

$$y = a^x$$

is $\mathbb{R}$, and its range is positive number interval $(0, \infty)$.

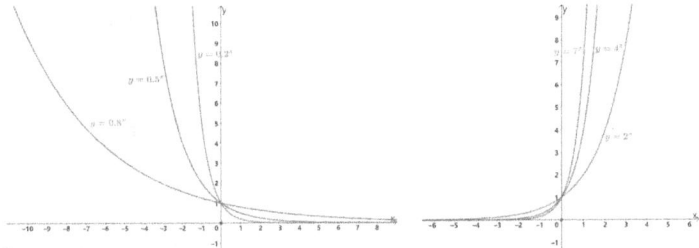

Figure 3.7: Graphs of exponential functions

For $a > 1$, we will show that $y = a^x$ is a strictly increasing function.

First, let us see the following example.

Example 3.17 Let a, b be two positive numbers. For any two positive integers m, n, prove

$$a > b \quad \text{if and only if} \quad a^{\frac{m}{n}} > b^{\frac{m}{n}}.$$

3.3 Exponential functions

Proof: We prove the statement via a few claims.

Claim 1: for $x > y > 0$, and any positive integer m, we have $x^m > y^m$.

We can use mathematical induction to prove the above claim. First, for $m = 1$, we know from the assumption that $x^1 > y^1$.

We now assume: for $m = k$, the inequality holds. That is
$$x^k > y^k.$$

Then, for $m = k + 1$
$$\begin{aligned} x^{k+1} &= x \cdot x^k \\ &> x \cdot y^k &&\text{(Induction assumption)} \\ &> y \cdot y^k &&\text{(Multiplication invariant)} \\ &= y^{k+1}. \end{aligned}$$

Claim 1 is proved.

Almost the same proof as the above (only replacing " $>$ " by " $\geq$ ") leads to the following

Claim 2: for $x \geq y > 0$, and any positive integer m, we have $x^m \geq y^m$.

Using Claim 2 we prove the next claim.

Claim 3: for $x > y > 0$, and any positive integer m, we have $x^{\frac{1}{n}} > y^{\frac{1}{n}}$.

We prove this claim by contradiction argument. If $x^{\frac{1}{n}} \leq y^{\frac{1}{n}}$, then, by Claim 2 we have $x \leq y$. Contradiction! Claim 3 is proved.

Combining Claim 1 and Claim 3, we have

Claim 4: Suppose x, y are two positive numbers. For any positive integer n,
$$x > y \Leftrightarrow x^n > y^n.$$

We now use Claim 4 to prove the statement in Example 3.17.

$$\begin{aligned} a > b &\Leftrightarrow a^{\frac{1}{n}} > b^{\frac{1}{n}} &&\text{(Using Claim 4)} \\ &\Leftrightarrow (a^{\frac{1}{n}})^m > (b^{\frac{1}{n}})^m &&\text{(Using Claim 4)} \\ &\Leftrightarrow a^{\frac{m}{n}} > b^{\frac{m}{n}}. &&\text{(Proposition 3.3)} \end{aligned}$$

□

From the above example, we know that for $a > 1$ and any positive rational number r,
$$a^r > 1.$$

Using the Fundamental Assumption of Elementary Math we know that the above in-

equality holds for any real number r. So, for any two real numbers $x_2 > x_1$,
$$\frac{a^{x_2}}{a^{x_1}} = a^{x_2-x_1} > 1.$$
Thus, for $a > 1$, $y = a^x$ is a strictly increasing function.

Exercise 3.3 Prove: for $a < 1$, $y = a^x$ is a strictly decreasing function. (**Hint:** you can either use the above result for $a > 1$ or use the same argument to prove this statement.)

We summarize the above discussion and the exercise into the following theorem.

> **Theorem 3.5. Monotonic property of exponential functions**
> For $a > 1$, $y = a^x$ is a strictly increasing function; For $a < 1$, $y = a^x$ is a strictly decreasing function.

One of the applications for exponentiation is to calculate the interests of deposits. We consider the so-called continuous compound interest: Once you deposit the money into your bank account, you can withdraw at anytime with interest (online banking makes the dream to become true!).

Example 3.18 Suppose that we deposit ten thousand dollars into a bank with an annual interest rate 2%. Then, after one year, we will have $10000 + 10000 \times 2\% = 10200$ dollar. The extra 200 dollars are the interest.

Now we are a bit greedy. Since the annual rate is 2%, so the half-year interest rate is 1%. So after half a year, we have $10000 + 10000 \times 1\% = 10100$ dollar. After another half year, we shall have $10100 + 10100 \times 1\% = 10201$ dollar! In this way, we earn extra one dollar!

Then, how about doing the above procedure every month, or every day? Will we get more and more money? Are we starting to worry that the bank may go bankrupt?

Well, we do not need to worry about the bankruptcy of a bank: we can show (after we learn calculus), that continuous compound interest has an upper bound. In fact, we can verify $(1 + \frac{1}{n})^n$ is increasing in n, but it is bounded. More specifically, we can prove (we usually learn this proof in an analysis course in college. See exercises in Section 8.2.4): as n tends to infinity, this number will tend to a number e. We write
$$(1 + \frac{1}{n})^n \to e.$$
e is called the natural constant or Euler constant.

We call the exponential function with base e "natural exponential function":
$$y = e^x.$$

Note *Euler constant was discovered by Swiss mathematician Jacob Bernoulli(1654-*

1705) when he tried to compute continuous compound interest. It is not an easy task to show that $(1 + \frac{1}{n})^n$ is increasing in n. After we learn derivative, we will come back to this question. We will also show that $(1 + \frac{1}{n})^n$ is bounded, in particular, we will see that $2 < e < 3$ (see exercises in Section 8.2.4). After we learn Taylor series, we will see a simpler formula (given by Euler) to compute e.

Example 3.19 Compare: which number is bigger between 10^9 and 9^{10}?

Solution: First, we observe
$$\frac{10^9}{9^{10}} = \frac{1}{9} \cdot \left(\frac{10}{9}\right)^9$$
$$= \frac{1}{9} \cdot \left(1 + \frac{1}{9}\right)^9.$$

Since
$$\left(1 + \frac{1}{9}\right)^9 < e < 3,$$

so
$$\frac{10^9}{9^{10}} < 3 \cdot \frac{1}{9} = \frac{1}{3}.$$

We conclude $10^9 < 9^{10}$.

3.3.1 Exercises after class

1. **Basic skills**. Graph functions
   (1) $y = 2^x$;   (2) $y - 3 = 2^{x+1}$;
   (3) $y = (\frac{1}{2})^x$;   (4) $y + 2 = (\frac{1}{2})^{x-1}$.

2. **Properties of exponential functions**.
   (1) Find the domain and range for the following functions:
   (a) $y = 3^{2-x}$;   (b) $y = e^{\frac{1}{x}}$.

   (2) Find the domain and range for the function:
   $$y = 2^x + 2^{-x}.$$

3. **Comprehensive questions**.
   (1) Prove: for any natural number n,
   $$2^n \geq n + 1.$$

   (2) Prove: for any positive integer n,
   $$2^{\frac{1}{n}} \leq 1 + \frac{1}{n}.$$

3.4 Logarithmic functions

For a given positive number $a \neq 1$, we consider the inverse function of the exponential function $y = a^x$.

We first solve x in terms of y. Using the definition of the logarithm, we know, for any positive y (in the range of the exponential function), that there is a x satisfying

$$x = \log_a y.$$

So, we define, for any positive number $a \neq 1$, the logarithmic function

$$f(x) = \log_a x. \tag{3.5}$$

The domain of this logarithmic function is $\mathbb{R}_+ = \{x \in \mathbb{R} \mid x > 0\}$, and its range is all real numbers. The graph of the function always passes through the point $(1, 0)$.

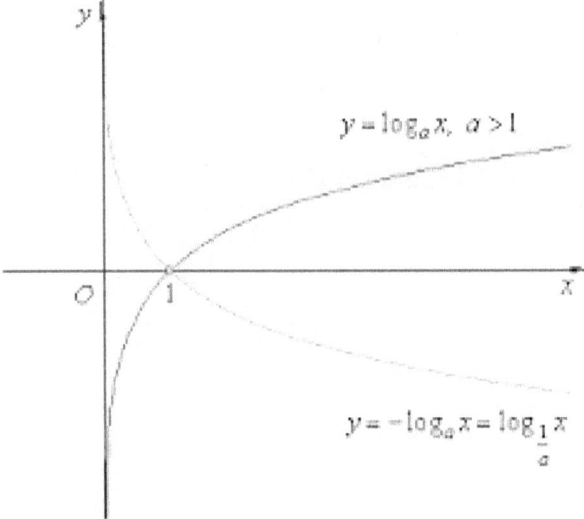

Figure 3.8: Graphs of logarithmic functions

In particular, if $a = e$, we call $f(x) = \log_e x$ the natural logarithmic function, and use the notation: $\log_e x = \ln x$.

Apparently, for fixed positive number $a \neq 1$, the exponential function and the logarithmic function are inverse functions to each other. We summarize this as the following proposition.

> **Proposition 3.4. Relation between exponential functions and logarithmic functions**
> For two positive numbers $a \neq 1$ and $x \neq 1$,
> $$a^{\log_a x} = x, \quad \log_a a^x = x.$$

Since the exponential function and the logarithmic function are inverse functions to each other, from Theorem 2.7 and Theorem 3.5 we have

> **Theorem 3.6. Monotonic properties for logarithmic functions**
>
> For $b > 1$,
> $$f(x) = \log_b x$$
> is a strictly increasing function in positive number interval $\mathbb{R}_+ = \{x \in \mathbb{R} \mid x > 0\}$.
>
> For $b < 1$,
> $$f(x) = \log_b x$$
> is a strictly decreasing function in positive number interval $\mathbb{R}_+ = \{x \in \mathbb{R} \mid x > 0\}$.

Example 3.20 Assume $a > 1$. By the definition, we know that $\log_a 1 = 0$. So, for $x > 1$,
$$\log_a x > \log_a 1 = 0.$$

□

Recall the following logarithmic operation rules:

> **Proposition 3.5. Rule for logarithmic operations**
>
> Suppose that A, B, C are three positive numbers that are not 1, and r is a real number. Then
> $$\log_C(AB) = \log_C A + \log_C B,$$
> $$\log_C A^r = r \log_C A,$$
> and
> $$\log_A B = \frac{\log_C B}{\log_C A}.$$

Example 3.21 From $9 < 10$, we know
$$\log_{10} 9 < 1.$$
So
$$2 \log_{10} 3 = \log_{10} 9 < 1.$$
Thus
$$\log_{10} 3 < \frac{1}{2}.$$

□

Example 3.22 Solve equations: (1) $\log_3(x^2 - 2x) = 1$; (2) $\log_6 x + \log_6(x+5) = 1$.

Solution: (1)
$$\log_3(x^2 - 2x) = 1 \Leftrightarrow x^2 - 2x = 3$$
$$\Leftrightarrow x^2 - 2x - 3 = 0$$
$$\Leftrightarrow x = 3 \text{ or } x = -1.$$

(2) If $x > 0$ and $x + 5 > 0$, then
$$\log_6 x + \log_6(x+5) = 1 \Leftrightarrow \log_6[x(x+5)] = 1$$
$$\Leftrightarrow x^2 + 5x - 6 = 0$$
$$\Leftrightarrow x = 1 \text{ or } x = -6.$$

Thus the solution to the equation is $x = 1$. ($x = -6$ is an "extraneous solution".)

$\square$

By now, we introduced all elementary functions except the trigonometric functions.

3.4.1 Exercises after class

1. **Basic skills.** Graph functions
   (1) $y = 2x$; (2) $y - 1 = 2(x + 2)$;
   (3) $y = 4x^2$; (4) $y - 1 = -4(x + 1)^2$;
   (5) $y = \frac{1}{x}$; (6) $y = \frac{1}{x+1}$.

2. **Properties of functions.**
   (1) Find the domain and range for the following functions:
   (a) $y = \log_2 \sqrt{x^2 - 2x + 5}$; (b) $y = \log_{10} \frac{1}{x^2 - 4x + 16}$.

   (2) Find the domain and range for the following functions:
   (a) $y = x^2 - 6x + 7$; (b) $y = \sqrt{x^2 - 3}$.

   (3) Assume $f(x) = e^x$; $g(x) = x^2 - 1$. Find
   (a) $f[g(x)]$; (b) $g[f(x)]$.

   (4) For $x \geq 0$, consider functions $f(x) = x \ln x$ and $g(x) = e^x$. Find
   (a) $f[g(x)]$; (b) $g[f(x)]$.

3. **Comprehensive questions.**
   (1) Solve equation: $\log_2(x - 1) = 2$.

   (2) Solve equation: (a) $2\log_2(x - 1) = 1$; (b) $\log_2(x - 1)^2 = 1$.

   (3) Solve equation: (a) $\ln[(x - 1)(x + 2)] = \ln 4$; (b) $\ln(x - 1) + \ln(x + 2) = \ln 4$.

   (4) Solve equation: (a) $\log_{x-1} 9 = 2$; (b) $\log_{(x-1)^2} 9 = 1$.

   (5) For any two positive number x, y, prove
   $$\ln \frac{x + y}{2} \geq \frac{\ln x + \ln y}{2}.$$

   (6) For any three positive number x, y, z, prove
   $$\ln \frac{x + y + z}{3} \geq \frac{\ln x + \ln y + \ln z}{3}.$$

3.5 Extra reading: Fundamental Theorem of Algebra for quadratic equations

We first show how solving quadratic equations inspires us to expand number fields.

Using the basic properties of equality, we can derive the following rule: For any two number x and y,
$$x^2 = y^2 \Leftrightarrow x = y \text{ or } x = -y.$$

This result essentially follows from the more fundamental rule:
$$xy = 0 \Leftrightarrow x = y \text{ o } rx = -y.$$

After completing the square, we can classify all quadratic equations into three types.

Type 1: Equations with only rational solutions.

Type 2: Equations with irrational solutions, requiring us to expand the number system from the rational numbers $\mathbb{Q}$ to the real numbers $\mathbb{R}$.

Type 3: Equations with complex solutions, requiring a further expansion from $\mathbb{R}$ to the complex numbers $\mathbb{C}$.

Type 1. Equations with only rational solutions.

Example 3.23 Solve the equation: $x^2 = 1$.

Solution: Using the rules above:
$$x^2 = 1 \Leftrightarrow x^2 = 1^2 \Leftrightarrow x = 1 \text{ or } x = -1,$$

We have $x = 1$, or $x = -1$.

$\square$

Type 2: Equations with irrational solutions.

Example 3.24 Solve the equation: $x^2 = 2$.

Before we learn how to solve this equation, we must ask: Are there any numbers whose square is 2? Mathematically, we can show that no integer or rational number (fraction) satisfies this condition. Therefore, to solve the equation, we must expand our number system by introducing a new number: the positive square root of 2, denoted $\sqrt{2}$, which satisfies
$$(\sqrt{2})^2 = 2.$$

Now we can solve the equation.

Solution:
$$x^2 = 2 \Leftrightarrow x^2 = (\sqrt{2})^2 \Leftrightarrow x = \sqrt{2} \text{ or } x = -\sqrt{2}.$$

Since $\sqrt{2}$ is not a fraction, we call it an irrational number. Solving this type of quadratic equations requires us to expand the number system from the rational numbers $\mathbb{Q}$ to the real numbers $\mathbb{R}$.

Type 3: Equations with complex solutions.

Example 3.25 Solve the equation: $x^2 = -1$.

Are there any real number whose square is -1?

By the properties of inequality, we know that there is no real number whose square is -1. However, this limitation inspires us to introduce a new symbol, called the imaginary unit (or imaginary symbol), denoted by i, with the defining property:
$$i^2 = -1.$$

Now we can solve the equation.

Solution:
$$x^2 = -1 \Leftrightarrow x^2 = (i)^2 \Leftrightarrow x = i \text{ or } x = -i.$$

□

In the same way, we can solve similar equations.

Example 3.26 Solve the equation: $x^2 = -4$.

Solution:
$$x^2 = -4 \Leftrightarrow x^2 = (2i)^2 \Leftrightarrow x = 2i \text{ or } x = -2i.$$

□

Any number of the form bi, where b is a real number, is called an imaginary number. So far, so good. But what happens in the next example?

Example 3.27 Solve the equation: $x^2 = i$.

We observe that there is no real number or purely imaginary number whose square is i. This leads us to introduce a broader class of numbers called complex numbers, of the form
$$a + bi,$$
where a, b are real numbers. As agreed at the beginning of our study of algebra, complex numbers must satisfy the three fundamental rules of algebra: (1) the commutative rule, (2) the associative rule, and (3) the distributive rule.

Using these rules, we can verify that
$$(\frac{\sqrt{2}}{2} + \frac{\sqrt{2}}{2}i)^2 = i.$$

With these preparations, we can solve the equation now.

Solution:
$$x^2 = i \Leftrightarrow x^2 = (\frac{\sqrt{2}}{2} + \frac{\sqrt{2}}{2}i)^2 \Leftrightarrow x = \frac{\sqrt{2}}{2} + \frac{\sqrt{2}}{2}i \text{ or } x = -\frac{\sqrt{2}}{2} - \frac{\sqrt{2}}{2}i.$$

□

For a genral quadratic equation:
$$x^2 + px + q = 0$$
After completing the square, we can reduce it to a simpler equation
$$x^2 = a + bi$$
where a, b are real numbers. We need to find two real numbers α and β, such that
$$(\alpha + \beta i)^2 = a + bi.$$
Expanding the left-hand side and equating real and imaginary parts, we get:

Real part: $\alpha^2 - \beta^2 = a$.

Imaginary part: $2\alpha\beta = b$.

In order to compute α and β we can also use the modulus of both sides:
$$\alpha^2 + \beta^2 = |\alpha + \beta i|^2 = |a + bi| = \sqrt{a^2 + b^2}.$$
From the above, we have
$$\alpha^2 = \frac{a + \sqrt{a^2 + b^2}}{2}, \quad \beta = \frac{-a + \sqrt{a^2 + b^2}}{2}.$$
Since $2\alpha\beta = b$, we can determine the signs of α and β accordingly (they have the same or opposite signs depending on the sign of b). So, for quadratic equations with complex coefficients, we can always find two complex solutions. This is exactly the Fundamental Theorem of Algebra for quadratic equations.

To wrap up, let us provide one more example to show how to solve for the roots of a quadratic equation with complex coefficients.

Example 3.28 Solve the equation: $x^2 + 2ix = 2 + i$

Solution: First, complete the square:
$$(x + i)^2 = 1 + i.$$
Let $z = x + i$, the equation is reduced to
$$z^2 = 1 + i.$$
We are looking for a complex number $\alpha + \beta i$ (where α, β are real numbers), such that
$$(\alpha + \beta i)^2 = 1 + i.$$

Expanding the left-hand side and equating real and imaginary parts, we get

Real part: $\alpha^2 - \beta^2 = 1$

Imaginary part: $2\alpha\beta = 1$

In order to compute α and β we can also use the modulus of both sides:
$$\alpha^2 + \beta^2 = |\alpha + \beta i|^2 = |1+i| = \sqrt{2}.$$

Thus,
$$\alpha^2 = \frac{\sqrt{2}+1}{2}, \quad \beta^2 = \frac{\sqrt{2}-1}{2}.$$

Since $2\alpha\beta = 1$, α and β will have the same sign. We obtain
$$\begin{cases} \alpha = \sqrt{\frac{\sqrt{2}+1}{2}} \\ \beta = \sqrt{\frac{\sqrt{2}-1}{2}}. \end{cases} \quad \text{or} \quad \begin{cases} \alpha = -\sqrt{\frac{\sqrt{2}+1}{2}} \\ \beta = -\sqrt{\frac{\sqrt{2}-1}{2}}. \end{cases}$$

That is
$$z = \sqrt{\frac{\sqrt{2}+1}{2}} + \sqrt{\frac{\sqrt{2}-1}{2}} \cdot i \quad \text{or} \quad z = -\sqrt{\frac{\sqrt{2}+1}{2}} - \sqrt{\frac{\sqrt{2}-1}{2}} \cdot i.$$

Thus the solutions to the original equation are
$$x = \sqrt{\frac{\sqrt{2}+1}{2}} + (\sqrt{\frac{\sqrt{2}-1}{2}} - 1) \cdot i \quad \text{or} \quad x = -\sqrt{\frac{\sqrt{2}+1}{2}} - (\sqrt{\frac{\sqrt{2}-1}{2}} + 1) \cdot i.$$

□

3.6 Chapter review and exercises

We covered all elementary functions (except the trigonometric functions) in this chapter.

1. Polynomial functions.

1.1. Monomials $y = x^n$

Symmetric functions: For positive integer n, $y = x^{2n}$ is an even function; $y = x^{2n+1}$ is an odd function.

Monotonic functions: For positive integer n, $y = x^n$ is a strictly increasing function on interval $(0, \infty)$.

1.2. Polynomial functions. The graph for the first-order polynomial function $y = ax + b$ is a straight line. Its slope is a, its x-interception is $-\frac{b}{a}$, y-interception is b.

The graph for the second-order polynomial function $y = ax^2 + bx + c$ is a parabola. Its intersections with x-axis are the solutions to the quadratic equation $ax^2 + bx + c = 0$ (there could be two, could be one, or could be no solutions). The graph can be used to solve quadratic inequalities.

Remainder Theorem (Theorem 3.1): Let R be the remainder for polynomial function $f(x) = c_n x^n + c_{n-1} x^{n-1} + \cdots c_1 x + c_0$ divided by the first degree polynomial $x - a$, then

$$R = f(a).$$

Fundamental Theorem of Algebra (Theorem 3.2) Consider n-th order ($n \geq 1$) polynomial equation

$$f(x) = c_n x^n + c_{n-1} x^{n-1} + \cdots c_1 x + c_0 = 0$$

where the leading coefficient is not zero (that is: $c_n \neq 0$), and all coefficients are complex numbers. There is a complex number r, such that $f(r) = 0$. That is: there is always a complex solution to the above equation.

Factorization of polynomial (Theorem 3.3) Consider polynomial equation

$$f(x) = c_n x^n + c_{n-1} x^{n-1} + \cdots c_1 x + c_0 = 0.$$

If the leading coefficient is not zero (that is: $c_n \neq 0$), and all coefficients are complex numbers, then there are n complex numbers $x_1, x_2, \cdots, x_n$, such that

$$f(x) = c_n x^n + c_{n-1} x^{n-1} + \cdots c_1 x + c_0 = c_n (x - x_1)(x - x_2) \cdots (x - x_n).$$

1.3. Binomial formula. For any natural number n,
$$(x+1)^n = \sum_{k=0}^{n} C_n^{n-k} x^{n-k} = x^n + C_n^{n-1} x^{n-1} + \cdots + C_n^1 x + C_n^0.$$

2. Power function

First, we need to get familiar with the graphs of power functions.

2.1. Integer exponent power functions. If n is a positive integer, $y = x^n$ is a strictly increasing function on interval $(0, \infty)$.

2.2. Fraction exponent power functions. If n is a positive integer, $y = x^{\frac{1}{n}}$ is a strictly increasing function on interval $(0, \infty)$ (inverse functions of increasing functions).

2.3. Radical expressions. Simplify radical expressions and solve radical equations.

3. Exponent and exponential equations

3.1. Exponential operation rules. Assume a, b are two positive numbers, r and s are two real numbers. Then
$$a^r \cdot a^s = a^{r+s},$$
$$(a^r)^s = a^{rs}$$
and
$$(ab)^r = a^r b^r.$$

3.2. Exponential functions. Graph of function $y = a^x$.

Monotonic property for $y = a^x$: if $a > 1$, $y = a^x$ is a strictly increasing function if $a < 1$, $y = a^x$ is a strictly decreasing function.

3.3. Euler constant e. Definition of e.

$(1 + \frac{1}{n})^n$ is increasing in natural number n.

4. Logarithm and logarithmic functions

4.1. Logarithmic operation rules. Suppose that A, B, C are three positive numbers that are not 1, and r is a real number. Then
$$\log_C(AB) = \log_C A + \log_C B,$$
$$\log_C A^r = r \log_C A,$$
and
$$\log_A B = \frac{\log_C B}{\log_C A}.$$

4.2. Properties of logarithmic functions.

4.2.1. Relation between exponent and logarithm: For any two positive numbers $a \neq 1$ and $x \neq 1$,
$$a^{\log_a x} = x, \quad \log_a a^x = a.$$

4.2.2. Monotonic properties of logarithmic functions (as the inverse functions of exponential functions):

For $b > 1$,
$$f(x) = \log_b x$$
is a strictly increasing function on $\mathbb{R}_+ = \{x \in \mathbb{R} \mid x > 0\}$.

For $b < 1$,
$$f(x) = \log_b x$$
is a strictly decreasing function on $\mathbb{R}_+ = \{x \in \mathbb{R} \mid x > 0\}$.

3.6.1 Chapter 3 test

1. Properties on elementary functions.

    (1) Graph functions:

    (1.1) $y = x^2$; (1.2) $y = x^2 - 4x + 3$;

    (1.3) $y = 3^x$; (1.4) $y = \log_3(x+1)$.

    (2) Find the domain and the range for the following functions:

    (2.1) $y = \sqrt{2x - x^2}$; (2.2) $y = 2^{x^2-x}$;

    (2.3) $y = \log_2(x^2 - 3x + 2)$; (2.4) $y = \log_3(2^x + 1)$.

    (3) Compute and simplify your answer.

    (3.1) $(x^2 + 1)(x^4 - x^2 + 1)$; (3.2) $(x^5 + 5x^4 + 5x^3 + 1) \div (x+1)^2$;

    (3.3) $e^{3\ln 5}$; (3.4) $\log_8 6 - \log_8 3 + \log_8 2$.

    (3.5) $\log_2(x^2 - 3x + 2) - \log_2(x - 1)$; (3.6) $\log_2(x^2 + 2) \cdot \log_{x^2+2} 4$.

2. More exercises on functions.

    (1) Solve equations:

    (1.1) $x^2 - 3x + 2 = 0$; (1.2) $x^2 - 5x + 2 = 0$;

    (1.3) $x^4 - 13x^2 + 36 = 0$; (1.4) $x^3 - 3x + 2 = 0$;

    (1.5) $\sqrt{2x+3} = x$; (1.6) $x - 3\sqrt{x+1} + 3 = 0$;

    (1.7) $2^{3x-4} = 16$; (1.8) $3^{x^2-4x} = \frac{1}{27}$;

    (1.9) $3^{2x} - 10 \times 3^x + 9 = 0$; (1.10) $\log_2(x - 1) + \log_2(x + 4) = 5$.

    (2) Solve inequalities:

    (2.1) $x^2 - 3x + 2 > 0$; (2.2) $\frac{x-\sqrt{2}}{x+\sqrt{5}} \geq 0$;

    (2.3) $x^4 - 13x^2 + 36 < 0$; (2.4) $x^3 - 3x + 2 \geq 0$;

    (2.5) $2^{3x-4} < 16$; (2.6) $3^{x^2-4x} \geq \frac{1}{27}$;

    (2.7) $\log_2(x-1) + \log_2(x+5) \leq 4$.

3. Comprehensive questions

    (1) Suppose $y = kx + b$ passes through two points $(1, 2)$ and $(-2, 1)$, then what are k, b? Does the line pass through the point $(2, 3)$?

    (2) Solve equation:
    $$\frac{7x - 2}{x^2 - 4} - \frac{2}{x - 2} = \frac{1}{x - 1}$$

(3) Solve the equation:
$$x^2 + \frac{1}{x^2} - 13 \cdot (x + \frac{1}{x}) + 38 = 0.$$

(4) If $3x^2 - mx + m = 0$ has real roots, find the range for m.

(5) Solve the equation:
$$x^{2x} - 31x^x + 108 = 0.$$

Chapter 4 Trigonometric functions

Introduction

- Sine function
- Cosine function
- Tangent function
- Cotangent function
- Pythagorean Theorem
- Addition and subtraction formulas
- Product-to-Sum formula
- Sum-to-Product formula
- Double angle formulas
- Half angle formulas
- de Moivre formula
- Euler formula
- General exponential functions
- Sine Theorem
- Cosine Theorem
- Heron-Qin's formula

4.1 Definition and graph

In the book "Introductory Geometry and Proofs", we defined all trigonometric functions via the unit circle in the Cartesian coordinate system.

Let $P(x, y)$ be a point on the unit circle, whose coordinates are (x, y), see Figure 4.1. Then
$$x^2 + y^2 = 1.$$

We define
$$\sin \alpha = y, \quad \cos \alpha = x, \quad \tan \alpha = \frac{y}{x} \text{ (for } x \neq 0\text{)} \quad \text{and} \quad \cot \alpha = \frac{x}{y} \text{ (for } y \neq 0\text{)}.$$

Due to the properties of similar triangles, we sure can introduce all trigonometric functions in any circle. For any point $P(x, y)$ which is not the origin, we define them in the following.

> **Definition 4.1. Definition of trigonometric function**
>
> *For a given point $P(x, y)$ which is not the origin, let $\angle POX = \alpha$. We define*
> $$\sin \alpha = \frac{y}{\sqrt{x^2 + y^2}}, \quad \cos \alpha = \frac{x}{\sqrt{x^2 + y^2}},$$
> $$\tan \alpha = \frac{y}{x} \text{ (for } x \neq 0\text{)} \quad \text{and} \quad \cot \alpha = \frac{x}{y} \text{ (for } y \neq 0\text{)}.$$

4.1 Definition and graph

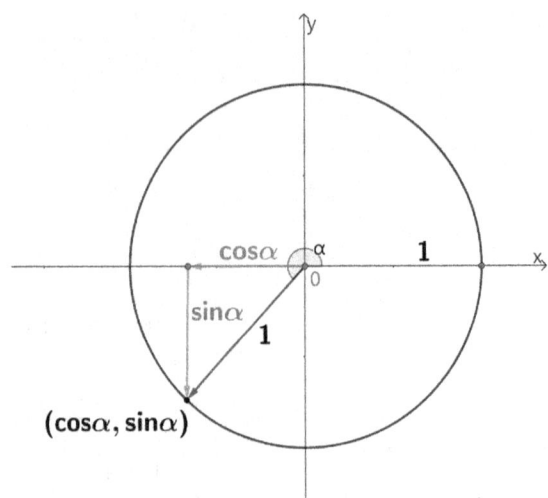

Figure 4.1: Trigonometric functions in the unit circle

From the definition, we can obtain some basic properties for trigonometric functions.

For sine function $y = \sin x$, we have

Proposition 4.1. Basic properties for sine function

Function $y = \sin x$ has the real number set as its domain. Its range is $[-1, 1]$. It is an odd function, and it is a periodic function with period 2π: $\sin(x + 2\pi) = \sin x$.

Apparently, for any radian x, we can define its sine value. So the domain of $y = \sin x$ is $\mathbb{R}$. On the circle, the sine value of a radian is the opposite of the sine value of the opposite radian. So sine function is an odd function (it is symmetric with respect to the origin). Easy to see from the definition that, the sine function has a period 2π. Finally, from inequality

$$-\sqrt{x^2 + y^2} \leq y \leq \sqrt{x^2 + y^2},$$

we know: the range of $y = \sin x$ is $[-1, 1]$. Moreover, for any integer n, $\sin(2n\pi + \frac{\pi}{2}) = 1$, $\sin(2n\pi - \frac{\pi}{2}) = -1$.

All the properties listed above for the sine function can be easily seen from its graph: Figure 4.2

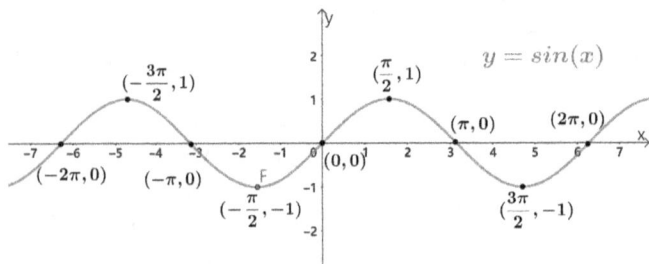

Figure 4.2: Graph of $y = \sin x$

Though $y = \sin x$ is not monotonic in the whole domain, for $x \in [-\frac{\pi}{2}, \frac{\pi}{2}]$, the sine function is a strictly increasing function.

Example 4.1 Compare the values: $\sin \frac{\pi}{3}$ and $\sin \frac{3\pi}{4}$.

Solution: By the definition, we know $\sin \frac{3\pi}{4} = \sin \frac{\pi}{4}$. Thus $\sin \frac{\pi}{3} > \sin \frac{\pi}{4} = \sin \frac{3\pi}{4}$.

$\square$

For cosine function $y = \cos x$, we have

> **Proposition 4.2. Basic properties for cosine function**
>
> Function $y = \cos x$ has the real number set as its domain. Its range is $[-1, 1]$. It is an even function, and it is a periodic function with period 2π: $\cos(x + 2\pi) = \cos x$.

Similar to the sine function, for any radian x, we can define its cosine value. So the domain of $y = \cos x$ is $\mathbb{R}$. On the circle, the cosine value of a radian is the same as the cosine value of the opposite radian. So the cosine function is an even function (it is symmetric with respect to the x-axis). Easy to see from the definition that, the cosine function has a period 2π. Finally, from inequality

$$-\sqrt{x^2 + y^2} \leq y \leq \sqrt{x^2 + y^2},$$

we know the range for $y = \cos x$ is $[-1, 1]$. For any integer n, $\cos(2n\pi) = 1$, $\sin(2n\pi + \pi) = -1$.

All the properties listed above for the cosine function can be easily seen from its graph: Figure 4.3.

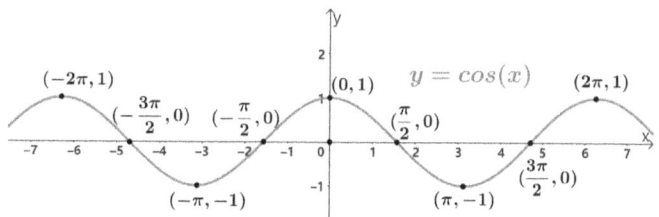

Figure 4.3: Graph of $y = \cos x$

Though $y = \cos x$ is not monotonic in the whole domain, for $x \in [0, \pi]$, the cosine function is a strictly decreasing function.

Example 4.2 Compare the values: $\cos \frac{\pi}{3}$ and $\cos(-\frac{\pi}{4})$.

Solution: Since cosine is an even function, we know $\cos(-\frac{\pi}{4}) = \cos \frac{\pi}{4}$. So $\cos(-\frac{\pi}{4}) = \cos \frac{\pi}{4} > \cos \frac{\pi}{3}$.

$\square$

For tangent function $y = \tan x$, we have

Proposition 4.3. Basic properties for tangent functions

The domain of $y = \tan x$ is $\mathbb{R} \setminus \{n\pi + \frac{\pi}{2} : n \in \mathbb{Z}\}$, the range is $(-\infty, +\infty)$. It is an odd function, as well as a periodic function with period π: $\tan(x + \pi) = \tan x$.

Observe
$$\tan x = \frac{\sin x}{\cos x},$$

and for integer $n \in \mathbb{Z}$, $\cos(n\pi + \frac{\pi}{2}) = 0$. So the domain of $y = \tan x$ is $\mathbb{R} \setminus \{n\pi + \frac{\pi}{2} : n \in \mathbb{Z}\}$. Since the product of an odd function and an even function is an odd function, so Tangent function is odd. Carefully examining the definition of the tangent function $y = \tan x$, we can see that it is a π-period function — We will give a rigorous proof of this fact via the properties of sine and cosine functions (See example 4.9 later with the help of trigonometric identity). As x gets close to $\frac{\pi}{2}$, $\tan x$ becomes larger and larger — we say the value "tends to infinity"; As x gets close to $-\frac{\pi}{2}$, $\tan x$ becomes negatively larger and larger — we say the value "tends to negative infinity". So the range of $y = \tan x$ is $(-\infty, +\infty)$.

All the properties listed above for the tangent function can be easily seen from its graph: Figure 4.4.

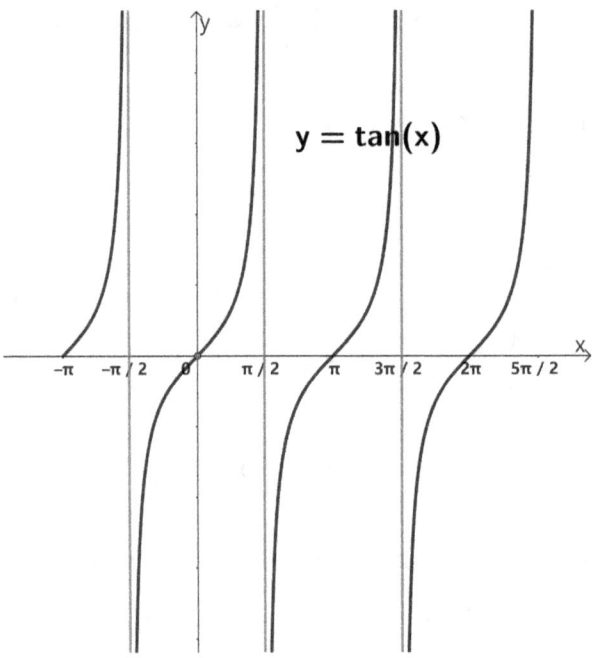

Figure 4.4: Graph of $y = \tan x$

For cotangent function $y = \cot x$, we have

4.1 Definition and graph

> **Proposition 4.4. Basic properties for cotangent functions**
>
> The domain of $y = \cot x$ is $\mathbb{R} \setminus \{n\pi : n \in \mathbb{Z}\}$, the range is $(-\infty, +\infty)$. It is an odd function, as well as a periodic function with period π: $\cot(x + \pi) = \cot x$.

Observe
$$\cot x = \frac{\cos x}{\sin x},$$
and for integer $n \in \mathbb{Z}$, $\sin(n\pi) = 0$. So the domain of $y = \cot x$ is $\mathbb{R} \setminus \{n\pi : n \in \mathbb{Z}\}$. Since the product of an odd function and an even function is an odd function, so cotangent function is odd. Carefully examining the definition of the cotangent function $y = \cot x$, we can see that it is a π-period function — Again, we will give a rigorous proof of this fact via the properties of sine and cosine functions (See example 4.9 later with the help of trigonometric identity). As x gets close to 0, $\cot x$ becomes larger and larger — we say the value "tends to infinity"; As x gets close to π, $\cot x$ becomes negatively larger and larger — we say the value "tends to negative infinity". So the range of $y = \cot x$ is $(-\infty, +\infty)$.

All the properties listed above for the cotangent function can be easily seen from its graph: Figure 4.5.

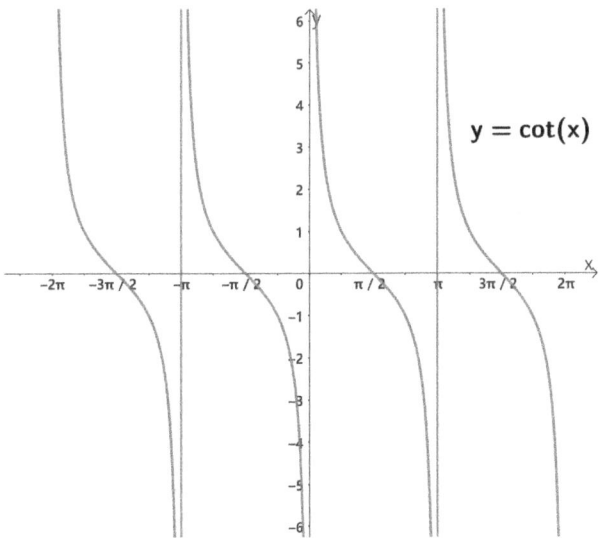

Figure 4.5: Graph of $y = \cot x$

Note Since the domains of tangent and cotangent functions are different, the tangent function is NOT the reciprocal function of the cotangent function. So we can not prove the period of the cotangent function is the same as that of the tangent function.

Example 4.3 Show that $T = \frac{2\pi}{3}$ is a period for the function $f(x) = \sin 3x$.

Solution:

$$\begin{aligned} f(x+\tfrac{2\pi}{3}) &= \sin[3(x+\tfrac{2\pi}{3})] \\ &= \sin[3x+2\pi] \\ &= \sin[3x] \\ &= f(x). \end{aligned}$$

So, from the definition, we know that $\frac{2\pi}{3}$ is a period of $f(x) = \sin 3x$.

$\square$

The signs for sine values, cosine values and tan values are displayed in the following diagram.

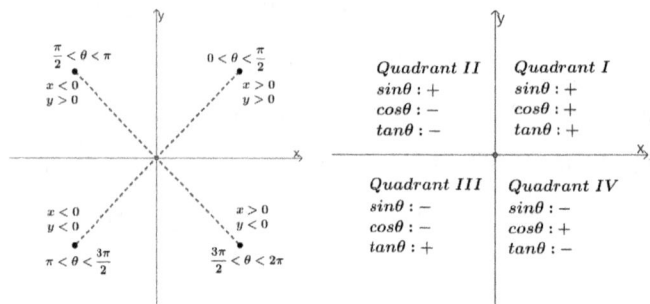

Figure 4.6: Signs of trigonometric functions in different quadrants.

4.1.1 Exercises after class

1. **Basic skills**. Find the values
   (1) $\sin \frac{2\pi}{3}$; (2) $\cos \frac{2\pi}{3}$;
   (3) $\sin \frac{31\pi}{3}$; (4) $\cos \frac{29\pi}{3}$;
   (5) $\tan(4\pi - \frac{\pi}{3})$; (6) $\cot(\frac{2\pi}{3} - 3\pi)$.

2. **Properties on trigonometric functions**.
   (1) Find the period for the following function:
   (i) $y = \sin(\frac{x}{2})$; (ii) $y = \cos(\frac{3}{4}x)$;
   (iii) $y = \tan(\frac{x}{2} + \frac{\pi}{5})$; (iv) $y = \sin x - \cos 2x$.

   (2) Check whether the following functions are odd, or even, or not symmetric at all.
   (i) $y = \sin 4x$; (ii) $y = \sin(x^2 - 4)$;
   (iii) $y = \sin x \cos x$; (iv) $y = \sin x^2 \cos x^3$.

3. **Comprehensive questions**.
   (1) Using the definitions of trigonometric functions to verify: for any radian θ,
   $$\sin^2 \theta + \cos^2 \theta = 1.$$

   (2) Prove: for any two numbers x, y,
   $$x^2 + y^2 \geq \frac{(x+y)^2}{2}.$$

   (3) Using the results obtained in part (2), can you find the range for the function $y = \sin x + \cos x$?

4.2 Trigonometric identities

We will learn various trigonometric identities in this section.

4.2.1 General Pythagorean identity

First, we have

Theorem 4.1. General Pythagorean identity
For any angle θ,
$$\sin^2\theta + \cos^2\theta = 1.$$

Proof: From the definitions for sine and cosine functions, we have
$$\sin^2\theta = \frac{y^2}{x^2+y^2},$$
and
$$\cos^2\theta = \frac{x^2}{x^2+y^2}.$$
So,
$$\sin^2\theta + \cos^2\theta = \frac{y^2+x^2}{x^2+y^2} = 1.$$

$\square$

We need to point out here: one shall not prove the classical Pythagorean identity (for angles less than $\frac{\pi}{2}$) from the above General Pythagorean Identity (for any angles). In fact, we need to use the classical Pythagorean identity to define sine and cosine functions.

Example 4.4 Assume $\theta \in (\frac{\pi}{2}, \pi)$, and $\sin\theta = \frac{\sqrt{5}}{5}$. Find the value for $\cos\theta$.

Solution: By the General Pythagorean Identity, we have
$$\cos^2\theta = 1 - \sin^2\theta$$
$$= 1 - \frac{5}{25}$$
$$= \frac{20}{25}.$$

Since angle $\theta \in (\frac{\pi}{2}, \pi)$, it is in the second quadrant, thus the cosine value for the angle is negative. We have
$$\cos\theta = -\frac{\sqrt{20}}{5} = -\frac{2\sqrt{5}}{5}.$$

$\square$

Exercise 4.1 Assume $\theta \in (\frac{3\pi}{2}, 2\pi)$, and $\cos\theta = \frac{\sqrt{3}}{3}$. Find the value for $\sin\theta$.

4.2.2 Complimentary angle identities

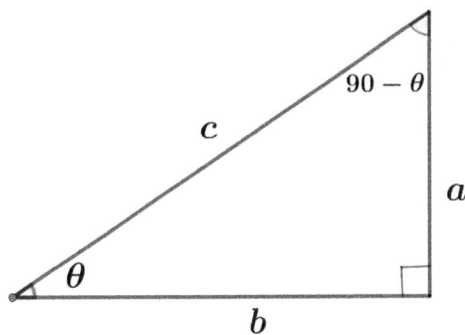

Figure 4.7: Complimentary angle identities for sine and cosine

If θ is an acute angle, we can observe from a right triangle, see Figure 4.7, that $\sin\theta = \frac{a}{c}$, $\cos\theta = \frac{b}{c}$, $\sin(90^o - \theta) = \frac{b}{c}$, and $\cos(90^o - \theta) = \frac{a}{c}$. Thus

$$\cos(\frac{\pi}{2} - \theta) = \sin\theta, \quad \sin(\frac{\pi}{2} - \theta) = \cos\theta. \tag{4.1}$$

For θ in any quadrant, we can also verify the above statement based on the location of $\frac{\pi}{2} - \theta$ in each quadrant.

Exercise 4.2 For θ in the second quadrant, prove: $\cos(\frac{\pi}{2} - \theta) = \sin\theta$.

Example 4.5 Prove: for $\theta \neq n\pi + \frac{\pi}{2}$, where $n \in \mathbb{Z}$,

$$\tan\theta + \frac{\cos\theta}{1 + \sin\theta} = \frac{1}{\cos\theta}.$$

Proof: We have

$$\tan\theta + \frac{\cos\theta}{1 + \sin\theta} = \frac{\sin\theta}{\cos\theta} + \frac{\cos\theta}{1 + \sin\theta}$$
$$= \frac{\sin\theta + \sin^2\theta + \cos^2\theta}{\cos\theta(1 + \sin\theta)}$$
$$= \frac{\sin\theta + 1}{\cos\theta(1 + \sin\theta)}$$
$$= \frac{1}{\cos\theta}.$$

$\square$

If θ is an acute angle, we can observe from a right triangle, see fig 4.7, that $\tan\theta = \frac{a}{b}$, $\cot\theta = \frac{b}{a}$, $\tan(90° - \theta) = \frac{b}{a}$, and $\cot(90° - \theta) = \frac{a}{b}$. Thus

$$\cot(\frac{\pi}{2} - \theta) = \tan\theta, \quad \tan(\frac{\pi}{2} - \theta) = \cot\theta. \tag{4.2}$$

Exercise 4.3 Derive formula (4.2) from formula (4.1).

4.2.3 Addition and subtraction formulas

The following elegant, yet hard to understand at the beginning, Addition and subtraction formulas [1] hold for sine and cosine functions.

> **Theorem 4.2. Addition and subtraction formulas-1**
> For any angles α, β, we have
> $$\sin(\alpha \pm \beta) = \sin\alpha\cos\beta \pm \cos\alpha\sin\beta, \qquad (4.3)$$
> and
> $$\cos(\alpha \pm \beta) = \cos\alpha\cos\beta \mp \sin\alpha\sin\beta. \qquad (4.4)$$

Addition and subtraction formulas are basic ones. Many other formulas can be derived from them. For instance, we will use it to derive the General Pythagorean Identity.

Example 4.6 Use addition and subtraction formulas to prove the General Pythagorean Identity.

Prove: Since $\cos 0 = 1$, we have
$$\begin{aligned} 1 = \cos 0 &= \cos(x - x) \\ &= \cos x \cos x + \sin x \sin x. \\ &= \cos^2 x + \sin^2 x. \end{aligned}$$
$\square$

We can also use addition and subtraction formulas to derive the complimentary angle identity (4.1).

Example 4.7 Use addition and subtraction formulas to derive the complimentary angle identity (4.1).

Prove
$$\begin{aligned} \cos(\frac{\pi}{2} - \theta) &= \cos\frac{\pi}{2}\cos x + \sin\frac{\pi}{2}\sin x \\ &= \sin x, \end{aligned}$$
and
$$\begin{aligned} \sin(\frac{\pi}{2} - \theta) &= \sin\frac{\pi}{2}\cos x - \cos\frac{\pi}{2}\sin x \\ &= \cos x, \end{aligned}$$

---

[1] It is not clear where this Addition formula comes from. However, the famous Ptolemy Theorem of Cyclic Quadrilateral implies this formula. We thus prefer to call this addition formula the Ptolemy addition formula.

Next, we show how to prove the addition formula (4.3) for acute angles α, β.

In a unit circle, let $\angle BOA = \alpha$ and $\angle AOC = \beta$. Passing through the point C, we draw a line that is perpendicular to line AO and intersects the AO line at point D. Passing through the point C, we draw another straight line that is perpendicular to the x-axis and intersects x-axis at point E. Finally, passing through the point D, we draw another straight line that is perpendicular to line CE and intersects CE at point F.

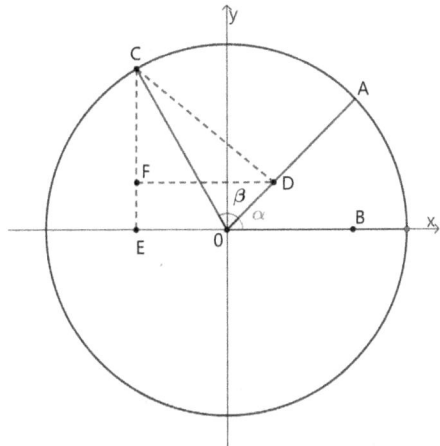

Figure 4.8: Addition formula for sine function

From the definition of trigonometric functions, we know
$$|CD| = \sin\beta, \quad |OD| = \cos\beta.$$
We use $|\cdot|$ to represent the length of a line segment. Using the property for the rectangle, we know the distance from point D to the x-axis is the same as $|FE|$. So
$$|FE| = |OD| \cdot \sin\alpha, \quad |CF| = |CD| \cdot \sin\angle CDF = |CD| \cdot \sin(\frac{\pi}{2} - \alpha) = |CD| \cdot \cos\alpha.$$
Again, from the definition of trigonometric functions, we have $\sin(\alpha + \beta) = |CE|$, thus
$$\sin(\alpha + \beta) = |CE| = |FE| + |CF|$$
$$= |OD| \cdot \sin\alpha + |CD| \cdot \cos\alpha$$
$$= \sin\alpha \cos\beta + \cos\alpha \sin\beta.$$
We thus obtain the addition formula for the sine functions with acute angles.

Example 4.8 Calculate $\cos 15°$.

4.2 Trigonometric identities

Solution: Using $\cos 45° = \sin 45° = \frac{\sqrt{2}}{2}$, $\sin 30° = \frac{1}{2}$, and $\cos 30° = \frac{\sqrt{3}}{2}$, we have

$$\cos 15° = \cos(45° - 30°)$$
$$= \cos 45° \cos 30° + \sin 45° \sin 30°$$
$$= \frac{\sqrt{2}}{2} \cdot \frac{\sqrt{3}}{2} + \frac{\sqrt{2}}{2} \cdot \frac{1}{2}$$
$$= \frac{\sqrt{6} + \sqrt{2}}{4}.$$

□

Using the substitution method, we can derive the subtraction formula from the addition formula:

$$\sin(\alpha - \beta) = \sin(\alpha + \gamma) \qquad \text{(Substitution } \gamma = -\beta\text{)}$$
$$= \sin \alpha \cos \gamma + \cos \alpha \sin \gamma \qquad \text{(Addition formula)}$$
$$= \sin \alpha \cos(-\beta) + \cos \alpha \sin(-\beta) \qquad \text{(Substitution)}$$
$$= \sin \alpha \cos \beta - \cos \alpha \sin \beta. \qquad \text{(Even and odd function properties)}$$

We can also use the addition formula for sine function and the complementary formula (4.1) to derive the addition and subtraction formulas for cosine function.

Exercise 4.4 Use formula (4.3) and formula (4.1) to prove formula (4.4).

From the addition and subtraction formula for sine and cosine, we can derive the addition and subtraction formulas for tangent and cotangent functions:

Theorem 4.3. Addition and subtraction formulas-2

For any two angles α, β,

(1) If $\alpha, \beta, \alpha \pm \beta \neq \frac{\pi}{2} + n\pi$, where $n \in \mathbb{Z}$, we have
$$\tan(\alpha \pm \beta) = \frac{\tan \alpha \pm \tan \alpha}{1 \mp \tan \alpha \tan \beta};$$

(2) If $\alpha, \beta, \alpha \pm \beta \neq n\pi$, where $n \in \mathbb{Z}$, we have
$$\cot(\alpha \pm \beta) = \frac{\cot \alpha \cot \beta \mp 1}{\cot \beta \pm \cot \alpha}.$$

Proof: (1) If $\alpha, \beta, \alpha \pm \beta \neq \frac{\pi}{2} + n\pi$,

$$\tan(\alpha \pm \beta) = \frac{\sin(\alpha \pm \beta)}{\cos(\alpha \pm \beta)}$$
$$= \frac{\sin \alpha \cos \beta \pm \cos \alpha \sin \beta}{\cos \alpha \cos \beta \mp \sin \alpha \sin \beta}$$
$$= \frac{\tan \alpha \pm \tan \beta}{1 \mp \tan \alpha \tan \beta}.$$

(2) If $\alpha, \beta, \alpha \pm \beta \neq n\pi$,

$$\cot(\alpha \pm \beta) = \frac{\cos(\alpha \pm \beta)}{\sin(\alpha \pm \beta)}$$
$$= \frac{\cos\alpha\cos\beta \mp \sin\alpha\sin\beta}{\sin\alpha\cos\beta \pm \cos\alpha\sin\beta}$$
$$= \frac{\cot\alpha\cot\beta \mp 1}{\cot\beta \pm \cot\alpha}$$

□

Note *Apparently, we can not use Theorem 4.3 to prove the complementary formula for tangent formula (4.2). However, since we already proved that* $\cos(\frac{\pi}{2} - \theta) = \sin x$ *and* $\sin(\frac{\pi}{2} - \theta) = \cos x$. *So using the definition we obtain*

$$\tan(\frac{\pi}{2} - \theta) = \frac{\sin(\pi/2 - \theta)}{\cos(\pi/2 - \theta)} = \frac{\cos\theta}{\sin\theta} = \cot\theta$$
$$\cot(\frac{\pi}{2} - \theta) = \frac{\sin(\pi/2 - \theta)}{\cos(\pi/2 - \theta)} = \frac{\sin\theta}{\cos\theta} = \tan\theta.$$

Example 4.9 Prove that π is a period for functions $y = \tan x$ and $y = \cot x$.

Proof Since $x \ne \frac{\pi}{2} + n\pi$,

$$\tan(\pi + x) = \frac{\tan\pi + \tan x}{1 - \tan\pi\tan x}$$
$$= \tan x.$$

and for $x \ne \frac{\pi}{2} + n\pi$, as well as $x \ne n\pi$,

$$\cot(\pi + x) = \frac{1}{\tan(\pi + x)}$$
$$= \frac{1}{\tan x}$$
$$= \cot x.$$

And for $x \ne \frac{\pi}{2} + n\pi$, $\cot x = \cot(\pi + x) = 0$.

□

4.2.4 Sum-to-Product, Product-to-Sum formulas

In this section, we will learn Product-to-Sum, Sum-to-Product formulas. These formulas are very useful in the future (especially important for the study of the Fourier series, the fundamental theory for the modern digital world).

We first learn the Sum-to Product formulas and their applications.

Theorem 4.4. Sum-to-Product formulas

For any angles A, B, we have

$$\sin A + \sin B = 2\sin(\frac{A + B}{2})\cos(\frac{A - B}{2}), \qquad (4.5)$$

$$\sin A - \sin B = 2\sin(\frac{A-B}{2})\cos(\frac{A+B}{2}), \quad (4.6)$$

and

$$\cos A + \cos B = 2\cos(\frac{A+B}{2})\cos(\frac{A-B}{2}), \quad (4.7)$$

$$\cos A - \cos B = -2\sin(\frac{A+B}{2})\sin(\frac{A-B}{2}), \quad (4.8)$$

Proof: Since

$$\sin A = \sin(\frac{A+B}{2} + \frac{A-B}{2}) = \sin\frac{A+B}{2}\cos\frac{A-B}{2} + \cos\frac{A+B}{2}\sin\frac{A-B}{2},$$

and

$$\sin B = \sin(\frac{A+B}{2} - \frac{A-B}{2}) = \sin\frac{A+B}{2}\cos\frac{A-B}{2} - \cos\frac{A+B}{2}\sin\frac{A-B}{2}.$$

We add both identities to get formula (4.5); We subtract one identity from another one to get formula (4.6).

Similarly,

$$\cos A = \cos(\frac{A+B}{2} + \frac{A-B}{2}) = \cos\frac{A+B}{2}\cos\frac{A-B}{2} - \sin\frac{A+B}{2}\sin\frac{A-B}{2},$$

and

$$\cos B = \cos(\frac{A+B}{2} - \frac{A-B}{2}) = \cos\frac{A+B}{2}\cos\frac{A-B}{2} + \sin\frac{A+B}{2}\sin\frac{A-B}{2}.$$

We add both identities to get formula (4.7); We subtract one identity from another one to get formula (4.8).

□

Example 4.10 Prove:
$$\frac{\sin 3x - \sin x}{\cos 3x + \cos x} = \tan x.$$

Proof:

$$\frac{\sin 3x - \sin x}{\cos 3x + \cos x} = \frac{2\sin x \cos 2x}{2\cos 2x \cos x}$$
$$= \frac{\sin x}{\cos x}$$
$$= \tan x.$$

□

Note *In the study of trigonometric functions, students will do a lot of practice on proofs, just like in the above example. We suggest that students spend a few minutes thinking over the questions. After a certain amount of training, students will more naturally work*

4.2 Trigonometric identities

on the proofs.

Example 4.11 Find the maximum and minimum values for function $f(x) = \sin x + \sin(x + \frac{\pi}{3})$.

Solution: First, let us spend five minutes reading and thinking about the question. The maximum value for each function $\sin x$ and $\sin(x + \frac{\pi}{3})$ is 1, but they are achieved at different values of variables! So this is not a simple addition problem.

After some thought, we present two methods to compute the maximum and the minimum values.

Method 1:

$$\begin{aligned}
\sin x + \sin(x + \frac{\pi}{3}) &= \sin x + \sin x \cos \frac{\pi}{3} + \cos x \sin \frac{\pi}{3} \\
&= \sin x + \frac{1}{2} \sin x + \frac{\sqrt{3}}{2} \cos x \\
&= \frac{3}{2} \sin x + \frac{\sqrt{3}}{2} \cos x \\
&= \sqrt{3}\{\frac{\sqrt{3}}{2} \sin x + \frac{1}{2} \cos x\} \\
&= \sqrt{3}\{\sin x \cos \frac{\pi}{6} + \cos x \sin \frac{\pi}{6}\} \\
&= \sqrt{3} \sin(x + \frac{\pi}{6}).
\end{aligned}$$

So the maximum value for $\sin x + \sin(x + \frac{\pi}{3})$ is $\sqrt{3}$, the minimum value is $-\sqrt{3}$.

Method 2:

$$\begin{aligned}
\sin x + \sin(x + \frac{\pi}{3}) &= 2 \sin(x + \frac{\pi}{6}) \cos(-\frac{\pi}{6}) \\
&= \sqrt{3} \sin(x + \frac{\pi}{6}).
\end{aligned}$$

So the maximum value for $\sin x + \sin(x + \frac{\pi}{3})$ is $\sqrt{3}$, the minimum value is $-\sqrt{3}$.

□

In the book "Introductory Algebra", we did some exercises on "Product-To-Difference". Recall this example:

Example 4.12 Compute

$$1 \times \frac{1}{2} + \frac{1}{2} \times \frac{1}{3} + \cdots + \frac{1}{2021} \times \frac{1}{2022}.$$

Solution: Observe:
$$1 \times \frac{1}{2} = 1 - \frac{1}{2}$$
$$\frac{1}{2} \times \frac{1}{3} = \frac{1}{2} - \frac{1}{3}$$
$$\ldots\ldots$$
$$\frac{1}{2021} \times \frac{1}{2022} = \frac{1}{2021} - \frac{1}{2022}.$$

We add all the above equations together:
$$1 \times \frac{1}{2} + \frac{1}{2} \times \frac{1}{3} + \cdots + \frac{1}{2021} \times \frac{1}{2022} = 1 - \frac{1}{2022} = \frac{2021}{2022}.$$

$\square$

The above approach is referred to as "partial fraction". It indicates that the products of fractions are harder to handle than the summation/difference. The partial fraction method will be used later in the study of integrals.

> **Theorem 4.5. Product-to-Sum, Product-to-Subtraction formulas**
>
> *For all angles A, B, we have*
>
> $$\sin A \cos B = \frac{1}{2}\{\sin(A+B) + \sin(A-B)\}, \qquad (4.9)$$
>
> $$\sin A \sin B = \frac{1}{2}\{\cos(A-B) - \cos(A+B)\}, \qquad (4.10)$$
>
> *and*
>
> $$\cos A \cos B = \frac{1}{2}\{\cos(A+B) + \cos(A-B)\}, \qquad (4.11)$$

Proof: Since
$$\sin(A+B) = \sin A \cos B + \cos A \sin B,$$
and
$$\sin(A-B) = \sin A \cos B - \cos A \sin B.$$

We add the above two equations to get the formula (4.9).

Similarly,
$$\cos(A+B) = \cos A \cos B - \sin A \sin B,$$
and
$$\cos(A-B) = \cos A \cos B + \sin A \sin B,$$

We add the above two equations to get the formula (4.11). We subtract one equation from another one to get the formula (4.10).

Example 4.13 Evaluate: (a) $\sin \frac{\pi}{12} \cos \frac{5\pi}{12}$; (b) $\cos \frac{\pi}{12} \sin \frac{5\pi}{12}$.

Solution:

(a).
$$\sin \frac{\pi}{12} \cos \frac{5\pi}{12} = \frac{1}{2}\{\sin \frac{\pi}{2} + \sin(-\frac{\pi}{3})\}$$
$$= \frac{1}{2}(1 - \frac{\sqrt{3}}{2})$$
$$= \frac{2-\sqrt{3}}{4}.$$

(b).
$$\cos \frac{\pi}{12} \sin \frac{5\pi}{12} = \frac{1}{2}\{\sin \frac{\pi}{2} + \sin \frac{\pi}{3}\}$$
$$= \frac{1}{2}(1 + \frac{\sqrt{3}}{2})$$
$$= \frac{2+\sqrt{3}}{4}.$$

4.2.5 Double angle, half angle formulas

In formula (4.3), if two angles are the same, we obtain the double angle formulas.

Theorem 4.6. Double angle formulas

For any angle A, we have
$$\sin 2A = 2 \sin A \cos A, \tag{4.12}$$

and
$$\cos 2A = \cos^2 A - \sin^2 A = 1 - 2\sin^2 A = 2\cos^2 A - 1. \tag{4.13}$$

From formula (4.13), we can derive the following half-angle formulas.

Theorem 4.7. Half angle formulas

For any angle A, we have
$$\sin^2 \frac{A}{2} = \frac{1 - \cos A}{2}, \tag{4.14}$$

and
$$\cos^2 \frac{A}{2} = \frac{1 + \cos A}{2}. \tag{4.15}$$

4.2.6 Trigonometric functions and complex numbers

In the book "Introductory Geometry and Proofs", we introduce the history of imaginary numbers and complex numbers. The imaginary numbers were proposed by the Greek mathematician Hero of Alexandria as a square root of negative numbers in the first century. It was kind of rediscovered by Girolamo Cardano (September 24, 1501—September 21, 1976) when he was deriving the solutions to third-order polynomial equations. The broad applications of complex numbers follow from the discovery of de Moivre formula and Euler formula. The discovery of de Moivre formula and the Euler formula benefits from the deep understanding of trigonometric functions.

Example 4.14 Find the maximum and minimum values of the function $f(x) = 5\cos x + 12\sin x$.

Five-minute thought: The maximum value for functions $\cos x$ and $\sin x$ are 1. So the maximum value of $f(x)$ will be less than or equal to $5 + 12 = 17$. Since two functions achieve the maximum at different values, it is not an easy question.

Solution: First, we observe

$$5\cos x + 12\sin x = \sqrt{5^2 + 12^2} \cdot \left(\frac{5}{\sqrt{5^2 + 12^2}} \cos x + \frac{12}{\sqrt{5^2 + 12^2}} \sin x\right)$$
$$= 13 \cdot \left(\frac{5}{13} \cos x + \frac{12}{13} \sin x\right).$$

Since $(\frac{5}{13}, \frac{12}{13})$ is a point on the unit circle, there is an angle x_0, such that

$$\sin x_0 = \frac{5}{13}, \quad \cos x_0 = \frac{12}{13}.$$

So,

$$5\cos x + 12\sin x = 13 \cdot \left(\frac{5}{13} \cos x + \frac{12}{13} \sin x\right)$$
$$= 13(\sin x_0 \cos x + \cos x_0 \sin x)$$
$$= 13\sin(x_0 + x).$$

(Even though we do not know the value of x_0.) So, $f(x) = 5\cos x + 12\sin x = 13\sin(x_0 + x)$. Its maximum value is 13, its minimum value is -13.

□

In fact, it is not easy to see from the graph that $f(x) = 5\cos x + 12\sin x = 13\sin(x_0 + x)$. See Figure 4.9.

We now re-exam complex numbers.

Suppose $z = a + bi$, where a, b are two real numbers. Then,

$$z = a + bi = \sqrt{a^2 + b^2} \cdot \left(\frac{a}{\sqrt{a^2 + b^2}} + \frac{bi}{\sqrt{a^2 + b^2}}\right)$$

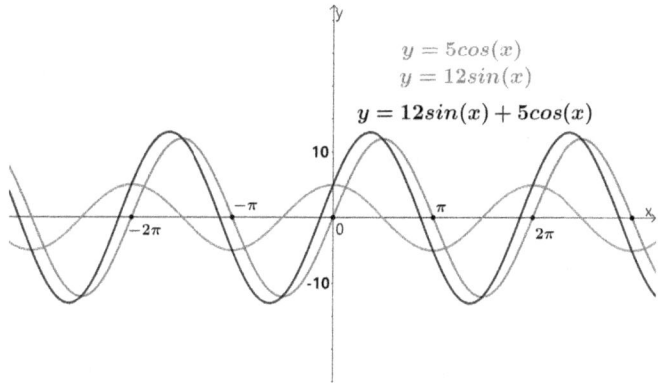

Figure 4.9: $f(x) = 5\cos x + 12\sin x = 13\sin(x_0 + x)$

We call $\sqrt{a^2 + b^2}$ the "modulus" of complex number z, and use notation $|z|$. Let

$$r = |z|, \quad \cos\theta = \frac{a}{\sqrt{a^2 + b^2}} \quad \text{and} \quad \sin\theta = \frac{b}{\sqrt{a^2 + b^2}}$$

Here, we restrict $\theta \in [0, 2\pi)$, and call θ the principal phase angle of complex number z. In this way, we identify the complex number $z = a + bi$ with the point (a, b) on the complex plane.

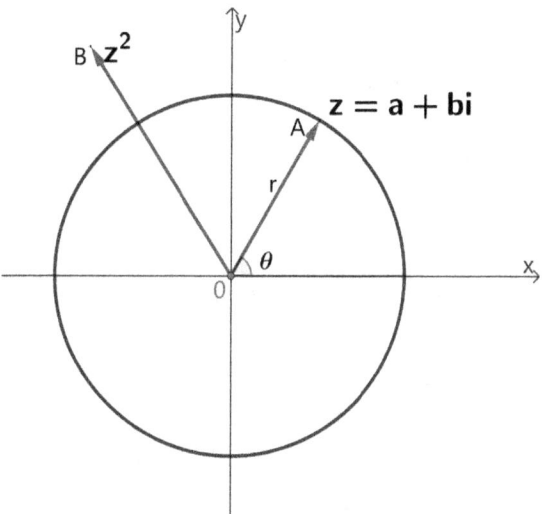

Figure 4.10: Complex plane

Using addition and subtraction formulas, we derive the following de Moivre formula.[2]

> **Proposition 4.5. de Moivre formula**
>
> *For any two angles α and β,*
>
> $$(\cos\alpha + i\sin\alpha)(\cos\beta + i\sin\beta) = \cos(\alpha + \beta) + i\sin(\alpha + \beta).$$

---

[2]Historically, when $\alpha = \beta$, this formula is called de Moivre formula. When $\alpha \neq \beta$, it can be derived from Ptolemy addition formula.

4.2 Trigonometric identities

Proof:

$$(\cos\alpha + i\sin\alpha)(\cos\beta + i\sin\beta) = \cos\alpha\cos\beta - \sin\alpha\sin\beta + i(\sin\alpha\cos\beta + \cos\alpha\sin\beta)$$
$$= \cos(\alpha + \beta) + i\sin(\alpha + \beta).$$

(Addition and Subtraction formulas)

The proof is completed.

□

With a deep understanding of the natural exponential function $y = e^x$ (see deep reading at the end of this chapter), we can derive the following Euler formula:

Proposition 4.6. Euler formula

For any radian θ (Caution: the angle θ is in radian),

$$e^{i\theta} = \cos\theta + i\sin\theta.$$

So

$$z = a + bi = re^{i\theta}, \qquad (4.16)$$

where

$$r = \sqrt{a^2 + b^2}, \quad \cos\theta = \frac{a}{\sqrt{a^2 + b^2}}, \quad \text{and} \quad \sin\theta = \frac{b}{\sqrt{a^2 + b^2}}.$$

Example 4.15 Taking $\theta = \pi$ in the Euler formula, we have

$$e^{i\pi} = -1. \qquad (4.17)$$

Usually, we call the above identity "Euler identity". In Euler identity, two transcendental numbers are magically linked[3].

□

Using the Euler formula, we can write de Moivre formula in the following form: Suppose α, β are two real numbers, then

$$e^{\alpha i} \cdot e^{\beta i} = e^{(\alpha+\beta)i}. \qquad (4.18)$$

Generally, we define, for any given complex number $z = a + bi$, that

$$e^{a+bi} = e^a \cdot e^{ib} = e^a \cos b + i e^a \sin b.$$

Recall, in the book "Introductory Geometry and Proofs", we learned that, with Euler and de Moivre formula, the following exponential operations for complex exponents.

---

[3] Transcendental numbers are the irrational numbers which are not solutions to rational coefficient polynomial equations.

> **Proposition 4.7. Exponential operations for complex exponents.**
> For any two complex numbers z and w, we have
> $$e^z \cdot e^w = e^{z+w}.$$

Proof: Assume that $z = a + bi$, $w = c + di$, where a, b, c, d are all real numbers. Then
$$e^{z+w} = e^{(a+c)+(b+d)i}.$$

On the other hand,

$$\begin{aligned}
e^z \cdot e^w &= e^a \cdot e^{bi} \cdot e^c \cdot e^{di} & \text{(Definition)} \\
&= e^a \cdot e^c \cdot e^{bi} \cdot e^{di} & \text{(Commutative rule)} \\
&= e^{a+c} \cdot e^{(b+d)i} & \text{(De Moivre formula (4.18))} \\
&= e^{(a+c)+(b+d)i} & \text{(Definition)}
\end{aligned}$$

The proof is completed. □

We now try to extend exponential operations for general exponents (with other positive bases other than e). First of all, by Theorem 1.4 we have: for any positive number a and any real number x:
$$\ln a^x = x \ln a.$$
That is
$$a^x = e^{x \ln a}.$$

We extend the above formula for complex exponents. We define:

> **Definition 4.2. General exponent**
> For any positive number $a \neq 1$ and any complex number x, we define
> $$a^x = e^{x \ln a}.$$

From this definition, we can obtain exponential operation rules for complex exponents.

> **Proposition 4.8. Exponent rules for complex exponents**
> Assume $a \neq 1$, $b \neq 1$ are two positive numbers, x and y are two complex numbers, then
> $$a^x \cdot a^y = a^{x+y} \qquad (4.19)$$
> and
> $$(ab)^y = a^y b^y. \qquad (4.20)$$

Proof: Let $x = r + ui$, $y = v + wi$, where r, u, v, w are real numbers. Then
$$a^x = e^{(r+ui)\ln a} = e^{r\ln a} \cdot e^{(u\ln a)i},$$
$$a^y = e^{(v+wi)\ln a} = e^{v\ln a} \cdot e^{(w\ln a)i},$$
$$b^y = e^{(v+wi)\ln b} = e^{u\ln b} \cdot e^{(w\ln b)i},$$

and
$$(ab)^y = e^{(v+wi)\ln(ab)} = e^{v\ln(ab)} \cdot e^{[w\ln(ab)]i}.$$

So
$$\begin{aligned} a^x \cdot a^y &= e^{r\ln a + (v\ln a)i} \cdot e^{u\ln a + (w\ln a)i} \\ &= e^{[(r+iu)+(v+iw)]\ln a} \\ &= e^{(x+y)\ln a} \\ &= a^{x+y}, \end{aligned}$$

and
$$\begin{aligned} (ab)^y &= e^{v\ln(ab)} \cdot e^{[w\ln(ab)]i} \\ &= e^{v\ln a + v\ln b} \cdot e^{[w\ln a + w\ln b]i} \\ &= e^{v\ln a + (w\ln a)i} \cdot e^{(v\ln b + (w\ln b)i)} \\ &= e^{(v+wi)\ln a} \cdot e^{(v+wi)\ln b} \\ &= a^y b^y. \end{aligned}$$

$\square$

Note Here, we warn: for positive number $a \neq 1$ and two complex numbers x and y, formula
$$\left(a^x\right)^y = a^{xy}$$

generally is NOT correct! The main reason is: $a^x = e^{x\ln a + 2n\pi i}$ holds for all integer n, thus $\left(a^x\right)^y$ would yield different values if the above formula held.

4.2.7 Complex solutions to polynomial equations

Let us recall how to find the n-th root for complex numbers with modulus 1 in Section 1.6.3, and how to apply this to solve polynomial equations in the complex number set.

Example 4.16 Find all complex solutions to polynomial equations:

(1).
$$x^4 + 1 = 0$$

(2).
$$x^8 = 1.$$

Solution: (1) Here we present two different methods.

Method 1: We first factorize $x^4 + 1$ in real number set.:

$$\begin{aligned}x^4 + 1 &= x^4 + 2x^2 + 1 - 2x^2 \\ &= (x^2 + 1)^2 - (\sqrt{2}x)^2 \\ &= (x^2 - \sqrt{2}x + 1)(x^2 + \sqrt{2}x + 1).\end{aligned} \quad (4.21)$$

So
$$x^4 + 1 = 0 \Leftrightarrow (x^2 - \sqrt{2}x + 1)(x^2 + \sqrt{2}x + 1) = 0$$
$$\Rightarrow x^2 - \sqrt{2}x + 1 = 0 \text{ or } x^2 + \sqrt{2}x + 1 = 0.$$

Using the quadratic formula or directly completing the square, we have:
$$x^2 - \sqrt{2}x + 1 = 0 \Leftrightarrow (x - \frac{\sqrt{2}}{2})^2 = -\frac{1}{2}$$
$$\Rightarrow x = \frac{\sqrt{2}}{2} + \frac{\sqrt{2}}{2}i \text{ or } x = \frac{\sqrt{2}}{2} - \frac{\sqrt{2}}{2}i.$$

Similarly,
$$x^2 + \sqrt{2}x + 1 = 0 \Leftrightarrow (x + \frac{\sqrt{2}}{2})^2 = -\frac{1}{2}$$
$$\Rightarrow x = -\frac{\sqrt{2}}{2} + \frac{\sqrt{2}}{2}i \text{ or } x = -\frac{\sqrt{2}}{2} - \frac{\sqrt{2}}{2}i.$$

Thus, we obtain for solutions:
$$x_1 = \frac{\sqrt{2}}{2} + \frac{\sqrt{2}}{2}i \text{ or } x_2 = \frac{\sqrt{2}}{2} - \frac{\sqrt{2}}{2}i$$
$$\text{or } x_3 = -\frac{\sqrt{2}}{2} + \frac{\sqrt{2}}{2}i \text{ or } x_4 = -\frac{\sqrt{2}}{2} - \frac{\sqrt{2}}{2}i.$$

Remark In the beginning, some students have difficulty understanding the purpose of factorizing the polynomial in the real number set (the purpose to get identity (4.21)), since the factorization does not yield any solutions. Yes, the factorization does not yield the solutions, but it reduces a complicated equation (here it is a 4th-order equation) into a few relatively easier equations (here it reduces into two second-order equations).

4.2 Trigonometric identities

Method 2: We recalled from Section 1.6.3 how to use the representations of complex numbers to solve the equation.

Observe
$$-1 = e^{2\pi k i + \pi i}, \quad \text{for all } k = 0, 1, 2, 3.$$

So
$$x_k = e^{\frac{2\pi k}{4}i + \frac{\pi}{4}i}, \quad \text{for all } k = 0, 1, 2, 3$$

solves $x^4 = -1$. That is
$$x_1 = e^{\frac{\pi}{4}i} = \frac{\sqrt{2}}{2} + \frac{\sqrt{2}}{2}i, \quad \text{(Principal 4-th root to -1)}$$

$$x_2 = e^{\frac{2\pi}{4}i + \frac{\pi}{4}i} = -\frac{\sqrt{2}}{2} + \frac{\sqrt{2}}{2}i,$$

$$x_3 = e^{\frac{4\pi}{4}i + \frac{\pi}{4}i} = -\frac{\sqrt{2}}{2} - \frac{\sqrt{2}}{2}i,$$

$$x_4 = e^{\frac{6\pi}{4}i + \frac{\pi}{4}i} = \frac{\sqrt{2}}{2} - \frac{\sqrt{2}}{2}i.$$

We actually can not see the 4-th principal root to -1 by using method 1.

(2). Observe
$$1 = e^{2\pi k i}, \quad \text{for } k = 0, 1, \cdots, 7.$$

So
$$x_k = e^{\frac{2\pi k}{8}i}, \quad \text{for } k = 0, 1, \cdot, 7$$

all solve $x^8 = 1$. That is:
$$x_1 = e^{\frac{0}{4}i} = 1, \quad \textbf{(The principal 8-th root of 1)}$$

$$x_2 = e^{\frac{2\pi}{8}i} = \frac{\sqrt{2}}{2} + \frac{\sqrt{2}}{2}i,$$

$$x_3 = e^{\frac{4\pi}{8}i} = i,$$

$$x_4 = e^{\frac{6\pi}{8}i} = -\frac{\sqrt{2}}{2} + \frac{\sqrt{2}}{2}i.$$

$$x_5 = e^{\frac{8\pi}{8}i} = -1,$$

$$x_6 = e^{\frac{10\pi}{8}i} = -\frac{\sqrt{2}}{2} - \frac{\sqrt{2}}{2}i,$$

$$x_7 = e^{\frac{12\pi}{8}i} = -i,$$

$$x_8 = e^{\frac{14\pi}{8}i} = \frac{\sqrt{2}}{2} - \frac{\sqrt{2}}{2}i.$$

□

4.2.8 Exercises after class

1. **Simplify**.

    (1) $\dfrac{\sin x}{\tan x}$; (2) $\dfrac{\cot \theta}{\cos \theta}$;

    (3) $\sin^3 y + \sin y \cos^2 y$; (4) $\cos^4 x - \sin^4 x + \sin^2 x$;

    (5) $(1 + \tan^2 x) \cdot \cos^2 x$; (6) $\dfrac{1}{1+\cot^2 x} + \cos^2 x$.

2. **Prove the following trigonometric identities**.

    (1) $\cos(-x) - \sin(-x) = \sin x + \cos x$;

    (2) $\dfrac{1}{1-\cos^2 x} = 1 + \cot^2 x$;

    (3) $\cos(\dfrac{\pi}{2} + x) = -\sin x$;

    (4) $\dfrac{1+\tan \theta}{1-\tan \theta} = \tan(\dfrac{\pi}{4} + x)$;

    (5) $\dfrac{\sin(x+y)}{\sin x \cos y} = \tan x \cot y + 1$;

    (6) $\dfrac{\cos(x-y)}{\sin x \cos y} = \cot x + \tan y$;

    (7) $\cos 3\theta = 4\cos^3 \theta - 3\cos \theta$;

    (8) $\sin 3\alpha = -4\sin^3 \alpha + 3\sin \alpha$;

    (9) $\cos^2 3\alpha - \sin^2 3\alpha = \cos 6\alpha$;

    (10) $\cos^4 5\alpha - \sin^4 5\alpha = \cos 10\alpha$;

    (11) $\dfrac{\sin 4x}{\sin x} = 4\cos x \cos 2x$;

    (12) $\dfrac{\sin 6x}{\sin 5x + \sin x} = \dfrac{\cos 3x}{\cos 2x}$.

3. **Comprehensive questions**.

    (1) Evaluate

    (i) $\sin \dfrac{\pi}{12}$; (ii) $\cos \dfrac{7\pi}{12}$.

    (2) Solve equations

    (i) $x^2 - 2x + 3 = 0$; (ii) $x^8 - 15x^4 - 16 = 0$.

    (3) Prove:

    $$\sin^4 x + \cos^4 x \geq \dfrac{1}{2}$$

    (4) Find the range for function $y = \sin^4 x + \cos^4 x$.

4.3 Trigonometric functions in triangles

In the book "Introductory Geometry and Proofs", we have learned sine law and cosine law.

Theorem 4.8. Sine Law
In a nontrivial triangle $\triangle ABC$ (the lengths of all three sides are larger than zero), there holds the following identities:
$$\frac{\sin \angle CAB}{BC} = \frac{\sin \angle CBA}{AC} = \frac{\sin \angle ACB}{AB}.$$

Example 4.17 Prove: in a triangle, the largest interior angle is opposite the longest side.

Proof. If all three angles in $\triangle ABC$ are less than or equal to $\frac{\pi}{2}$ then we obtain the above result from the Sine Law and the fact that sine function is an increasing function in the interval $(0, \frac{\pi}{2}]$.

If the largest angle in $\triangle ABC$, say for example $\angle A$ is large than $\frac{\pi}{2}$, then by the exterior angle formula we know that $180^0 - \angle A = \angle B + \angle C$. So $\sin \angle A = \sin(180^0 - \angle A) > \sin \angle B$, and $\sin \angle A = \sin(180^0 - \angle A) > \sin \angle C$. From the Sine Law we know that the side BC, which is opposite to angle $\angle A$, is the longest side.

$\square$

Theorem 4.9. Cosine Law
In a triangle $\triangle ABC$, assume $BC = a$, $AC = b$ and the angle between AC and BC is θ, then the length c of the side AB is given by
$$c^2 = a^2 + b^2 - 2ab\cos\theta.$$

As an application of trigonometric identities, we now derive the famous Heron-Qin's formula

Theorem 4.10. Heron-Qin's formula
In a triangle $\triangle ABC$, assume $BC = a$, $AC = b$ and $AB = c$. Then the area $Area_{\triangle ABC}$ of the triangle is given by
$$Area_{\triangle ABC} = \sqrt{s(s-a)(s-b)(s-c)}$$
where $s = (a+b+c)/2$ is half of the circumference of the triangle.

Proof: First of all, we know the area of the triangle is given by
$$Area_{\triangle ABC} = \frac{1}{2}ab\sin \angle C.$$

So,
$$(Area_{\triangle ABC})^2 = \frac{1}{4}a^2b^2 \sin^2 \angle C$$
$$= \frac{1}{4}a^2b^2(1 - \cos^2 \angle C)$$
$$= \frac{1}{4}a^2b^2(1 - \cos \angle C)(1 + \cos \angle C).$$

By Cosine Law, we know
$$\cos \angle C = \frac{a^2 + b^2 - c^2}{2ab}.$$

Thus
$$1 + \cos \angle C = \frac{a^2 + b^2 - c^2 + 2ab}{2ab}$$
$$= \frac{(a+b)^2 - c^2}{2ab}$$
$$= \frac{(a+b+c)(a+b-c)}{2ab}.$$

and
$$1 - \cos \angle C = \frac{2ab - a^2 - b^2 + c^2}{2ab}$$
$$= \frac{-(a-b)^2 + c^2}{2ab}$$
$$= \frac{(-a+b+c)(a-b+c)}{2ab}.$$

Combining the above identities, we reach
$$(Area_{\triangle ABC})^2 = \frac{(a+b+c)(a+b-c)(-a+b+c)(a-b+c)}{2^4}$$
$$= \frac{a+b+c}{2} \cdot \frac{a+b+c-2c}{2} \cdot \frac{a+b+c-2a}{2} \cdot \frac{a+b+c-2b}{2}$$
$$= s(s-c)(s-a)(s-b).$$

Taking a square root of both sides (area is a positive value), we obtain Heron-Qin's formula.

□

Note *Heron-Qin's formula is due to the Greek mathematician Hero of Alexandria (It may be known to Archimedes earlier) and to the Chinese mathematician Jiushao Qin (1208–1261).*

4.3.1 Exercises after class

1. **Basic skills.** In a triangle $\triangle ABC$, let $BC = a$, $AC = b$ and $AB = c$.

    (1) Assume $a = 5$, $b = 8$, $\angle A = 30^o$, find $\sin \angle B$.

    (2) Assume $a = 5$, $b = 8$, $\angle C = 30^o$, find c.

    (3) Assume $a = 5$, $b = 8$, $\angle C = 30^o$, find the area of $\triangle ABC$.

    (4) Assume $a = 5$, $b = 8$, $\angle A = 30^o$, find $\sin(\angle A + \angle C)$.

2. **Comprehensive questions.** In a triangle $\triangle ABC$, let $BC = a$, $AC = b$ and $AB = c$.

    (1) Assume $c = 5$, $b = 5\sqrt{3}$, $\angle C = 30^o$, find $\angle B$.

    (2) Assume $c = 5$, $b = 5\sqrt{3}$, $\angle C = 30^o$, find a.

    (3) Prove
    $$\sin \angle A + \sin \angle B + \sin \angle C = 4 \cos \frac{A}{2} \cos \frac{B}{2} \cos \frac{C}{2}.$$

    (4) Assume that $\triangle ABC$ is not a right triangle, prove
    $$\tan \frac{A}{2} \tan \frac{B}{2} + \tan \frac{A}{2} \tan \frac{C}{2} + \tan \frac{B}{2} \tan \frac{C}{2} = 1.$$

4.4 Extra reading: Euler formula*

In this section, we try to explain how can we derive the amazing Euler formula. Inevitably, we will use some formulas that have not been covered so far (but these formulas will be discussed in the last chapter).

In the last chapter, we will give other descriptions of all elementary functions covered in this book. This relies on the approximations of any given elementary function via polynomials (or infinite polynomials— which will be given precisely when we study the Taylor series). For example, we will prove: for any real number x,

$$e^x = \sum_{n=0}^{\infty} \frac{x^n}{n!}. \tag{4.22}$$

Here the series means the following: for $x \neq 0$,

$$\sum_{n=0}^{\infty} \frac{x^n}{n!} = \frac{x^0}{0!} + \frac{x^1}{1!} + \ldots + \underbrace{\frac{x^k}{k!}}_{\text{the k-th term}} + \cdots.$$

From this, we can define, for any complex number z, that

$$e^z = \sum_{n=0}^{\infty} \frac{z^n}{n!}. \tag{4.23}$$

Recall: we define $0! = 1$; and for positive integer n, we define $n! = n \cdot (n-1) \cdots 2 \cdot 1$. We will come back to discuss the series in detail later in Section 7.3. The so-called "repeated decimal" in fact is defined in a similar way:

$$0.333\cdots = \sum_{n=0}^{\infty} 3 \cdot 10^{n-1}.$$

And, we will prove, for any real number x, that

$$\sin x = \sum_{n=0}^{\infty} (-1)^n \frac{x^{2n+1}}{(2n+1)!},$$

and

$$\cos x = \sum_{n=0}^{\infty} (-1)^n \frac{x^{2n}}{(2n)!}.$$

For a real number θ, we can compute:

$$\begin{aligned}
e^{i\theta} &= \frac{\theta^0}{0!} + \frac{\theta^1 i}{1!} + \frac{\theta^2 i^2}{2!} + \frac{\theta^3 i^3}{3!} + \frac{\theta^4 i^4}{4!} + \cdots \\
&= \frac{\theta^0}{0!} + \frac{\theta^1}{1!}i - \frac{\theta^2}{2!} - \frac{\theta^3}{3!}i + \frac{\theta^4}{4!} + \cdots \\
&= \frac{\theta^0}{0!} - \frac{\theta^2}{2!} + \frac{\theta^4}{4!} - \frac{\theta^6}{6!} + \cdots \\
&\quad + (\frac{\theta^1}{1!} - \frac{\theta^3}{3!} + \frac{\theta^5}{5!} - \frac{\theta^7}{7!} \cdots)i \\
&= \cos\theta + i\sin\theta.
\end{aligned}$$

We thus obtain the Euler formula!

4.5 Chapter review and exercises

We study four trigonometric functions and various trigonometric identities.

1. Four trigonometric functions.

   1.1. Sine function $y = \sin x$

   Symmetry: $y = \sin x$ is an odd function defined on $\mathbb{R}$.

   Period: $y = \sin x$ is a 2π periodic function.

   Monotone: Since $y = \sin x$ is a 2π periodic function, it can not be a monotone function in the whole domain. But on interval $(-\frac{\pi}{2}, \frac{\pi}{2})$, $y = \sin x$ is a strictly increasing function.

   1.2. Cosine function $y = \cos x$

   Symmetry: $y = \cos x$ is an even function defined on $\mathbb{R}$.

   Period: $y = \cos x$ is a 2π periodic function.

   Monotone: Since $y = \cos x$ is a 2π periodic function, it can not be a monotone function in the whole domain. But on interval $(0, \pi)$, $y = \cos x$ is a strictly decreasing function.

   1.3. Tangent function $y = \tan x$

   Symmetry and period: The domain of $y = \tan x$ is $\mathbb{R} \setminus \{n\pi + \frac{\pi}{2} : n \in \mathbb{Z}\}$. It is an odd function; it is also a periodic function with period π.

   Monotone: on interval $(-\frac{\pi}{2}, \frac{\pi}{2})$, $y = \tan x$ is a strictly increasing function.

   1.4. Cotangent function $y = \cot x$

   Symmetry and period: The domain of $y = \cot x$ is $\mathbb{R} \setminus \{n\pi : n \in \mathbb{Z}\}$. It is an odd function; it is also a periodic function with period π.

   Monotone: on interval $(0, \pi)$, $y = \cot x$ is a strictly decreasing function.

   We remind students: it will help students to remember the properties of trigonometric functions if they can memorize their graphs.

2. Trigonometric identities

   2.1. General Pythagorean Identity. For any angle θ,
   $$\sin^2 \theta + \cos^2 \theta = 1.$$

   2.2. Addition and subtraction formulas. For any angles α, β, we have
   $$\sin(\alpha \pm \beta) = \sin \alpha \cos \beta \pm \cos \alpha \sin \beta,$$
   and
   $$\cos(\alpha \pm \beta) = \cos \alpha \cos \beta \mp \sin \alpha \sin \beta.$$

Further, if $\alpha, \beta, \alpha \pm \beta \neq \frac{\pi}{2} + n\pi$, where $n \in \mathbb{Z}$, we have
$$\tan(\alpha \pm \beta) = \frac{\tan \alpha \pm \tan \alpha}{1 \mp \tan \alpha \tan \beta};$$

On the other hand, if $\alpha, \beta, \alpha \pm \beta \neq n\pi$, where $n \in \mathbb{Z}$, we have
$$\cot(\alpha \pm \beta) = \frac{\cot \alpha \cot \beta \mp 1}{\cot \beta \pm \cot \alpha}.$$

2.3. Sum-to-Product, Difference-to-Product formulas. For any angles A, B, we have
$$\sin A + \sin B = 2 \sin(\frac{A+B}{2}) \cos(\frac{A-B}{2}),$$

$$\sin A - \sin B = 2 \sin(\frac{A-B}{2}) \cos(\frac{A+B}{2}),$$

and
$$\cos A + \cos B = 2 \cos(\frac{A+B}{2}) \cos(\frac{A-B}{2}),$$

$$\cos A - \cos B = -2 \sin(\frac{A+B}{2}) \sin(\frac{A-B}{2}).$$

2.4. Product-to-Sum or Difference formulas. For any angles A, B, we have

$$\sin A \cos B = \frac{1}{2}\{\sin(A+B) + \sin(A-B)\},$$

$$\sin A \sin B = \frac{1}{2}\{\cos(A-B) - \cos(A+B)\},$$

and
$$\cos A \cos B = \frac{1}{2}\{\cos(A+B) + \cos(A-B)\}.$$

2.5. Double angle formula. For any angle A, we have
$$\sin 2A = 2 \sin A \cos A,$$
and
$$\cos 2A = \cos^2 A - \sin^2 A = 1 - 2\sin^2 A = 2\cos^2 A - 1.$$

2.6. Half angle formula. For any angle A, we have
$$\sin^2 \frac{A}{2} = \frac{1 - \cos A}{2},$$
and
$$\cos^2 \frac{A}{2} = \frac{1 + \cos A}{2}.$$

3. The applications of trigonometric identities

3.1. Operation on complex numbers

3.1.1. de Moivre formula. For any α and β,
$$(\cos\alpha + i\sin\alpha)(\cos\beta + i\sin\beta) = \cos(\alpha+\beta) + i\sin(\alpha+\beta).$$

3.1.2. Euler formula. For any radian θ,
$$e^{i\theta} = \cos\theta + i\sin\theta.$$

3.1.3. Exponential operation rules. Let $a \neq 1$, $b \neq 1$ be two positive numbers, and r and s be two complex numbers. We have
$$a^r \cdot a^s = a^{r+s}$$
and
$$(ab)^r = a^r b^r.$$

3.2. Trigonometric functions in triangles.

3.2.1. Sine Law. In a nontrivial triangle $\triangle ABC$, there hold the following identities
$$\frac{\sin\angle CAB}{BC} = \frac{\sin\angle CBA}{AC} = \frac{\sin\angle ACB}{AB}.$$

3.2.2. Cosine Law. In triangle $\triangle ABC$, assume $BC = a$, $AC = b$, and the angle between AC and BC is θ. Then the length c of side AB is given by
$$c^2 = a^2 + b^2 - 2ab\cos\theta.$$

3.2.3. Heron-Qin's formula In a triangle $\triangle ABC$, assume $BC = a$, $AC = b$ and $AB = c$. Then the area $Area_{\triangle ABC}$ of the triangle is given by
$$Area_{\triangle ABC} = \sqrt{s(s-a)(s-b)(s-c)}$$
where $s = (a+b+c)/2$ is half of the circumference of the triangle.

4.5.1 Chapter 4 test

1. **Basic skills-1.** Among the following functions, which one is odd, which one is even, and which one is periodic?

   (1) $y = \sin x \cos 2x$;   (2) $y = 2\sin x - \cos x$;

   (3) $y = \sin^2 x$;   (4) $y = \sin(x^2 + 1)$.

2. **Basic skills-2.** Find the domains and ranges for the following functions.

   (1) $y = \sin x + 2\cos x$;   (2) $y = \sin^2 x + 2\sin x + 2$.

   (3) $y = \sin x + \cos(x + \frac{\pi}{3})$;   (4) $y = \cos 2x + 2\sin x$.

3. **Basic skills-3.** Operation on complex numbers

   (1) Use two methods (direct computation, and Euler formula) to compute
   $$(\sqrt{3} + i)^4$$

   (2) Find all 5-th roots of -32, and indicate which one is the principal one.

4. **Comprehensive questions-1:** trigonometric identities. Prove the following trigonometric identities:

   (1)
   $$\frac{\sin(x+y)}{\cos(x-y)} = \frac{\tan x + \tan y}{1 + \tan x \tan y}$$

   (2)
   $$(\cos x - \sin x)^2 = 1 - \sin 2x$$

   (3)*
   $$\sin(x+y)\sin(x-y) = \sin^2 x - \sin^2 y$$

   (4)**
   $$\cos\frac{\pi}{5} - \cos\frac{2\pi}{5} = \frac{1}{2}$$

   (5)**
   $$\frac{1}{\cos 0^o \cos 1^o} + \frac{1}{\cos 1^o \cos 2^o} + \cdots + \frac{1}{\cos n^o \cos(n+1)^o} = \frac{\tan(n+1)^o}{\sin 1^o}$$

5. **Comprehensive questions-2:** Solve the following questions in the complex field.

   (1) (a). Factorize in real number set:
   $$x^4 + x^2 + 1.$$

(b). Solve equations in complex number set:
$$x^4 + x^2 + 1 = 0.$$

(2) (a). Factorize in real number set:
$$x^5 + x^4 + x^3 + x^2 + x + 1.$$

(b). Solve equations in complex number set:
$$x^5 + x^4 + x^3 + x^2 + x + 1 = 0.$$

(3)* Solve equations in complex number set without factorizing:
$$x^5 + x^4 + x^3 + x^2 + x + 1 = 0.$$

6. **Comprehensive questions-3:** Suppose $x \in [0, 2\pi)$, solve the following trigonometric equations.

(1)
$$4\sin^2 x = 3.$$

(2)
$$3\cos^2 x = \sin^2 x.$$

(3)
$$4\cos x = \sin 2x.$$

(4)
$$\sin 3x + \sin 2x + \sin x = 0.$$

(5)
$$4\sin^2 x + (2\sqrt{3} - 2)\sin x = \sqrt{3}.$$

Chapter 5 Inverse trigonometric functions

Introduction

- arcsin function
- arccos function
- arctan function
- arccot function
- Complementary properties
- Trigonometric equation

Since trigonometric functions are periodic functions, we can only consider their inverse functions in certain small intervals.

5.1 Definitions

5.1.1 $y = \arcsin x$: inverse of sine function

We first study the inverse of sine function.

Recall the graph of $y = \sin x$:

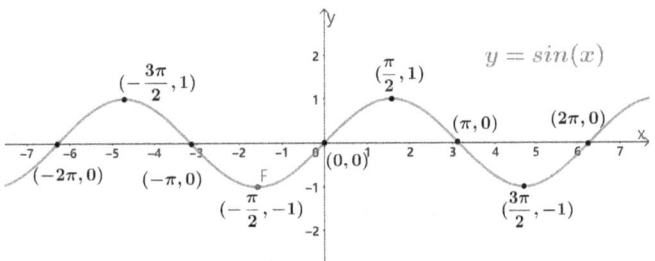

Figure 5.1: Graph of $y = \sin x$

From the graph, we see that: $y = \sin x$ is a periodic function defined on $\mathbb{R}$; its range is $[-1, 1]$.

Though $y = \sin x$ is not a monotonic function in the whole domain, for $x \in [-\frac{\pi}{2}, \frac{\pi}{2}]$, it is a strictly increasing function with function value increasing from -1 to 1.

> **Definition 5.1. Arcsine function**
> *The inverse function of $y = \sin x$ on the interval $[-\frac{\pi}{2}, \frac{\pi}{2}]$, denoted as $y = \arcsin x$, is a strictly increasing function with domain $[-1, 1]$, and the range $[-\frac{\pi}{2}, \frac{\pi}{2}]$.* ♣

The graph of $y = \arcsin x$ (with the comparison with sine function in the same figure) is given in Figure 5.2.

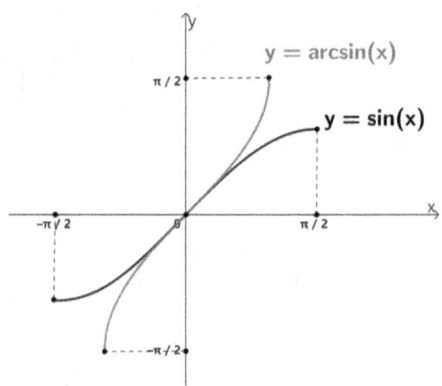

Figure 5.2: Graph of $y = \arcsin x$

By definition, we know that for $x \in [-1, 1]$, $\sin(\arcsin x) = x$; for $x \in [-\frac{\pi}{2}, \frac{\pi}{2}]$, $\arcsin(\sin x) = x$. Caution: if $x \notin [-\frac{\pi}{2}, \frac{\pi}{2}]$, by the range of the inverse sine function, we know that $\arcsin(\sin x) \neq x$.

Example 5.1 Evaluate $\arcsin(\sin \frac{3\pi}{4})$.

Solution: First of all, we caution: by the range of the inverse sine function, $\arcsin(\sin \frac{3\pi}{4}) \neq \frac{3\pi}{4}$.

In fact,
$$\arcsin(\sin \frac{3\pi}{4}) = \arcsin(\frac{\sqrt{2}}{2})$$
$$= \frac{\pi}{4}.$$

□

Example 5.2 Let x be an angle in the second quadrant. Evaluate $\arcsin(\sin x)$.

Solution: We first assume that $x \in [\frac{\pi}{2}, \pi]$. Then $x_0 = \pi - x \in [0, \frac{\pi}{2}]$, and
$$\sin x_0 = \sin(\pi - x) = \sin x.$$
Since $x_0 \in [0, \frac{\pi}{2}]$, by the definition, we have $\arcsin(\sin x_0) = x_0$. So
$$\arcsin(\sin x) = \arcsin(\sin x_0) = x_0 = \pi - x.$$

Then we consider the other case. Since x is in the second quadrant, we can write
$$x = x_1 + 2n\pi \quad \text{here n is an integer,} \quad x_1 \in [\frac{\pi}{2}, \pi].$$
Thus
$$\arcsin(\sin x) = \arcsin(\sin x_1) = \pi - x_1.$$

□

5.1.2 $y = \arccos x$: the inverse cosine function

We then study the inverse of cosine function.

5.1 Definitions

Recall the graph of $y = \cos x$:

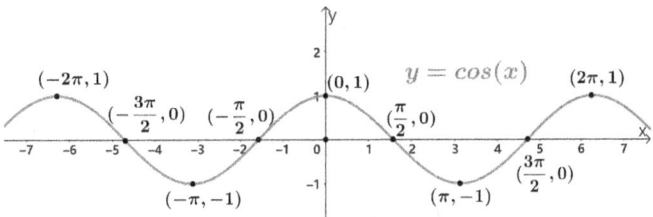

Figure 5.3: Graph of $y = \cos x$

From the graph, we see that $y = \cos x$ is a periodic function defined on $\mathbb{R}$; its range is $[-1, 1]$.

Though $y = \cos x$ is not a monotonic function in the whole domain, for $x \in [0, \pi]$, it is a strictly decreasing function with function value decreasing from 1 to -1.

Definition 5.2. Arccos function
The inverse function of $y = \cos x$ on the interval $[0, \pi]$, denoted as $y = \arccos x$, is a strictly decreasing function with domain $[-1, 1]$, and the range $[-\frac{\pi}{2}, \frac{\pi}{2}]$.

The graph of $y = \arccos x$ (with the comparison with cosine function in the same figure) is the following:

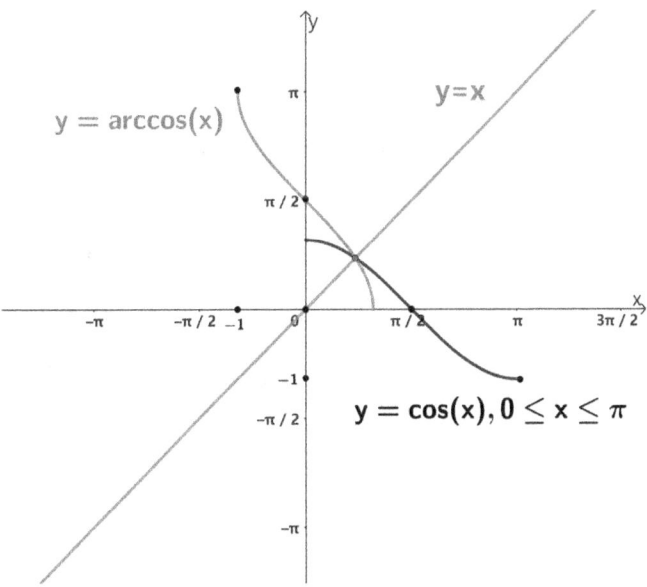

Figure 5.4: Graph of $y = \arccos x$

From the definition we know: for $x \in [-1, 1]$, $\cos(\arccos x) = x$; for $x \in [0, \pi]$, $\arccos(\cos x) = x$. Similarly, we caution: if $x \notin [0, \pi]$, then $\arccos(\cos x) \neq x$.

Example 5.3 Evaluate $\arccos(\cos \frac{5\pi}{4})$.

Solution: We first remind: due to the range of the inverse of cosine function, we know $\arccos(\cos \frac{5\pi}{3}) \neq \frac{5\pi}{3}$.

Let us compute

$$\arccos(\cos\frac{5\pi}{3}) = \arccos(\frac{1}{2})$$
$$= \frac{\pi}{3}.$$

□

Example 5.4 Let x be an angle in the fourth quadrant. Evaluate $\arccos(\cos x)$.

Solution: We first consider the case : $x \in [\frac{3\pi}{2}, 2\pi]$. Then $x_0 = 2\pi - x \in [0, \frac{\pi}{2}]$, and

$$\cos x_0 = \cos(2\pi - x) = \cos x.$$

Since $x_0 \in [0, \frac{\pi}{2}]$ from the definition we have $\arccos(\cos x_0) = x_0$. So

$$\arccos(\cos x) = \arccos(\cos x_0) = x_0 = 2\pi - x.$$

Then we consider the other case. Since x is in the fourth quadrant, we can write

$$x = x_1 + 2n\pi \quad \text{here n is an integer,} \quad x_1 \in [\frac{3\pi}{2}, 2\pi].$$

Thus

$$\arccos(\cos x) = \arccos(\cos x_1) = 2\pi - x_1.$$

□

5.1.3 $y = \arctan x$: inverse of tangent function

Finally, We study the inverse of tangent and cotangent functions.

Recall the graph of $y = \tan x$:

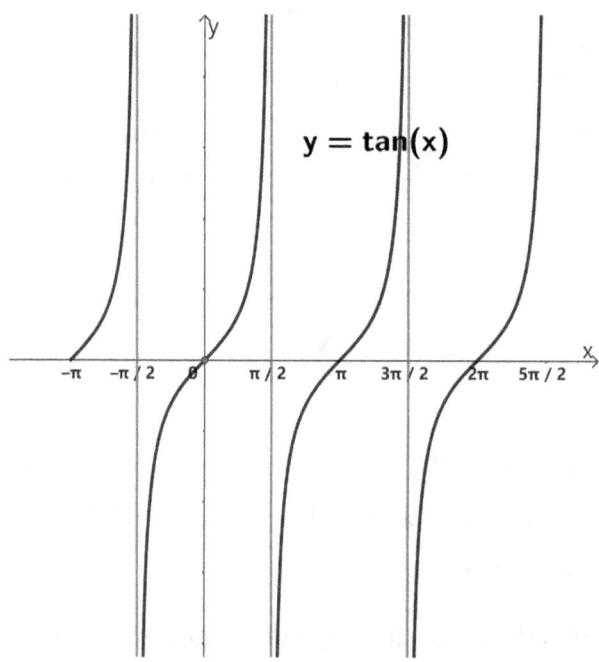

Figure 5.5: Graph of $y = \tan x$

We know $y = \tan x$ is a strictly increasing function on the interval $(-\frac{\pi}{2}, \frac{\pi}{2})$. Its range is $(-\infty, +\infty)$.

So, we introduce the inverse of the tangent function in the following.

Definition 5.3. Arctan function
The inverse function of $y = \tan x$ on the interval $(-\frac{\pi}{2}, \frac{\pi}{2})$, denoted as $y = \arctan x$, is a strictly increasing function with domain $(-\infty, \infty)$, and the range $(-\frac{\pi}{2}, \frac{\pi}{2})$.

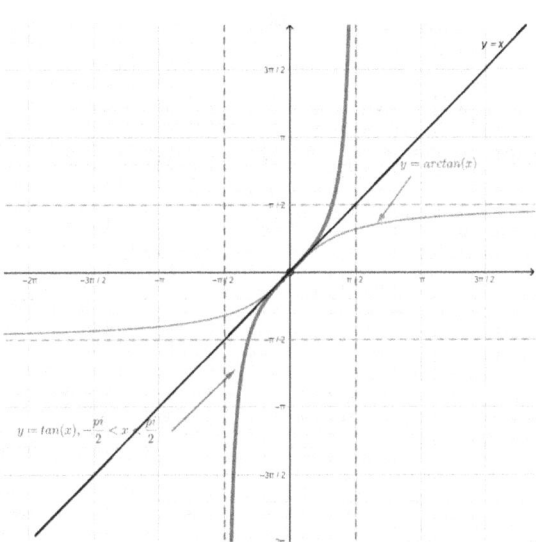

Figure 5.6: Graph of $y = \arctan x$

By definition, we know that for $x \in (-\infty, +\infty)$, $\tan(\arctan x) = x$; for $x \in (-\frac{\pi}{2}, \frac{\pi}{2})$, $\arctan(\tan x) = x$. Caution: for $x \notin (-\frac{\pi}{2}, \frac{\pi}{2})$, $\arctan(\tan x) \neq x$.

Example 5.5 Evaluate $\arctan(\tan \frac{5\pi}{4})$.

Solution: We compute
$$\arctan(\tan \frac{5\pi}{4}) = \arctan(-1)$$
$$= -\frac{\pi}{4}.$$

$\square$

Similarly, we know $y = \cot x$ is a strictly decreasing function on the interval $(0, \pi)$. Its range is $(-\infty, +\infty)$.

So, we introduce the inverse of the cotangent function in the following.

Definition 5.4. Inverse of cot function
The inverse function of $y = \cot x$ on the interval $(0, \pi)$, denoted as $y = \text{arccot} x$, is a strictly decreasing function with domain $(-\infty, \infty)$, and the range $(0, \pi)$.

Here is the graph of the function $y = arccotx$.

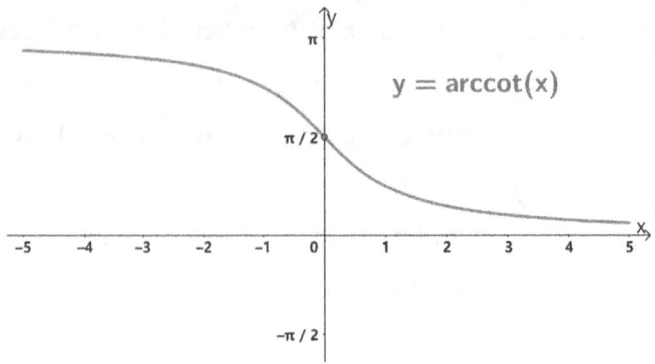

Figure 5.7: Graph of $y = arccotx$

By definition, we know that for $x \in (-\infty, +\infty)$, $\cot(arccotx) = x$; for $x \in (0, \pi)$, $arcot(\cot x) = x$. Caution: for $x \notin (0, \pi)$, $arccot(\cot x) \neq x$.

5.1.4 Exercises after class

1. **Basic skill**.

    (1) (a) Compute $\arcsin \frac{1}{2}$; (b) Solve equation: $\sin x = \frac{1}{2}$.

    (2) (a) Compute: $\arctan(-\frac{\sqrt{3}}{3})$; (b) Solve equation $\tan x = -\frac{\sqrt{3}}{3}$.

    (3) (a) Compute: $\sin(\arcsin \frac{1}{8})$; (b) Compute: $\arcsin[\sin(\frac{\pi}{8})]$.

    (4) (a) Compute: $\cos(\arcsin \frac{1}{8})$; (b) Compute: $\arcsin[\sin(-2\pi + \frac{\pi}{9})]$.

2. **Comprehensive questions**.

    (1) Find the domain of function $y = \ln(\arctan x)$;

    (2) Compute $\sin(2 \arcsin \frac{2}{3})$;

    (3) Compute $\cos(\arcsin \frac{1}{3} + \arccos \frac{1}{4})$;

    (4) Assume $x \in [-1, 0]$, compute

    $$\cos(\arccos x + \arcsin x).$$

5.2 Basic properties

5.2.1 Properties for inverse trigonometric functions

First, we do the following exercise.

Example 5.6 Compute
$$\sin(\arccos \frac{1}{2}).$$

Solution: Since $\arccos \frac{1}{2} \in [0, \pi]$, so its sine value is nonnegative.
$$\sin(\arccos \frac{1}{2}) = \sqrt{1 - \cos^2(\arccos \frac{1}{2})}$$
$$= \sqrt{1 - (\frac{1}{2})^2}$$
$$= \frac{\sqrt{3}}{2}.$$

□

Exercise 5.1 Compute $\cos(\arcsin \frac{\sqrt{2}}{2})$.

Generally, for $|x| \leq 1$, we can compute the following.

Example 5.7 If $|x| \leq 1$, compute
$$\sin(\arccos x).$$

Solution: Since $\arccos x \in [0, \pi]$, so its sine value is nonnegative.

$$\sin(\arccos x) = \sqrt{1 - \cos^2(\arccos x)}$$
$$= \sqrt{1 - x^2}.$$

□

Example 5.8 Compute
$$\arccos(\sin \frac{2\pi}{3}).$$

Solution:
$$\arccos(\sin \frac{2\pi}{3}) = \arccos(\frac{\sqrt{3}}{2})$$
$$= \frac{\pi}{6}.$$

□

Exercise 5.2 Suppose $x \in (\frac{3\pi}{2}, 2\pi)$. Find $\arccos(\sin x)$.

From the definitions of inverse trigonometric functions, we can obtain the following properties.

> **Proposition 5.1. Complementary properties**
>
> $$\arcsin x + \arccos x = \frac{\pi}{2}. \tag{5.1}$$
>
> and
>
> $$\arctan x + \mathrm{arccot}\, x = \frac{\pi}{2} \tag{5.2}$$

Proof: We only give the proof for (5.1), and leave the proof of (5.2) for exercise.

By the definition we have $\arcsin x \in [-\frac{\pi}{2}, \frac{\pi}{2}]$, and $\frac{\pi}{2} - \arccos x \in [-\frac{\pi}{2}, \frac{\pi}{2}]$. So, we only need to prove $\sin(\frac{\pi}{2} - \arccos x) = x$.

We can verify

$$\sin(\frac{\pi}{2} - \arccos x) = \cos(\arccos x) \quad \text{(Formula (4.1))}$$

$$= x.$$

$\square$

Exercise 5.3 Prove formula (5.2).

5.2.2 Trigonometric equations

Finally, let us study how to solve trigonometric equations. We list three typical examples.

Example 5.9 Solve equations:

(a). $\sin x + \frac{1}{2} = 0$;

(b). $\sin x + \sqrt{3} \cos x = -1$;

(c). $\cos^2 x + 2 \sin x - 2 = 0$.

Solution: (a). Viewing $\sin x$ as a variable, we obtain:

$$\sin x = -\frac{1}{2}.$$

For $x \in [0, 2\pi)$, $x = \frac{7\pi}{6}$ and $x = \frac{11\pi}{6}$ are two solutions. Since $y = \sin x$ is a 2π periodic function, we can obtain all solutions: for any integer n,

$$x = 2n\pi + \frac{7\pi}{6}, \quad \text{or} \quad x = 2n\pi + \frac{11\pi}{6}.$$

(b). The difficulty for this equation is: we have trouble finding the values for $\sin x$

or $\cos x$. We make the following transformation, based on trigonometric identities.

$$\sin x + \sqrt{3}\cos x = 1 \Leftrightarrow \frac{1}{2}\sin x + \frac{\sqrt{3}}{2}\cos x = -\frac{1}{2}$$
$$\Leftrightarrow \cos\frac{\pi}{3}\sin x + \sin\frac{\pi}{3}\cos x = -\frac{1}{2}$$
$$\Leftrightarrow \sin(x + \frac{\pi}{3}) = -\frac{1}{2}.$$

Using the results obtained in (a), we have: for any integer n,

$$x + \frac{\pi}{3} = 2n\pi - \frac{\pi}{6}, \quad \text{or} \quad x + \frac{\pi}{3} = 2n\pi + \frac{7\pi}{6}.$$

We thus obtain all solutions: for any integer n,

$$x = 2n\pi - \frac{\pi}{2}, \quad \text{or} \quad x = 2n\pi + \frac{5\pi}{6}.$$

(c). Again, we have trouble finding the values for $\sin x$ or $\cos x$. We make the following transformation:

$$\cos^2 x + 2\sin x - 2 = 0 \Leftrightarrow 1 - \sin^2 x + 2\sin x - 2 = 0$$
$$\Leftrightarrow \sin^2 x\, 2\sin x + 1 = 0$$
$$\Leftrightarrow (\sin x - 1)^2 = 0.$$

So,

$$\sin x = 1.$$

On interval $[0, 2\pi)$, it has one solution:

$$x = \arcsin 1 = \frac{\pi}{2}.$$

Again, using the periodic property, we obtain all solutions: for any integer n,

$$x = 2n\pi + \frac{\pi}{2}.$$

□

From the above three examples, we learn that it is usually a hard and comprehensive question to solve a trigonometric equation. The only way to gain the skills is to do more exercises.

5.2.3 Exercises after class

1. **Basic skills**.

    (1) Compute $\cos(\arcsin \frac{1}{4})$;

    (2) Compute: $\tan(\arcsin \frac{1}{4})$;

    (3) Compute: $\arcsin[\cos(\frac{5\pi}{2} - \frac{\pi}{3})]$;

    (4) Compute: $\cos(\arcsin \frac{1}{8} + \frac{\pi}{6})$.

2. **Comprehensive problems**.

    (1) Prove
    $$\arctan x + arccotx = \frac{\pi}{2}$$

    (2) Solve trigonometric equation:
    $$\sin x + \cos x = \cos 2x$$

    (3) Solve trigonometric equation:
    $$4\cos x \cos 2x = 1.$$

    (4) Solve trigonometric inequality:
    $$\arccos x > \arccos x^2.$$

5.3 Chapter review and exercises

In Chapter 5, we learned the basic properties of four inverse trigonometric functions.

1. Four inverse trigonometric functions.

1.1. Arcsin function $y = \arcsin x$

Its domain: $[-1, 1]$, range: $[-\frac{\pi}{2}, \frac{\pi}{2}]$.

Symmetry: $y = \arcsin x$ is an odd function defined on $[-1, 1]$.

Monotonicity: $y = \arcsin x$ is a strictly increasing function on $[-1, 1]$.

1.2. Arccos function $y = \arccos x$

Its domain: $[-1, 1]$, range: $[-\frac{\pi}{2}, \frac{\pi}{2}]$.

Symmetry: $y = \arccos x$ is not an even function defined on $[-1, 1]$.

Monotonicity: $y = \arccos x$ is a strictly decreasing function on $[-1, 1]$.

1.3. Arctan function $y = \arctan x$

Its domain: $(-\infty, \infty)$, range: $(-\frac{\pi}{2}, \frac{\pi}{2})$.

Symmetry: $y = \arctan x$ is an odd function defined on $(-\infty, \infty)$.

Monotonicity: $y = \arctan x$ is a strictly increasing function on $(-\infty, \infty)$.

1.4. Arccot function $y = arccotx$

Its domain: $(-\infty, \infty)$, range: $(0, \pi)$.

Symmetry: $y = arccotx$ is not an even function defined on $(-\infty, \infty)$.

Monotonicity: $y = arccotx$ is a strictly decreasing function on $(-\infty, \infty)$.

2. Properties of inverse trigonometric functions.

Complementary property of inverse trigonometric function:
$$\arcsin x + \arccos x = \frac{\pi}{2}$$
and
$$\arctan x + arccotx = \frac{\pi}{2}$$

5.3.1 Chapter 5 test

1. **Comprehensive question-1**. Which of the following functions is odd? Which is even? Which is increasing? Which is decreasing?

    (1) $y = \arcsin(x + \frac{1}{2})$;   (2) $y = \arcsin(x^2 + \frac{1}{2})$;

    (3) $y = \arcsin x - 2\arccos x$;   (4) $y = \arctan \frac{\pi x}{2}$.

2. **Comprehensive question-2**.

    (1) Suppose that $x = \arcsin(sinx)$. Find the range of x.

    (2) Suppose that $y = \cos(\arccos y)$. Find the range of y.

    (3) Solve the following equation:
    $$\arctan x + 2arccotx = \frac{2\pi}{3}.$$

    (4) Solve the following inequality:
    $$\arcsin x > \arcsin(1 - x).$$

Chapter 6 Geometric properties of functions

Introduction

- Cartesian coordinate system
- Straight line on the plane
- Curves on the plane
- Convex and concave functions
- Jensen inequality
- Polar coordinate system
- Vector
- Inner product and angle
- Cross product and area
- Line equations
- Plane equations
- Parametric equations
- Vector functions

6.1 Two-dimension and three-dimension coordinate systems

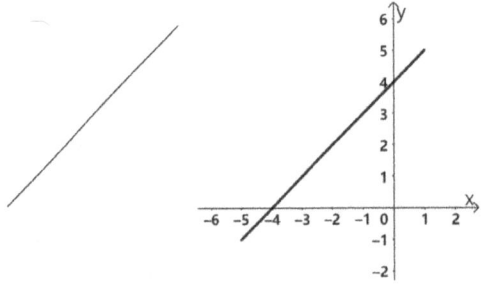

Figure 6.1: Line on plane and line in coordinate system

We use a straight line as an example to discuss the graphs of functions on the plane. For example, a straight line (see Figure 6.1) is nothing but a collection of points. Our eyes can quickly see whether a given point is on the line or not. If we want the machine to determine whether a point is on a straight line or not, we need to use a function to help - first, we need to put the graph in the coordinate system, as shown in Figure 6.1. Then we write the algebraic representation of the function — The common property satisfied by all points (represented by a pair of ordered numbers (x_0, y_0)). Then we can algebraically verify whether a given point (x, y) satisfies this common property or not.

For example, we can see below that the linear equation that passes through two points $(-4, 0)$ and $(0, 4)$ is
$$y - x = 4.$$

Now it is easy to verify whether another point $(-9, -7)$ is on this line or not. We calculate
$$-7 - (-9) = 2 \neq 4.$$

So we know that point $(-9, -7)$ is not on this line (even though we did not use our eyes to verify this. Remember: most machines by now have no "eyes".).

With the help of algebra in the study of plane geometry, lots of problems can be solved by a machine. For example, in general, measuring the distance between two points requires two people to pull a straight line. Now we put both points in a rectangular coordinate system and mark the coordinates (we can use satellites to locate them), then the machine can automatically calculate the distance using the following two-point distance formula.

Theorem 6.1. Distance formula of two points on the plane

The distance formula of two given points $A(x_1, y_1)$ and $B(x_2, y_2)$ in the rectangular coordinate system is
$$dist(A, B) = \sqrt{(x_2 - x_1)^2 + (y_2 - y_1)^2}.$$

Prove: See Chapter 6 in the book "Introductory Geometry and Proofs" book.

□

Since we live in a three-dimensional space (namely, any position in the space can be represented by an ordered three numbers), we often use a three-dimensional rectangular coordinate system. See Figure 6.2.

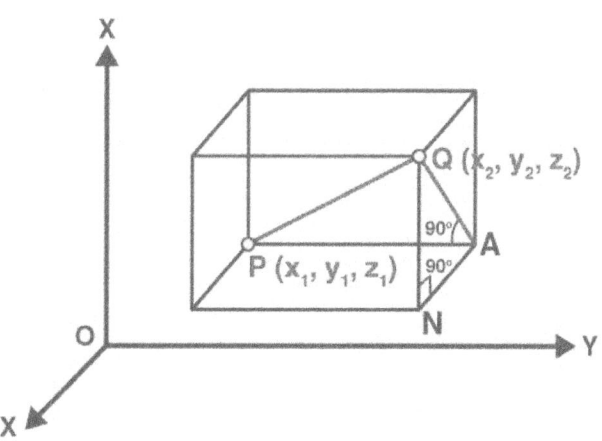

Figure 6.2: 3D distance map

Notice that the x-axis of our drawing points to us. The x, y and z axes satisfy the right-hand side rule, which is often called "positive orientation" in mathematics. In fact, GPS orientation is the positive orientation: the east rotates to the north, then to the west, and then to the south on the map counterclockwise.

Some geometric problems in three dimension, for example, the angle between two planes, and the distance between two lines that are not co-plane will be discussed after we introduce the inner product later(see, for example, Comprehensive problem 6 in this Chapter test).

The formula for the distance between two points in three-dimensional space is the following:

> **Theorem 6.2. The distance formula between two points in space**
> The distance formula of given two points $P(x_1, y_1, z_1)$ and $Q(x_2, y_2, z_2)$ in a rectangular coordinate system is
> $$dist(P, Q) = \sqrt{(x_2 - x_1)^2 + (y_2 - y_1)^2 + (z_2 - z_1)^2}.$$

In the later sections of this chapter, we give the distance formula of any two points in the n dimensional space via the inner product operation.

6.1.1 Exercises after class

1. **Basic skills**.

   (1) Draw a two-dimensional coordinate system and mark the point $A(-3, 2)$. Try to explain how to find the point $A(-3, 2)$ in the coordinate system.

   (2) Draw a three-dimensional coordinate system and mark the point $B(0, -3, 2)$. Try to explain how to find the point $B(0, -3, 2)$. What is the relationship between point A (in part (1))) and point B?

   (3) Draw a three-dimensional coordinate system and mark the point $C(1, -3, 2)$. Try to explain how to find the point $C(1, -3, 2)$. What is the relationship between point C and point B?

   (4) Calculate the distance from point C to point B.

2. **Comprehensive questions**.

   (1) Calculate the distance from point $A(1, 2, 3)$ to point $B(1, 2, 7)$. Can you "see" their distance[1].

   (2) Calculate the shortest distance from point $A(1, 2)$ to point $B(3, x)$.

   (3) Calculate the shortest distance from point $A(1, 2, 3)$ to point $B(3, 1, x)$.

---

[1] "Seeing" the result can not be used to derive the result mathematically. But it may stimulate interest and fun for students to do exercises.

6.2 Graphs of functions in Cartesian coordinate system

We will examine which propositions we can find by studying the graphs of functions in the rectangular coordinate system.

We first look at the graph represented by a two-variable linear equation $ax+by+c = 0$.

Example 6.1 Draw the graph of $2x + 3y - 6 = 0$.

Solution: We learned in the book "Introductory Geometry and Proofs": all points

$$\{(x,y) : ax + by + c = 0\}$$

together form a straight line. So, we just need to select two points, for example, point $A(0,2)$ and point $B(3,0)$, and then connect the two points to get a straight line.

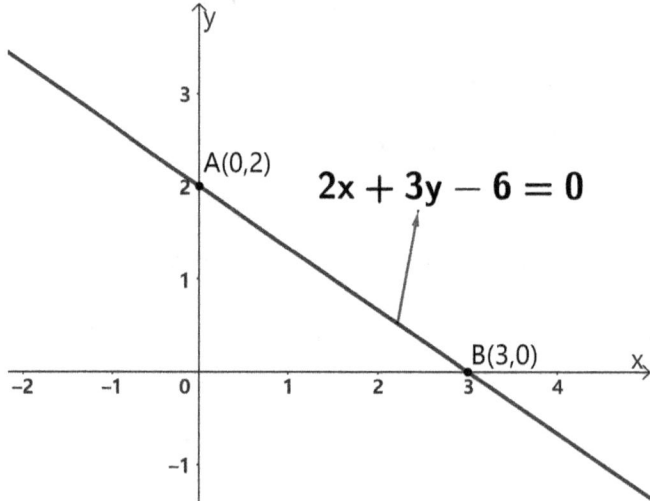

Figure 6.3: straight line $2x + 3y - 6 = 0$

□

Review: we call the x coordinate of the point where the line $ax+by+c = 0$ intersects the x-axis the "x-intercept"; Its y-intercept is the y-coordinate of the point where the line intersects the y-axis.

Example 6.2 (1). Find the x-intercept and y-intercept of $2x + 3y - 6 = 0$.

(2). If the x-intercept of a line is -1 and the y-intercept is 3, Find the linear equation.

Solution: (1). Let $y = 0$, from the equation $2x + 3y - 6 = 0$, we get

$$2x - 6 = 0 \implies x = 3.$$

So the x-intercept is 3.

Let $x = 0$, from the equation $2x + 3y - 6 = 0$, we get

$$3y - 6 = 0 \implies y = 2.$$

So the y-intercept is 2.

(2). From the two-point formula of the straight line, we have: the straight line equation is
$$\frac{y-0}{x-(-1)} = \frac{y-3}{x-0} \Longrightarrow xy = (x+1)(y-3)$$
$$\Longrightarrow y - 3x - 3 = 0.$$

□

The geometric meaning of the first-order inequality has been discussed in the early study of general functions, we shall not repeat it again.

To solve a 2×2 linear system, geometrically, is equivalent to finding the intersection of two straight lines.

Example 6.3 Draw $x+3y+6 = 0$ and $-3x+y+2.01 = 0$ in the same coordinate system. Find the intersection of these two lines.

Solution: We first draw the graphs of two straight lines:

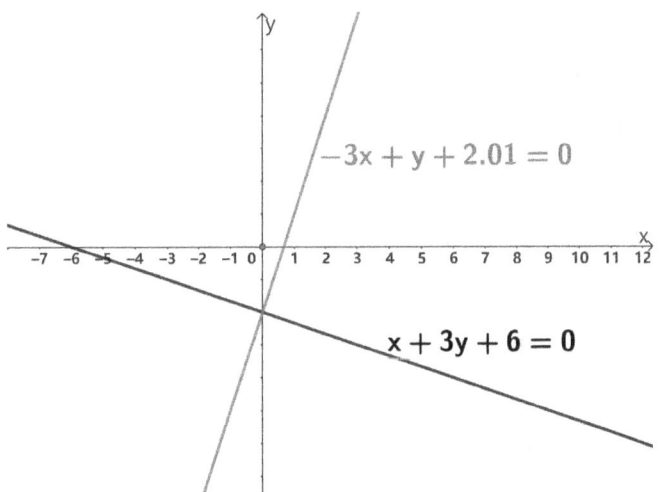

Figure 6.4: Intersection of two lines on the plane

It is difficult to read the small differences in the graph, so it is difficult to tell precisely the location of the intersection point.

Let us solve the equations:
$$\begin{cases} x + 3y + 6 = 0 \\ -3x + y + 2.01 = 0. \end{cases}$$
Using the elimination method, we get:
$$\begin{cases} x = 0.003 \\ y = 2.001. \end{cases}$$
Therefore, the intersection point of these two lines is $(0.003, 2.001)$.

□

In the book "Introductory Geometry and Proofs", we have learned that the graph of a two-variable quadratic equation

$$ax^2 + 2bxy + cy^2 + dx + ey + f = 0$$

is a conic curve, where, a, b, c, d, e, f are constants. Through a "linear transformation", the above equation can be transformed into

$$Ax^2 + By^2 + Cx + Dy + E = 0,$$

where, A, B, C, D, E are other constants[2]. We further write the above equation as (via completing the square)

$$A(x + \frac{C}{2A})^2 + B(y + \frac{D}{2B})^2 = -E + \frac{C^2}{2A^2} + \frac{D^2}{4B^2}.$$

There are three cases:

Case 1: $AB > 0$. The points satisfying the equation are all on an ellipse[3]. In the after-class exercise, we can also see that this case corresponds to $ac > b^2$.

Case 2: $AB < 0$. The points satisfying the equation are on a hyperbola. In the after-class exercise, we can also see that this corresponds to $ac < b^2$.

Case 3: $A = 0$ or $B = 0$. The points satisfying the equation are on a parabola.

We also learned in Chapter 3 that the graph of parabolas can help us solve quadratic inequalities.

We now study the graphs of linear functions and quadratic functions in three-dimensional space.

First, we study the graph of the three-variable linear equation $ax + by + cz = d$ in the three-dimensional coordinate system.

Example 6.4 Draw the graph of $x + 2y + 2z = 4$ in the three-dimensional coordinate system.

Solution: We will discuss in Section 6.5 below that, all points satisfying

$$\{(x, y, z) : ax + by + cz = d\}$$

will form a plane. So, we just need to select three points, for example, point $A(4, 0, 0)$ (on the x- axis), point $B(0, 2, 0)$ (on the y- axis), and point $C(0, 0, 2)$ (on the z- axis). Then connect these three points to get the plane.

□

Geometrically, finding the solution of a 3×3 linear system is equivalent to finding the intersection of three planes.

---

[2] We will learn such linear transformations in detail in a "linear algebra" course in college.

[3] Or it is only one point (if the right side of the equation is zero), or no point satisfies the equation (if the right side of the equation is negative).

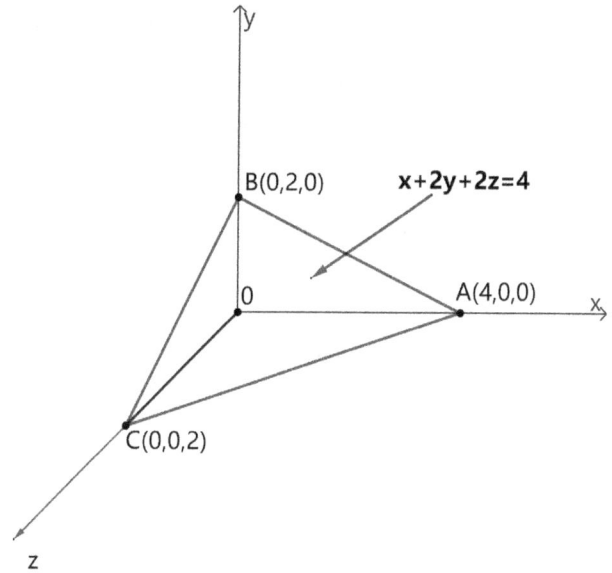

Figure 6.5: Plane graph in three-dimensional coordinate system

Example 6.5 Find the intersection point of the plane $x + y + z = 3$, $x - y + z = 1$ and $x + y - z = 1$.

Solution:

Let us solve the linear system:
$$\begin{cases} x + y + z = 3 & (1) \\ x - y + z = 1 & (2) \\ x + y - z = 1 & (3) \end{cases}$$

We use the elimination method. From (2)+(3), we get $x = 1$. Bringing this in (1) and (2), we get:
$$\begin{cases} y + z = 2 & (4) \\ -y + z = 0 & (5). \end{cases}$$

Again, from (4)+(5), we get $z = 1$; From (4)-(5), we get $y = 1$. Therefore, the intersection of the three planes is $(1, 1, 1)$.

□

The graph of a three-variable quadratic equation is much more complex. Here we only discuss the sphere equation in three-dimensional space.
$$x^2 + y^2 + z^2 + ax + by + cz + d = 0.$$

After completing the square, we have
$$(x + \frac{a}{2})^2 + (y + \frac{b}{2})^2 + (z + \frac{c}{2})^2 = -d + \frac{a^2}{4} + \frac{b^2}{4} + \frac{c^2}{4}.$$

If the right side of the equation is positive, then this is a sphere with the center at $(-\frac{a}{2}, -\frac{b}{2}, -\frac{c}{2})$ and the radius of $\sqrt{-d + \frac{a^2}{4} + \frac{b^2}{4} + \frac{c^2}{4}}$.

Finally, we use geometric intuition to understand Jensen's inequality[4]. First, we define convex and concave functions on interval I.

> **Definition 6.1. Definition of convex and concave functions**
>
> Consider a function $y = f(x)$ defined on an interval I. If for any two points $x_1, x_2 \in I$ and $\lambda \in [0, 1]$, we always have
> $$f[\lambda x_2 + (1 - \lambda)x_1] \leq \lambda f(x_2) + (1 - \lambda)f(x_1),$$
> then we call this function a convex function on I.
>
> Consider a function $y = f(x)$ defined on an interval I. If for any two points $x_1, x_2 \in I$ and $\lambda \in [0, 1]$, always have
> $$f[\lambda x_2 + (1 - \lambda)x_1] \geq \lambda f(x_2) + (1 - \lambda)f(x_1),$$
> then we call this function a concave function on I.

The geometric properties of convex and concave functions can be seen in the figure below. Later, when we study Calculus, we will find mathematical tools to verify the convexity and concavity of a function (see, for example, Chapter 8, Section 8.2).

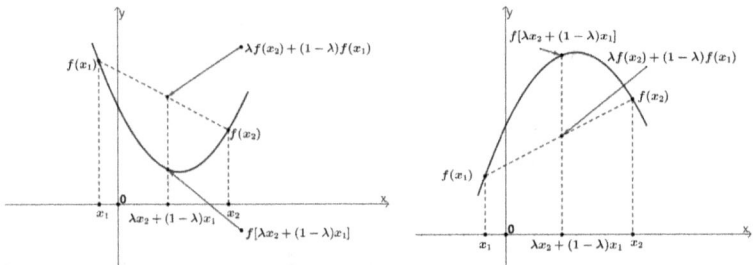

Figure 6.6: Convex and concave functions

Example 6.6 (**The square function is a convex function**) From the graph, one can see that the square function $y = x^2$ is a convex function.

Example 6.7 (**The natural logarithmic function is a concave function**) From the graph, we can see that the natural logarithm function $y = \ln x$ is a concave function.

As we can see from the graph, we give the following Jensen's inequality.

> **Theorem 6.3. (Jensen's Inequality)**
>
> Assuming that a function $y = f(x)$ on interval I is continuous and convex, then for k points $x_1, x_2, \cdots, x_k$ on interval I, we have
> $$f\left(\frac{x_1 + x_2 + \cdots + x_k}{k}\right) \leq \frac{f(x_1) + f(x_2) + \cdots + f(x_k)}{k}.$$
> Assuming that a function $y = g(x)$ on interval I is continuous and concave, then

---
[4]Jensen's inequality was proved by Danish mathematician Johan Jensen in 1906.

> for k points $x_1, x_2, \cdots, x_k$ on interval I, we have
> $$g(\frac{x_1+x_2+\cdots+x_k}{k}) \geq \frac{g(x_1)+g(x_2)+\cdots+g(x_k)}{k}.$$

Jensen's inequality is widely used in the proof of inequalities.

Example 6.8 Prove: For given k numbers $x_1, x_2, \cdots, x_k$,
$$(\frac{x_1+x_2+\cdots+x_k}{k})^2 \leq \frac{x_1^2+x_2^2+\cdots+x_k^2}{k}.$$

Proof: We provide two methods to prove this inequality.

Method 1: We use Jensen's inequality to prove it. Let $f(x) = x^2$. We know it is a convex function, so
$$f(\frac{x_1+x_2+\cdots+x_k}{k}) \leq \frac{f(x_1)+f(x_2)+\cdots+f(x_k)}{k}.$$

That is:
$$(\frac{x_1+x_2+\cdots+x_k}{k})^2 \leq \frac{x_1^2+x_2^2+\cdots+x_k^2}{k}.$$

Method 2: We use Cauchy-Schwarz inequality to prove it. Introduce two vectors
$$\mathbf{U} = (x_1, x_2, \cdots, x_k), \quad \mathbf{V} = (1, 1, \cdots, 1).$$

By Cauchy-Schwarz inequality $|\mathbf{U} \cdot \mathbf{V}| \leq |\mathbf{U}| \cdot |\mathbf{V}|$, we have
$$|x_1 + x_2 + \cdots + x_n| \leq \sqrt{x_1^2 + x_2^2 + \cdots + x_k^2} \cdot \sqrt{k}$$

Take a square of the above equation, and then divide by k^2 on both sides. We have
$$(\frac{x_1+x_2+\cdots+x_k}{k})^2 \leq \frac{x_1^2+x_2^2+\cdots+x_k^2}{k}.$$

$\square$

Example 6.9 Prove[5]: for any given k positive numbers $x_1, x_2, \cdots, x_k$,
$$\sqrt[k]{x_1 \cdot x_2 \cdots x_k} \leq \frac{x_1+x_2+\cdots+x_k}{k}.$$

Proof: We use Jensen's inequality to prove it. Let $f(x) = \ln x$. We know it is a concave function, so
$$f(\frac{x_1+x_2+\cdots+x_k}{k}) \geq \frac{f(x_1)+f(x_2)+\cdots+f(x_k)}{k}.$$

---

[5] We call $\sqrt[k]{x_1 \cdot x_2 \cdots x_k}$ the geometric mean of numbers $x_1, x_2, \cdots, x_k$; $\frac{x_1+x_2+\cdots+x_k}{k}$ is called the arithmetic mean of numbers $x_1, x_2, \cdots, x_k$.

That is:
$$\ln \frac{x_1 + x_2 + \cdots + x_k}{k} \geq \frac{\ln x_1 + \ln x_2 + \cdots + \ln x_k}{k}$$
$$= \frac{1}{k} \cdot \ln(x_1 x_2 \cdots x_k)$$
$$= \ln \sqrt[k]{x_1 \cdot x_2 \cdot \cdots \cdot x_k}.$$

Since $\ln x$ is an increasing function, we have
$$\sqrt[k]{x_1 \cdot x_2 \cdot \cdots \cdot x_k} \leq \frac{x_1 + x_2 + \cdots + x_k}{k}.$$

□

6.2.1 Exercises after class

1. **Basic skill.**

   (1) Find the linear equation passing through two points $A(1,2)$ and $B(2,1)$.

   (2) Find the x-interception and y-intercept of line $2x - 5y = 7$.

   (3) Find the intersection of two straight lines (if they intersect): (i) $3x + 4y = 5$ and $5x - 4y = 1$; (ii) $8y = 10x + 1$ and $5x - 4y = 1$.

   (4) Find the intersection of the following three planes: $3x+4y = 5, 5x-4y+2z = 1$ and $3x - 4y - 2z = 7$.

   (5) Prove the sum of two convex functions is also a convex function.

2. **Comprehensive questions.**

   (1) Prove:
   $$\log_\pi 3 + \log_3 \pi \geq 2.$$

   (2) Prove: for any three positive numbers a, b, c,
   $$\frac{a}{b} + \frac{b}{c} + \frac{c}{a} \geq 3.$$

   (3) Prove:

   (i) $2(1 + x^2) \geq (1 + x)^2$

   (ii) $3(1 + x^2 + x^4) \geq (1 + x + x^2)^2$

   (iii) Can you sum up a general inequality (including the previous two inequalities) and prove it?

   (4) Using mathematical induction to prove Theorem 6.3.

6.3 Polar coordinate system

In addition to using the Cartesian coordinate system, with the help of trigonometric functions, we can also use different methods to describe points and graphs in space. It should be pointed out that this new expressions vary with different dimensions.

6.3.1 Two dimensional polar coordinate system (r, θ)

Take a point O (we call it a pole) on the plane and the ray from O to the right $\vec{OX}$ (We call it the polar axis). For any given point P on the plane that is different from the pole, we can measure its distance to the pole $|\vec{OP}|$, recorded as r, and measure the angle between $\vec{OP}$ and $\vec{OX}$, denoted as θ. Then point P has a position representation (r, θ). We call (r, θ) the polar coordinates of point P. It should be pointed out that the polar coordinate representation of point P is obviously not unique: for any integer $n \in \mathbb{Z}$, polar coordinates (r, θ) and $(r, \theta + 2n\pi)$ represent the same point.

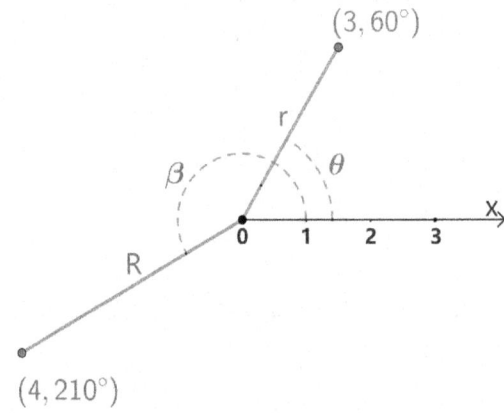

Figure 6.7: Polar plot

Using the definitions of trigonometric functions, we can easily convert the polar coordinates (r, θ) of a point P to its rectangular coordinates:

$$x = r \cos \theta$$
$$y = r \sin \theta.$$

Conversely, the rectangular coordinates (x, y) of a point P can also be converted to its polar coordinates:

$$r^2 = x^2 + y^2$$
$$\tan \theta = \frac{y}{x}.$$

Obviously, we also need to determine the quadrant of angle θ from the signs of $\cos\theta$ and $\sin\theta$.

Example 6.10 Draw the following points in the polar coordinate system and convert their polar coordinates into rectangular coordinates: (1) $A(2, \frac{\pi}{3})$; (2) $B(3, \frac{5\pi}{3})$; (3) $C(1, \frac{8\pi}{3})$; (4) $D(-1, \frac{\pi}{3})$.

Solution: (1). Cartesian coordinates of A point:
$$x = r\cos\theta = 2 \cdot \cos\frac{\pi}{3} = 1$$
$$y = r\sin\theta = 2 \cdot \sin\frac{\pi}{3} = \sqrt{3}.$$
That is, the rectangular coordinate of the A point is $(1, \sqrt{3})$.

(2). Cartesian coordinates of B point:
$$x = r\cos\theta = 3 \cdot \cos\frac{5\pi}{3} = \frac{3}{2}$$
$$y = r\sin\theta = 3 \cdot \sin\frac{5\pi}{3} = -\frac{3\sqrt{3}}{2}.$$
That is, the rectangular coordinates of the B point are $(\frac{3}{2}, -\frac{3\sqrt{3}}{2})$.

(3). Cartesian coordinates of the C point:
$$x = r\cos\theta = 1 \cdot \cos\frac{8\pi}{3} = -1$$
$$y = r\sin\theta = 1 \cdot \sin\frac{8\pi}{3} = \sqrt{3}.$$
That is, the rectangular coordinate of the A point is $(-1, \sqrt{3})$.

(4). Cartesian coordinates of D point:
$$x = r\cos\theta = -1 \cdot \cos\frac{\pi}{3} = -1$$
$$y = r\sin\theta = -1 \cdot \sin\frac{\pi}{3} = -\sqrt{3}.$$
That is, the rectangular coordinate of the A point is $(-1, -\sqrt{3})$.

□

Example 6.11 Draw the following points in the rectangular coordinate system and convert them into polar coordinates: (1) $A(\frac{1}{2}, \frac{\sqrt{3}}{2})$; (2) $B(-\frac{1}{2}, \frac{\sqrt{3}}{2})$; (3) $C(\frac{\sqrt{3}}{2}, \frac{1}{2})$; (4) $D(-\frac{\sqrt{3}}{2}, -\frac{1}{2})$.

Solution: (1). Let the polar coordinates of A point be (r, θ). We calculate
$$r = \sqrt{x^2 + y^2} = \sqrt{\frac{1}{4} + \frac{3}{4}} = 1$$
$$\tan\theta = \frac{y}{x} = \sqrt{3}.$$
Because point $A(\frac{1}{2}, \frac{\sqrt{3}}{2})$ is in the first quadrant, so $\theta = \frac{\pi}{3}$. Therefore, the polar coordinates of the A point are $(1, \frac{\pi}{3})$.

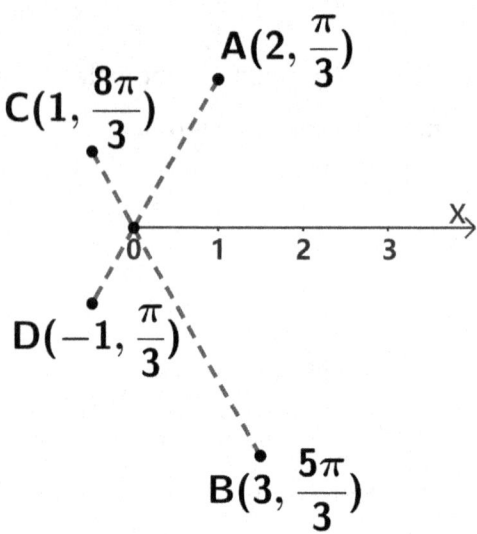

Figure 6.8: Points in polar coordinate system

(2). Let the polar coordinates of B point be (r, θ). We calculate
$$r = \sqrt{x^2 + y^2} = \sqrt{\frac{1}{4} + \frac{3}{4}} = 1$$
$$\tan \theta = \frac{y}{x} = -\sqrt{3}.$$
Because point $B(\frac{1}{2}, -\frac{\sqrt{3}}{2})$ is in the second quadrant, so $\theta = \frac{2\pi}{3}$. Therefore, the polar coordinates of the B point are $(1, \frac{2\pi}{3})$.

(3). Let the polar coordinates of C point be (r, θ). We calculate
$$r = \sqrt{x^2 + y^2} = \sqrt{\frac{3}{4} + \frac{1}{4}} = 1$$
$$\tan \theta = \frac{y}{x} = \frac{1}{\sqrt{3}}.$$
Because point $C(\frac{\sqrt{3}}{2}, \frac{1}{2})$ is in the first quadrant. So $\theta = \frac{\pi}{6}$. Therefore, the polar coordinates of the C point are $(1, \frac{\pi}{6})$.

(4). Let the polar coordinates of D point be (r, θ). We calculate
$$r = \sqrt{x^2 + y^2} = \sqrt{\frac{3}{4} + \frac{1}{4}} = 1$$
$$\tan \theta = \frac{y}{x} = \sqrt{3}.$$
Because point $D(-\frac{\sqrt{3}}{2}, -\frac{1}{2})$ is in the third quadrant, so $\theta = \frac{7\pi}{6}$. Therefore, the polar coordinates of the D point are $(1, \frac{7\pi}{6})$.

□

6.3 Polar coordinate system

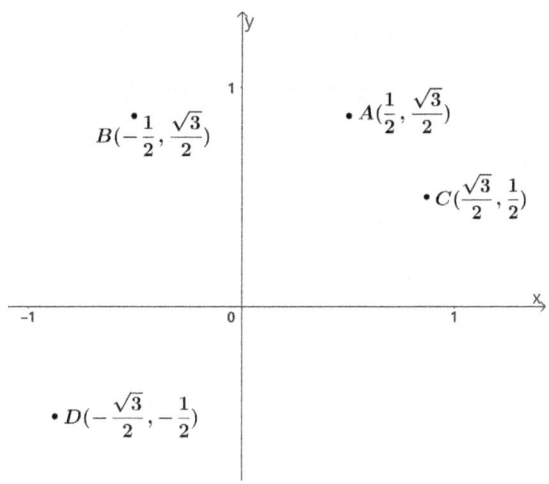

Figure 6.9: Points in rectangular coordinate system

6.3.2 Polar equation

We can also use polar coordinates to describe curves on a plane.

Example 6.12 In the polar coordinate system, draw the following diagrams of the polar equations:

(a) $r = 2$; (b) $\theta = \frac{2\pi}{3}$; (c) $r = 2\cos\theta$.

Solution: See Figure 6.10.

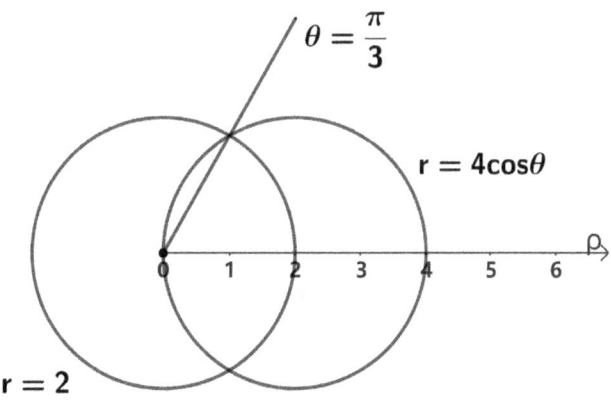

Figure 6.10: Example 6.12 Polar coordinate equations

□

Using the transformation formula from polar coordinates to rectangular coordinates, sometimes we can easily convert a polar equation into a rectangular coordinate equation (our familiar form).

Example 6.13 Convert the following polar equations into rectangular coordinate equations

(a) $r = 1$; (b) $\theta = \frac{\pi}{4}$; (c) $r = 2\cos\theta + 2\sin\theta$.

Solution: (a). $r = 1 \Longrightarrow r^2 = 1$, so the rectangular coordinate equation is
$$x^2 + y^2 = 1.$$
This is the equation of a circle with center at the origin and a radius of 1.

(b). $\theta = \frac{\pi}{4} \Longrightarrow \frac{y}{x} = 1$. So the rectangular coordinate equation is
$$y = x.$$
This is a straight line equation with a slope of 1, passing through the origin.

(c).
$$r = 2\cos\theta + 2\sin\theta \Longrightarrow r^2 = 2r\cos\theta + 2r\sin\theta$$
$$\Longrightarrow x^2 + y^2 = 2x + 2y$$
$$\Longrightarrow (x-1)^2 + (y-1)^2 = 2.$$
This is the equation of a circle with center $(1, 1)$ and a radius of $\sqrt{2}$.

□

Not all polar equations can be converted into some "easy" rectangular coordinate equations. For example, polar equation $r = 2 + 2\cos\theta$ (we call it Cardioid equation since its diagram looks like a heart, see Figure 6.11): its rectangular coordinate equation is not simple (see Exercise 2 (2) in this section).

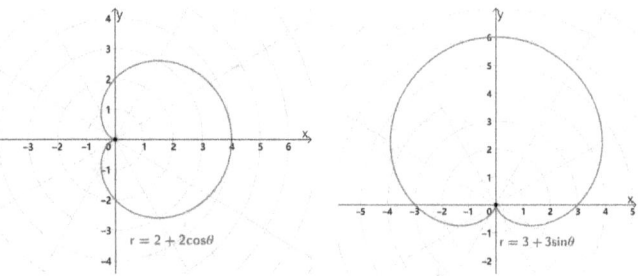

Figure 6.11: Cardioid pattern

Amazingly, there is a uniform representation of conic curves using polar equations. See Comprehensive Question-5 at the end of this chapter.

6.3.3 Exercises after class

1. **Basic skills**.

   (1) Draw the following points in the polar coordinate system and convert their polar coordinates into rectangular coordinates: (i) $A(1, \frac{2\pi}{3})$; (ii) $B(2, \frac{5\pi}{4})$; (iii) $C(-1, \frac{8\pi}{3})$; (iv) $D(-1, -\frac{\pi}{3})$.

   (2) Draw the following points in the rectangular coordinate system and convert them into polar coordinates: (i) $A(\frac{\sqrt{2}}{2}, \frac{\sqrt{2}}{2})$; (ii) $B(-\frac{1}{2}, -\frac{\sqrt{3}}{2})$; (iii) $C(\frac{\sqrt{3}}{2}, -\frac{1}{2})$; (iv) $D(-\frac{\sqrt{2}}{2}, -\frac{\sqrt{2}}{2})$.

   (3) In the polar coordinate system, draw the following diagram of the polar equation: (i) $r = \cos\theta$; (ii) $r = 4\sin\theta$.

   (4) Draw the diagram of $r = 2\cos\theta$ in polar coordinate system.

2. **Comprehensive questions**.

   (1) Calculate the distance between two points in the polar coordinate system: (i) $A(2, \frac{2\pi}{3})$ and $B(5, \frac{2\pi}{2})$; (ii) $A(8, \frac{5\pi}{3})$ and $B(3, \frac{2\pi}{3})$; (ii) $A(3, \frac{5\pi}{3})$ and $B(3, \frac{7\pi}{6})$; (iv) $A(8, \frac{\pi}{2})$ and $B(3, 0)$.

   (2) Find the Cartesian equation of the polar equation $r = 3 + 3\sin\theta$.

   (3) If two non-trivial circles $r = a$ and $r = b\sin\theta$ are tangent. Find the algebraic relationship between a and b.

6.4 Operations on vectors

We first review the addition, scalar multiplication, and inner product of vectors (we have learned all these in the book "Introductory Geometry and Proofs" (Chapter 6)).

6.4.1 Addition, scalar multiplication and inner product of vectors

For two n-dimensional vectors $\vec{u} = (u_1, u_2, \cdots, u_n)$ and $\vec{v} = (v_1, v_2, \cdots, v_n)$, we define the addition of vectors.

> **Definition 6.2. Addition of Vectors**
> For two vectors $\vec{u} = (u_1, u_2, \cdots, u_n)$ and $\vec{v} = (v_1, v_2, \cdots, v_n)$, we define their sum as
> $$\vec{u} + \vec{v} = (u_1 + v_1, u_2 + v_2, \cdots, u_n + v_n)$$

In short, vector addition is the addition of corresponding components (the first component plus the first component, $\cdots$, the nth component plus the nth component). Similar to the addition of numbers, vector addition also obeys the commutative rule and the associative rule (we have verified these rules for the two-dimensional vectors in the "Introductory Geometry and Proofs" book).

> **Proposition 6.1. Commutative rule for the addition of vectors**
> For two vectors $\vec{u} = (u_1, u_2, \cdots, u_n)$ and $\vec{v} = (v_1, v_2, \cdots, v_n)$, we have
> $$\vec{u} + \vec{v} = \vec{v} + \vec{u}.$$

> **Proposition 6.2. Associative rule for the addition of vectors**
> For three vectors $\vec{u} = (u_1, u_2, \cdots, u_n)$, $\vec{v} = (v_1, v_2, \cdots, v_n)$ and $\vec{w} = (w_1, w_2, \cdots, w_n)$, we have
> $$(\vec{u} + \vec{v}) + \vec{w} = \vec{u} + (\vec{v} + \vec{w}).$$

✍ Exercise 6.1 Use the associative rule for addition to prove the associative rule for the addition of vectors.

Adding the same vector multiple times leads us to introduce the concept of a scalar product.

> **Definition 6.3. Scalar product and subtraction of vectors**
> For any real number λ and any vector $\vec{u} = (u_1, u_2, \cdots, u_n)$, we define
> $$\lambda \vec{u} = (\lambda u_1, \lambda u_2, \cdots, \lambda u_n)$$

If $\vec{v} = (v_1, v_2, \cdots, v_n)$ is another arbitrary vector, we define the subtraction for vectors as:
$$\vec{u} - \vec{v} = \vec{u} + (-1) \cdot \vec{v} = (u_1 - v_1, u_2 - v_2, \cdots, u_n - v_n).$$

From the definition, we get a more general formula (which we usually call the linear combination of two vectors): for any two real numbers α, β and two vectors $\vec{u} = (u_1, u_2, \cdots, u_n)$, $\vec{v} = (v_1, v_2, \cdots, v_n)$,
$$\alpha \vec{u} + \beta \vec{v} = (\alpha u_1 + \beta v_1, \alpha u_2 + \beta v_2, \cdots, \alpha u_n + \beta v_n). \tag{6.1}$$

The geometric significance of vector addition, subtraction, and scalar multiplication has been discussed in Chapter 6 of the book "Introductory Geometry and Proofs". We shall not repeat it again here.

The products for vectors are relatively complicated. Let us first review the inner product of two vectors introduced in the book "Introductory Geometry and Proofs". Here we introduce the inner product in a general n dimensional space.

Definition 6.4. inner product

For any two vectors $\vec{u} = (u_1, u_2, \cdots, u_n)$, $\vec{v} = (v_1, v_2, \cdots, v_n)$, we define their inner product as
$$\vec{u} \cdot \vec{v} = u_1 v_1 + u_2 v_2 + \cdots + u_n v_n. \tag{6.2}$$

From the definition of the inner product, we can verify that the inner product satisfies the following algebraic operation rules.

Proposition 6.3. Operation rules on inner product

For any two real numbers α, β and three vectors $\vec{u}, \vec{v}$ and $\vec{w}$, we have
$$\vec{u} \cdot \vec{v} = \vec{v} \cdot \vec{u} \quad (Commutative\ rule)$$
and
$$\vec{u} \cdot (\alpha \vec{v} + \beta \vec{w}) = \alpha \vec{u} \cdot \vec{v} + \beta \vec{u} \cdot \vec{w}. \quad (Distribution\ rule)$$

Exercise 6.2 Prove the distribution rule for the inner product.

Note *Think: Is there any associative rule for the inner product?*

For a given vector $\vec{u} = (u_1, u_2, \cdots, u_n)$, it can be represented by a directional line segment from the origin $(0, 0, \cdots, 0)$ to the point $(u_1, u_2, \cdots, u_n))$ in the coordinate system (the length of this line segment is called the length of the vector and is denoted as $|\vec{u}|$). Using Pythagorean identity and the definition of the inner product, we have the

following formula to calculate the length of a vector.
$$|\vec{u}| = \sqrt{u_1^2 + u_2^2 + \cdots + u_n^2} = \sqrt{\vec{u} \cdot \vec{u}}. \tag{6.3}$$

If we use two points $A(x_1, x_2, \cdots, x_n)$ and $B(y_1, y_2, \cdots, y_n)$ in the coordinate system to represent a vector $\vec{AB}$, then
$$\vec{AB} = \vec{OB} - \vec{OA} = (y_1 - x_1, y_2 - x_2, \cdots, y_n - x_n).$$

From this and the above formula (6.3), we obtain the formula to calculate the distance between two points in any dimensional space.

Proposition 6.4. Distance formula

The distance from point $A(x_1, x_2, \cdots, x_n)$ to point $B(y_1, y_2, \cdots, y_n)$, denoted as $d(A, B)$, is given by the following formula:
$$d(A, B) = \sqrt{(y_1 - x_1)^2 + (y_2 - x_2)^2 + \cdots + (y_n - x_n)^2}. \tag{6.4}$$

One of the greatest applications of the inner product is that: human beings can use the algebraic operation of the inner product to "calculate" (not measure!) the angle between two vectors. This idea enables us to "teach" machines how to "read" the angle between two vectors: people only need to design a program that makes machines calculate the inner product of two vectors!

Theorem 6.4. The angle between two vectors

Suppose the angle between two nontrival vectors vectors $\vec{u}$ and $\vec{v}$ is $\theta \in [0, \pi]$, then
$$\cos \theta = \frac{\vec{u} \cdot \vec{v}}{|\vec{u}||\vec{v}|}. \tag{6.5}$$

Proof: See the proof of Theorem 6.1 in the book "Introductory Geometry and Proofs".

$\square$

Theorem 6.4 has the following simple but very useful corollary.

Corollary 6.1. Orthogonal of two vectors

The necessary and sufficient condition for two nontrivial vectors $\vec{u}$ and $\vec{v}$ to be orthogonal to each other is: $\vec{u} \cdot \vec{v} = 0$.

Example 6.14 Calculate the angle θ between vectors $\vec{u} = (1, 0, 3, 4)$ and $\vec{v} = (2, -1, -2, 4)$.

Solution:

$$\cos\theta = \frac{\vec{u}\cdot\vec{v}}{|\vec{u}|\cdot|\vec{v}|}$$
$$= \frac{12}{\sqrt{26}\cdot\sqrt{26}}$$
$$= \frac{6}{13}$$

So:

$$\theta = \arccos\frac{6}{13}.$$

6.4.2 More general inner product*

Different people have different viewpoints. There are also different inner products in space $\mathbb{R}^n = \{(x_1, x_2, \cdots, x_n) : x_i \in \mathbb{R} \text{ for } i = 1, 2, \cdots, n\}$. Here we introduce a definition of a general inner product.

6.4.2.1 General inner product

Definition 6.5. General inner product

The inner product operation on $\mathbb{R}^n$ is a mapping, denoted as $[\cdot,\cdot]$. It maps two vectors to a real number and satisfies the following three properties:

(1) Linear property: *If α, β are two real numbers and $\vec{u}, \vec{v}, \vec{w}$ are three vectors, then*

$$[\alpha\vec{u} + \beta\vec{v}, \vec{w}] = \alpha[\vec{u}, \vec{w}] + \beta[\vec{v}, \vec{w}].$$

(2) Symmetry: *For any two vectors $\vec{u}$ and $\vec{v}$,*

$$[\vec{u}, \vec{v}] = [\vec{v}, \vec{u}]$$

(3) Positivity: *For any vector $\vec{u}$,*

$$[\vec{u}, \vec{u}] \geq 0, \quad \text{and } [\vec{u}, \vec{u}] = 0 \quad \text{if and only if } \vec{u} = \vec{0} = (0, \cdots, 0).$$

For the general inner product, we define the length of the vector as:

$$|\vec{u}| = \sqrt{[\vec{u}, \vec{u}]}.$$

Let us prove the following theorem.

Theorem 6.5. Cauchy-Schwarz inequality

For any two vectors $\vec{u}$ and $\vec{v}$, we have

$$|[\vec{u}, \vec{v}]| \leq |\vec{u}||\vec{v}|,$$

> or
> $$-|\vec{u}||\vec{v}| \leq [\vec{u}, \vec{v}] \leq |\vec{u}||\vec{v}|.$$

Solution: For any two vectors $\vec{u}$ and $\vec{v}$, we introduce a function $f(t)$ of t variable:
$$f(t) = [\vec{u} + t\vec{v}, \vec{u} + t\vec{v}].$$
From the definition of the inner product, we know that $f(t) \geq 0$.

On the other hand, we calculate:
$$\begin{aligned} f(t) &= [\vec{u} + t\vec{v}, \vec{u} + t\vec{v}] \\ &= [\vec{u}, \vec{u} + t\vec{v}] + t[\vec{v}, \vec{u} + t\vec{v}] \\ &= [\vec{u}, \vec{u}] + 2t[\vec{u}, \vec{v}] + t^2[\vec{v}, \vec{v}] \\ &= |\vec{u}|^2 + 2t[\vec{u}, \vec{v}] + t^2|\vec{v}|^2. \end{aligned}$$

Since $f(t)$ is a non-negative second-order polynomial of t, $f(t) = 0$ has a duplicate real number solution or no solution. So its discriminant
$$\Delta = \{2[\vec{u}, \vec{v}]\}^2 - 4|\vec{u}|^2|\vec{v}|^2 \leq 0.$$

That is:
$$|[\vec{u}, \vec{v}]| \leq |\vec{u}||\vec{v}|.$$

□

Due to the Cauchy-Schwarz inequality, we can measure the angle between two nontrivial vectors by using any inner product:

> **Definition 6.6. Angle between two vectors**
> We define the angle $\theta \in [0, \pi]$ between two nontrivial vectors $\vec{u}$ and $\vec{v}$ as
> $$\theta = \arccos \frac{[\vec{u}, \vec{v}]}{|\vec{u}||\vec{v}|}.$$

Example 6.15 For any two vectors $\vec{u} = (u_1, u_2)$ and $\vec{v} = (v_1, v_2)$, we define a new inner product
$$<\vec{u}, \vec{v}>_{\text{new}} = u_1 v_1 + 2 u_2 v_2.$$
For two given vectors $\vec{U} = (\frac{\sqrt{2}}{2}, \frac{\sqrt{2}}{2})$ and $\vec{V} = (\frac{\sqrt{2}}{2}, -\frac{\sqrt{2}}{2})$,

(1) Use the new inner product to calculate their length;

(2) Use the new inner product to calculate the angle θ between these two vectors.

Solution: (1) By definition, we calculate

$$|\vec{U}|_{new} = \sqrt{<\vec{U}, \vec{U}>_{new}}$$

$$= \sqrt{\frac{\sqrt{2}}{2} \cdot \frac{\sqrt{2}}{2} + 2 \cdot \frac{\sqrt{2}}{2} \cdot \frac{\sqrt{2}}{2}}$$

$$= \sqrt{\frac{6}{4}}$$

$$= \frac{\sqrt{6}}{2},$$

and

$$|\vec{V}|_{new} = \sqrt{<\vec{V}, \vec{V}>_{new}}$$

$$= \sqrt{\frac{\sqrt{2}}{2} \cdot \frac{\sqrt{2}}{2} + 2 \cdot (-\frac{\sqrt{2}}{2}) \cdot (-\frac{\sqrt{2}}{2})}$$

$$= \sqrt{\frac{6}{4}}$$

$$= \frac{\sqrt{6}}{2}.$$

(2). By definition, we have

$$\cos\theta = <\frac{1}{|\vec{U}|_{new}}\vec{U}, \frac{1}{|\vec{V}|_{new}}\vec{V}>_{new}$$

$$= \frac{1}{|\vec{U}|_{new} \cdot |\vec{V}|_{new}} \cdot <\vec{U}, \vec{V}>_{new}$$

$$= \frac{1}{\frac{6}{4}} \cdot [\frac{\sqrt{2}}{2} \cdot \frac{\sqrt{2}}{2} + 2 \cdot \frac{\sqrt{2}}{2} \cdot (-\frac{\sqrt{2}}{2})]$$

$$= -\frac{1}{3}$$

So

$$\theta = \arccos(-\frac{1}{3}).$$

Remark Using the standard inner product on the plane, we know that

$$\vec{U} \cdot \vec{V} = \frac{\sqrt{2}}{2} \cdot \frac{\sqrt{2}}{2} + \frac{\sqrt{2}}{2} \cdot (-\frac{\sqrt{2}}{2}) = 0!$$

That is, the two vectors are perpendicular to each other. Under the new inner product above, these two vectors are not perpendicular.

We will look at one practical application of these inner products in the next section.

6.4.2.2 Relevance and irrelevance*

1. A prose by Martin Niemöller

"First they came ..." is the poetic form of a 1946 postwar confessional prose by the German Lutheran pastor Martin Niemöller (1892—1984). It is about the silence of German intellectuals and certain clergy—-including, by his own admission, Niemöller himself—following the Nazis' rise to power and subsequent incremental purging of their chosen targets, group after group. It deals with themes of persecution, guilt, repentance, and personal responsibility. —— from wikipedia

First they came...

First they came for the Communists
And I did not speak out
Because I was not a Communist

Then they came for the Socialists
And I did not speak out
Because I was not a Socialist

Then they came for the trade unionists
And I did not speak out
Because I was not a trade unionist

Then they came for the Jews
And I did not speak out
Because I was not a Jew

Then they came for me
And there was no one left
To speak out for me

The prose indicates that those irrelevant events (Nazis persecuting Communists, Socialist Jews, etc.), in the end, are closely related to the clergy's fate!

How could it happen? Here, we try to use data analysis to explain it.

2. Relevant or irrelevant

For simplicity, we use two-dimensional vectors to represent certain social events.

For two given vectors $\vec{U} = (u_1, u_2)$ and $\vec{V} = (v_1, v_2)$. In a given society, the relation between two vectors can be measured by the angle between them, which can be calculated via the inner product of these two vectors. If the angle is $0°$ or $180°$, we say two vectors are completely relevant (to be more precise: they are linearly relevant); if the angle is $90°$, we say they are completely irrelevant. Naturally, we call two vectors closely related if the angle between them is close to $0°$ or $180°$, and call two vectors almost irrelevant if the angle between them is close to $90°$.

In fact, if we use θ to denote the angle between $\vec{U}$ and $\vec{V}$, we can use the inner product $<\vec{U},\vec{V}>$ to calculate θ:

$$\cos\theta = <\frac{1}{|\vec{U}|}\vec{U}, \frac{1}{|\vec{V}|}\vec{V}>$$

where $|\vec{U}| = \sqrt{<\vec{U},\vec{U}>}$ is the length of vector $\vec{U}$ and $|\vec{V}| = \sqrt{<\vec{V},\vec{U}>}$ is the length of vector $\vec{V}$. In this section, we only talk about non-trivial vectors—those vectors with lengths larger than 0.

From Cauchy-Schwarz inequality, we know that for any two vectors $\vec{u}$ and $\vec{v}$,

$$|<\frac{1}{|\vec{u}|}\vec{u}, \frac{1}{|\vec{v}|}\vec{v}>| \leq 1,$$

and equality holds if and only if $\vec{u} \parallel \vec{v}$. So, if two vectors are completely relevant (thus they are parallel to each other), they are always completely relevant — whatever the inner product you use to measure it.

We now examine the irrelevant relation.

We use the following example. Consider two vectors $\vec{F} = (\frac{\sqrt{2}}{2}, \frac{\sqrt{2}}{2})$ and $\vec{W} = (\frac{\sqrt{2}}{2}, -\frac{\sqrt{2}}{2})$.

Using the standard inner product on the plane: For $\vec{U} = (u_1, u_2)$ and $\vec{V} = (v_1, v_2)$,

$$\vec{U} \cdot \vec{V} = u_1 v_1 + u_2 v_2,$$

we get:

$$\vec{F} \cdot \vec{W} = \frac{1}{2} - \frac{1}{2} = 0.$$

They are completely irrelevant!

We will check the relation between $\vec{F}$ and $\vec{W}$ under a new inner product. For any two vectors $\vec{U} = (u_1, u_2)$ and $\vec{V} = (v_1, v_2)$, we define a new inner product

$$<\vec{U}, \vec{V}>_{\text{new}} = u_1 v_1 + 2 u_2 v_2.$$

Using the new inner product, we calculate the length of vectors $\vec{F}$ and $\vec{W}$:

$$|\vec{F}|_{\text{new}} = \frac{\sqrt{6}}{2}, \quad |\vec{W}|_{\text{new}} = \frac{\sqrt{6}}{2}.$$

Thus

$$\cos\theta = <\frac{1}{|\vec{F}|_{\text{new}}}\vec{F}, \frac{1}{|\vec{W}|_{\text{new}}}\vec{W}>_{\text{new}} = -\frac{1}{3} \neq 0!$$

In layman's terms: from a different viewpoint, two completely irrelevant events may NOT be completely irrelevant!

If we go a bit further, for any positive integer number k, we introduce a new inner

product
$$< \vec{U}, \vec{V} >_k = k u_1 v_1 + u_2 v_2.$$

Using this new inner product, the lengths of $\vec{F}$ and $\vec{W}$ are:
$$|\vec{F}|_k = \sqrt{\frac{k+1}{2}}, \quad |\vec{W}|_k = \sqrt{\frac{k+1}{2}}.$$

Thus
$$< \frac{1}{|\vec{F}|_k}\vec{F}, \frac{1}{|\vec{W}|_k}\vec{W} >_k = \frac{k}{k+1} - \frac{1}{k+1} = \frac{k-1}{k+1}.$$

One can see that as k becomes large, the angle between $\vec{F}$ and $\vec{W}$ is getting close to $0°$. So, two completely irrelevant events are getting more and more relevant due to the change of viewpoints (the increasing of parameter k)!

6.4.3 Cross product*

Any vector $\vec{u} = (u_1, u_2, u_3)$ in the three-dimensional vector space $\mathbb{R}^3$ can be expressed as
$$(u_1, u_2, u_3) = u_1(1, 0, 0) + u_2(0, 1, 0) + u_3(0, 0, 1)$$

We call $\vec{e}_1 = (1, 0, 0)$, $\vec{e}_2 = (0, 1, 0)$, $\vec{e}_3 = (0, 0, 1)$ standard basis vectors of three-dimensional vector space. We also use $\mathbf{i} = \vec{e}_1$, $\mathbf{j} = \vec{e}_2$ and $\mathbf{k} = \vec{e}_3$ to represent the standard basis vectors of three-dimensional vector space

Let us define a product operation that satisfies the following properties (we call it exterior product, or cross product):

(1) **For the standard basis vectors**:
$$\vec{e}_1 \times \vec{e}_2 = \vec{e}_3; \quad \vec{e}_2 \times \vec{e}_3 = \vec{e}_1; \quad \vec{e}_3 \times \vec{e}_1 = \vec{e}_2.$$
$$\vec{e}_1 \times \vec{e}_1 = \vec{0}; \quad \vec{e}_2 \times \vec{e}_2 = \vec{0}; \quad \vec{e}_3 \times \vec{e}_3 = \vec{0}.$$

(2) **Skew commutative rule**: for any two vectors $\vec{u}$, $\vec{v}$,
$$\vec{u} \times \vec{v} = -\vec{v} \times \vec{u}.$$

(3) **Distributive rule**: For any three vectors $\vec{u}$, $\vec{v}$, $\vec{w}$ and two real numbers a, b
$$(a\vec{u} + b\vec{v}) \times \vec{w} = a\vec{u} \times \vec{w} + b\vec{v} \times \vec{w},$$

and
$$\vec{w} \times (a\vec{u} + b\vec{v}) = a\vec{w} \times \vec{u} + b\vec{w} \times \vec{v}.$$

It is easy to see from the definition that the commutative rule does not hold for the cross product "×"!

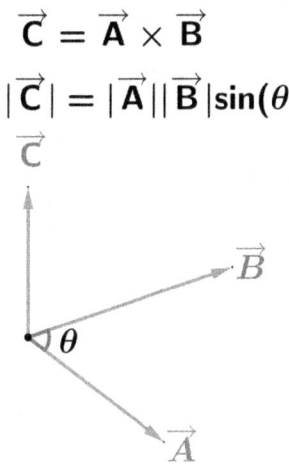

Figure 6.12: Cross product diagram

We have the following theorem.

Theorem 6.6. Cross product formula

For two vectors $\vec{u} = (u_1, u_2, u_3)$ and $\vec{v} = (v_1, v_2, v_3)$, we have

$$\vec{u} \times \vec{v} = \begin{bmatrix} \mathbf{i} & \mathbf{j} & \mathbf{k} \\ u_1 & u_2 & u_3 \\ v_1 & v_2 & v_3 \end{bmatrix}. \qquad (6.6)$$

Proof: From the definition and the operation rules for the cross product, we have:

$$\vec{u} \times \vec{v} = (u_1\mathbf{i} + u_2\mathbf{j} + u_3\mathbf{k}) \times (v_1\mathbf{i} + v_2\mathbf{j} + v_3\mathbf{k})$$
$$= (u_1v_2 - u_2v_1)\mathbf{i} \times \mathbf{j} + (u_1v_3 - u_3v_1)\mathbf{i} \times \mathbf{k} + (u_2v_3 - u_3v_2)\mathbf{j} \times \mathbf{k}$$
$$= (u_1v_2 - u_2v_1)\mathbf{k} - (u_1v_3 - u_3v_1)\mathbf{j} + (u_2v_3 - u_3v_2)\mathbf{i}.$$

On the other hand, we can also calculate the determinant

$$\begin{bmatrix} \mathbf{i} & \mathbf{j} & \mathbf{k} \\ u_1 & u_2 & u_3 \\ v_1 & v_2 & v_3 \end{bmatrix} = (u_1v_2 - u_2v_1)\mathbf{k} - (u_1v_3 - u_3v_1)\mathbf{j} + (u_2v_3 - u_3v_2)\mathbf{i}.$$

We thus obtain the proof of theorem 6.6 by comparing the above two identities.

□

Geometric meaning of cross-product is as follows:

Theorem 6.7. Cross product and area

Suppose the angle between two nontrivial vectors $\vec{u} = (u_1, u_2, u_3)$ and $\vec{v} =$

6.4 Operations on vectors

> (v_1, v_2, v_3) is $\theta \in [0, 180°]$, then
> $$|\vec{u} \times \vec{v}| = |\vec{u}| \cdot |\vec{v}| \sin \theta. \tag{6.7}$$
> That is, the length of the cross product is the area of the parallelogram formed by two vectors $\vec{u} = (u_1, u_2, u_3)$ and $\vec{v} = (v_1, v_2, v_3)$.

Proof: The proof of this theorem can not be seen immediately. So, let us first do the following calculation:

$$
\begin{aligned}
|\vec{u} \times \vec{v}|^2 &= (u_1^2 + u_2^2 + u_3^2)(v_1^2 + v_2^2 + v_3^2) - (u_1 v_1 + u_2 v_2 + u_3 v_3)^2 && \text{(Algebraic operation)} \\
&= |\vec{u}|^2 \cdot |\vec{v}|^2 - (\vec{u} \cdot \vec{v})^2 && \text{(Definition)} \\
&= |\vec{u}|^2 \cdot |\vec{v}|^2 - |\vec{u}|^2 \cdot |\vec{v}|^2 \cos^2 \theta && \text{(Definition for angle)} \\
&= |\vec{u}|^2 \cdot |\vec{v}|^2 (1 - \cos^2 \theta) \\
&= |\vec{u}|^2 \cdot |\vec{v}|^2 \sin^2 \theta. && \text{(Pythogorean identity)}
\end{aligned}
$$

Since $\sin \theta \geq 0$, we take a square root of both sides of the above identity and obtain the theorem.

$\square$

Example 6.16 (1) Find the area of a parallelogram whose vertices are points $A(1, 2)$, $B(2, 5)$, $C(6, 4)$, and $D(5, 1)$.

(2) Find the area of the triangle whose vertices are points $A(1, 2)$, $B(2, 5)$, and $D(5, 1)$.

Solution: First we calculate the vectors

$$\vec{AB} = (1, 3), \quad \vec{AD} = (4, -1).$$

View them as three-dimensional vectors: $\vec{AB}_n = (1, 3, 0)$, $\vec{AD}_n = (4, -1, 0)$. Then we calculate

$$\vec{AB}_n \times \vec{AD}_n = \begin{bmatrix} \mathbf{i} & \mathbf{j} & \mathbf{k} \\ 1 & 3 & 0 \\ 4 & -1 & 0 \end{bmatrix}$$

$$= 0\mathbf{i} + 0\mathbf{j} - 13\mathbf{k}$$

$$= (0, 0, -13).$$

So

(1)

$$\text{Area of parallelogram} = |(0, 0, -13)| = 13.$$

(2) The area of a triangle is half that of a parallelogram. So it is: $13/2$.

Consider three nontrivial vectors $\vec{u} = (u_1, u_2, u_3)$, $\vec{v} = (v_1, v_2, v_3)$ and $\vec{w} = (w_1, w_2, w_3)$. First, cross product $\vec{v} \times \vec{w}$ yields a vector. Then taking an inner product of this new vector with $\vec{u}$, we obtain a scalar number. We call $\vec{u} \cdot (\vec{v} \times \vec{w})$ the mixed product of $\vec{u}, \vec{v}, \vec{w}$, and use the notation $\vec{u} \cdot \vec{v} \times \vec{w}$.

> **Theorem 6.8. Mixed product and volume**
>
> *Consider three nontrivial vectors $\vec{u} = (u_1, u_2, u_3)$, $\vec{v} = (v_1, v_2, v_3)$ and $\vec{w} = (w_1, w_2, w_3)$. Their mixed product is*
>
> $$\vec{u} \cdot \vec{v} \times \vec{w} = \begin{vmatrix} u_1 & u_2 & u_3 \\ v_1 & v_2 & v_3 \\ w_1 & w_2 & w_3 \end{vmatrix}. \tag{6.8}$$

We will illustrate step by step how to prove the above theorem in the exercises after class.

> **Corollary 6.2. Direction of cross product vector**
>
> *Let $\vec{u} = (u_1, u_2, u_3)$ and $\vec{v} = (v_1, v_2, v_3)$ be two nontrivial vectors. Then*
>
> $$\vec{u} \times \vec{v} \perp \vec{u}, \quad \vec{u} \times \vec{v} \perp \vec{v}.$$

Exercise 6.3 Prove Corollary 6.2 from Theorem 6.8.

Note that
$$\vec{U} = \frac{\vec{u} \times \vec{v}}{|\vec{u} \times \vec{v}|}$$
is the unit vector of $\vec{u} \times \vec{v}$. It can be seen that
$$\vec{u} \cdot \vec{v} \times \vec{w} = |\vec{u} \times \vec{v}| \vec{U} \cdot \vec{w}$$
$$= |\vec{u} \times \vec{v}| |\vec{w}| \cos\theta$$
where θ is the angle between vector $\vec{U}$ and $\vec{w}$. So we have

> **Corollary 6.3. Volume of parallelepiped**
>
> *Consider a parallelepiped formed by three nontrivial vectors $\vec{u} = (u_1, u_2, u_3)$, $\vec{v} = (v_1, v_2, v_3)$ and $\vec{w} = (w_1, w_2, w_3)$. Its volume is given by*
>
> $$\textbf{Volume} = |\vec{u} \cdot \vec{v} \times \vec{w}| = \text{Absolute value of } \begin{vmatrix} u_1 & u_2 & u_3 \\ v_1 & v_2 & v_3 \\ w_1 & w_2 & w_3 \end{vmatrix}.$$

Using the above corollary, we can train the machine to determine whether any four points are co-planar or not.

Example 6.17 Consider four points $A(1,2,3)$, $B(-1,2,5)$, $C(1,-1,3)$ and $D(0,0,0)$. Check whether these four points are on the same plane or not.

Solution: First calculate the vector

$$\vec{DA} = (1,2,3), \quad \vec{DB} = (-1,2,5)) \quad \vec{DC} = (1,-1,3).$$

We calculate the volume of the parallelepiped formed by three vectors $\vec{DA}$, $\vec{DB}$, and $\vec{DC}$.

$$\textbf{Volume} = \vec{u} \cdot \vec{v} \times \vec{w}$$

$$= \text{Absolute value of } \begin{bmatrix} 1 & 2 & 3 \\ -1 & 2 & 5 \\ 1 & -1 & 3 \end{bmatrix}$$

$$= |1 \times (6+5) + 1 \times (6+3) + 1 \times (10-6)|$$

$$= 24$$

$$\neq 0.$$

So these four points are not on the same plane. □

6.4.4 Exercises after class

1. **Basic exercises**.

    (1) Calculate

    (a)
    $$(2,5) + (4,-2) =$$

    (b)
    $$3(1,2) - 2(3,5) =$$

    (2) If vector $\bar{u} = (0,4)$, $\bar{v} = (3,0)$, calculate

    (a) $\bar{u} \cdot \bar{v}$

    (b) Length of vector $\bar{u} - \bar{v}$.

    (c) $(0,4,0) \times (3,0,0)$

2. **Comprehensive questions**.

    (1) For any two vectors $\bar{u} = (u_1, u_2)$, $\bar{v} = (v_1, v_2)$, define operations
    $$\bar{u} \cdot \bar{v} = u_1 v_1 + 2u_2 v_2.$$

    (a). Verify that this is an inner product operation.

    (b). Prove: for any two numbers x, y,
    $$x + 2y \leq \sqrt{3(x^2 + 2y^2)}.$$

    (2) Proof: for any three numbers x, y, z,
    $$x + 16y + 64z \leq 9\sqrt{x^2 + 16y^2 + 64z^2}$$

    (a) It is difficult (perhaps very difficult) to use algebraic operations directly.

    (b) Prove it by using Cauchy-Schwartz inequality.(**Hint**: $1^2 + 4^2 + 8^2 = 9^2$).

    (3) Consider three nontrivial vectors $\vec{u} = (u_1, u_2, u_3)$, $\vec{v} = (v_1, v_2, v_3)$ and $\vec{w} = (w_1, w_2, w_3)$.

    (a) Calculate the cross product $\vec{u} \times \vec{v}$.

    (b) Write $\vec{F} = \vec{u} \times \vec{v}$. Calculate the inner product $\vec{F} \cdot \vec{w}$.

    (c) Calculate the determinant
    $$\begin{bmatrix} u_1 & u_2 & u_3 \\ v_1 & v_2 & v_3 \\ w_1 & w_2 & w_3 \end{bmatrix}.$$

    Verify: the determinant is the same as $\vec{F} \cdot \vec{w}$.

6.5 Equations for lines and planes

In this section, we learn how to use the inner product to derive algebraic equations for lines and planes in two-dimensional and three-dimensional space, and reveal their common properties.

6.5.1 Linear equations on the plane

Let us first recall that the first-order equation $y = ax + b$ represents a straight line in the coordinate system.

Then we recall the equation for the straight line that passes through two points $A(x_1, y_1)$ and $B(x_2, y_2)$ in the coordinate system: When $x \neq x_1$ and $x \neq x_2$, we get the two-point formula of the linear equation.

$$\frac{y - y_1}{x - x_1} = \frac{y - y_2}{x - x_2}. \tag{6.9}$$

If $x_1 = x_2$ but $y_1 \neq y_2$, we have a straight line parallel to the y-axis

$$x = x_1. \tag{6.10}$$

If $x_1 \neq x_2$ but $y_1 = y_2$, we get a straight line parallel to the x-axis

$$y = y_1. \tag{6.11}$$

Recall again: the slope of a two-dimensional straight line as follows:

> **Definition 6.7. Slope**
>
> Take two different points (x_1, y_1) and (x_2, y_2) on a line that is not parallel to the y-axis. Define the slope m of this line as
>
> $$m = \frac{y_2 - y_1}{x_2 - x_1}.$$

If we know that a straight line passes through the point (x_1, y_1) and its slope is m, then we can get from formula (6.9) that

$$y = m(x - x_1) + y_1. \tag{6.12}$$

Formula (6.12) is called the point-slope formula of the linear equation.

Exercise 6.4 Try to derive formula (6.12) from formula (6.9).

Another way to represent a line is to use the normal vector $\vec{N}$ of the line.

> **Definition 6.8. Normal vector**
>
> If a nontrivial vector $\vec{N} = (a, b)$ is perpendicular to a vector formed by any two different points on a line, we call $\vec{N}$ the normal vector of the line. ♣

If we know that the normal vector of a line is $\vec{N} = (a, b)$, and the line passes through a point (x_1, y_1). Then, for any point (x, y) on the line, we have

$$(x - x_1, y - y_1) \cdot (a, b) = 0.$$

Multiplying out, we have

$$a(x - x_1) + b(y - y_1) = 0. \tag{6.13}$$

We call the above equation the "normal form formula" for the line equation. If the second component b of the normal vector $\vec{N} = (a, b)$ of a straight line is not zero ($b \neq 0$), then its slope $m = -\frac{a}{b}$; If $b = 0$, we can see that this is a vertical line parallel to the y-axis. Obviously, the normal form is more general: no matter whether the straight line is vertical or horizontal, it has a normal form equation. (The same idea will be used to derive the linear equations for the high-dimensional planes below).

We now study the algebraic relationship between two perpendicular lines. If two lines are perpendicular to each other, their normal vectors are perpendicular to each other. It is easy to get:

> **Proposition 6.5. Formula for the slopes of two perpendicular lines**
>
> Consider two straight lines which are not parallel to x- or y-axis, whose slopes are m_1 and m_2. The necessary and sufficient conditions for them to be perpendicular to each other are
>
> $$m_1 \cdot m_2 = -1.$$
> ♣

Finally, we study the vector form of the straight line. Assuming that the line is parallel to vector $\vec{v} = (v_1, v_2)$ and passes through point $P_0(a, b)$, then for any point $P(x, y)$ on the line, we have $\vec{P_0P} // \vec{v}$. That is:

$$(x, y) = (a, b) + t(v_1, v_2) \quad \text{for all } t \in \mathbb{R}. \tag{6.14}$$

We call equation (6.14) the "vector form of the line" or the "parametric equation of the line" (the variable t is called the parameter).

6.5.2 Line equations in space

Line equations in space are very different from the line equations in the plane. For example, there is no definition of the slope for a straight line in space, and there are many normal vectors. The relatively easy way to obtain a straight-line equation is to use the intersection of two space planes, which will be discussed in the next section after the study of the plane equations. Another way to represent a line in the space is to use the parameter form of a line. We will discuss it in the last section of this chapter.

6.5.3 Plane equations in three-dimensional space

Let us look at the first-order equation with three unknown variables x, y, z

$$ax + by + cz + d = 0. \tag{6.15}$$

In the three-dimensional coordinate system, if we mark enough points of the following set:

$$\{(x, y, z) \mid ax + by + cz + d = 0\}$$

we can see that these points form a plane. We say the first-order equation $ax+by+cz+d = 0$ represents a plane in the coordinate system.

Suppose (x_0, y_0, z_0) is a point on the plane, that is

$$ax_0 + by_0 + cz_0 + d = 0.$$

Then we can rewrite equation (6.15) as

$$a(x - x_0) + b(y - y_0) + c(z - z_0) = 0. \tag{6.16}$$

We call equation (6.15) the general form of the plane equation, and call equation (6.16) the standard form of the plane equation.

Let us look at equation (6.16) closely: when (x, y, z) and (x_0, y_0, z_0) are two different points on the plane, $(x - x_0, y - y_0, z - z_0)$ is a vector on the plane. We rewrite equation (6.16) as

$$(a, b, c) \cdot (x - x_0, y - y_0, z - z_0) = 0.$$

So vector (a, b, c) is perpendicular to any vector on the plane! We call it the normal vector of the plane. From the perspective of the normal vector, the plane in 3D space is the same as the line in 2D space[6]. In fact, we can define any hyperplane in any higher dimensional space (with dimension larger than or equal to three) in future studies.

Example 6.18 Find the plane equation whose normal vector is $\vec{v} = (2, -1, 3)$, containing point $P_0(1, 3, 2)$.

---

[6]In fact, we can rigorously define a line in the plane and a plane in the space here. (1) For a given point P_0 and a nontrivial vector $\vec{N}$ in the plane, all points on the plane, from whom the line connects to p_0 are orthogonal to vector $\vec{N}$, form a straight line. (2) For a given point P_0 and a nontrivial vector $\vec{N}$ in the space, all points in the space, from whom the line connects to p_0 are orthogonal to vector $\vec{N}$, form a plane in the space.

Solution: Let $P(x, y, z)$ be any point on the plane. Then $\vec{P_0P} \perp \vec{v}$. Therefore

$$(x - 1, y - 3, z - 2) \cdot (2, -1, 3) = 0.$$

After simplification:

$$2x - y + 3z = 5.$$

□

Example 6.19 Find the plane equation containing three points: $A(1, 2, 3)$, $B(-1, 2, 5)$, $C(1, -1, 3)$

Solution: Let's first calculate two vectors on the plane:

$$\vec{AB} = (-2, 0, 2), \quad \vec{AC} = (0, -3, 0).$$

So we can get the normal vector:

$$\vec{v} = \vec{AB} \times \vec{AC} = \begin{bmatrix} \mathbf{i} & \mathbf{j} & \mathbf{k} \\ -2 & 0 & 2 \\ 0 & -3 & 0 \end{bmatrix}$$

$$= 6\mathbf{i} + 0\mathbf{j} + 6\mathbf{k}$$

$$= (6, 0, 6).$$

Let $P(x, y, z)$ be any point on the plane, then $\vec{AP} \perp \vec{v}$. Therefore

$$(x - 1, y - 2, z - 3) \cdot (6, 0, 6) = 0.$$

After simplification, we have

$$x + z = 4.$$

□

Now we learn how to use algebra to study the geometric relation between two planes (Can we train the robot dog to read the angle between two planes?).

> **Definition 6.9. Angle between planes**
>
> We define the angle between two planes as the angle between their normal vectors in interval $[0, \frac{\pi}{2}]$.

Example 6.20 Check whether the following two pairs of planes are perpendicular to each other or not, are parallel to each other or not. (a). $x + 2y + 3z - 1 = 0$ and $5x + 2y - 3z = 1$; (b). $x + 2y + 3z - 1 = 0$ and $3x + 6y = -9z + 1$.

Solution: (a). The normal vector of plane $x + 2y + 3z - 1 = 0$ is $\vec{v}_1 = (1, 2, 3)$, and the

6.5 Equations for lines and planes

normal vector of $5x + 2y - 3z = 1$ is $\vec{v}_2 = (5, 2, -3)$.

$$\vec{v}_1 \cdot \vec{v}_2 = (1, 2, 3) \cdot (5, 2, -3)$$
$$= 5 + 4 - 9$$
$$= 0.$$

So these two planes are perpendicular to each other.

(b). The normal vector of plane $x + 2y + 3z - 1 = 0$ is $\vec{v}_1 = (1, 2, 3)$, and the normal vector of $3x + 6y = -9z + 1$ is $\vec{v}_2 = (3, 6, 9)$.

$$\vec{v}_1 \cdot \vec{v}_2 = (1, 2, 3) \cdot (3, 6, 9)$$
$$= 3 + 12 + 27$$
$$= -42 \neq 0.$$

So these two planes are not perpendicular to each other.

On the other hand,

$$\vec{v}_1 \times \vec{v}_2 = \begin{bmatrix} \mathbf{i} & \mathbf{j} & \mathbf{k} \\ 1 & 2 & 3 \\ 3 & 6 & 9 \end{bmatrix}$$
$$= 0\mathbf{i} + 0\mathbf{j} + 0\mathbf{k}$$
$$= (0, 0, 0).$$

So these two planes are parallel to each other.

□

Note *For part (b) in the above example, if we can observe that $\vec{v}_2 = 3\vec{v}_2$, we immediately conclude that two planes are parallel to each other without other computation.*

Example 6.21 Find the intersection of two planes $x + 2y + 3z - 1 = 0$ and $3x + 6y = 9z + 1$, and find the unit vector $\vec{v}$ parallel to the line.

Solution: In fact, we can simply use:

$$\begin{cases} x + 2y + 3z - 1 = 0 \\ 3x + 6y = 9z + 1 \end{cases}$$

to represent this intersection line.

To find the unit vector $\vec{v}$ parallel to the line, We observe that $\vec{v}$ is perpendicular to vectors $\vec{n}_1 = (1, 2, 3)$ and $\vec{n}_2 = (3, 6, -9)$. Thus, $\vec{v} \parallel \vec{n}_1 \times \vec{n}_2$.

Write $\vec{w} = \vec{n}_1 \times \vec{n}_2$. We compute

$$\vec{w} = \vec{n}_1 \times \vec{n}_2 = \begin{bmatrix} \mathbf{i} & \mathbf{j} & \mathbf{k} \\ 1 & 2 & 3 \\ 3 & 6 & -9 \end{bmatrix}$$

$$= -36\mathbf{i} + 18\mathbf{j} + 0\mathbf{k}$$

$$= (-36, 18, 0).$$

So

$$\vec{v} = (-\frac{2\sqrt{5}}{5}, \frac{\sqrt{5}}{5}, 0) \quad \text{or} \quad \vec{v} = (\frac{2\sqrt{5}}{5}, -\frac{\sqrt{5}}{5}, 0).$$

□

6.5.4 Exercises after class

1. **Basic skills**.

   (1) Find the equation for the following line:

   (i) The line passing through point $A(1,3)$ and point $B(3,1)$;

   (ii) A straight line with a slope of 2, passing through point $A(1,3)$;

   (iii) A line passing through point $A(\frac{1}{2},\frac{2}{3})$ and parallel to the line $2x+3y=5$;

   (iv) A line passing through point $A(-\frac{3}{2},\frac{1}{2})$ and perpendicular to the line $x+3y=4$.

   (2) Find the plane equation:

   (i) The plane passing through point $A(1,3,2)$, point $B(0,3,1)$ and point $C(0,0,0)$;

   (ii) A plane whose normal vector is $\vec{n}=(2,3,1)$, and containing point $A(1,3,2)$;

   (iii) A plane passing through point $A(\frac{1}{2},\frac{2}{3},\frac{4}{3})$ and parallel to the plane $2x+3y-z=5$;

   (iv) A plane passing through point $A(-\frac{3}{2},\frac{1}{2},0)$ and perpendicular to the plane $x+3y+2z=4$.

   (3) Calculate

   (i) Is point $A(2,9))$ on the line $5x-6=7$?

   (ii) Find the angle between the line $2y-3x=1$ and $3x-2y=2$;

   (iii) Find the angle between the plane $x+2y-3z=1$ and $3x-2y+4z=2$.

2. **Comprehensive questions**.

   (1) Check whether three points $A(1,2)$, $B(2,-1)$ and $C(-9,-2)$ are col-linear or not.

   (2) Check whether four points $A(3,2,-1)$, $B(2,-1,-2)$, $C(-5,0,-2)$ and $D(-2,3,0)$ are co-planar or not.

   (3) Find the vector form for the intersection line of two planes $x+2y+3z-1=0$ and $3x+6y=9z+1$.

6.6 Parametric equations and vector functions

Another expression of a curve is its parametric equation. For example, if we know the coordinates of the curve at any time, we (or even the machine) can draw the curve.

6.6.1 Parametric equations in two-dimensional plane

For a curve on a two-dimensional plane, we can use

$$\begin{cases} x = x(t), & \alpha \le t \le \beta \\ y = y(t), & \alpha \le t \le \beta \end{cases}$$

to represent it.

First, consider the line passing through point (x_1, y_1) and parallel to vector (a, b). If (x, y) is a point on the line, then vector $(x - x_1, y - y_1)$ is parallel to vector (a, b). That is, there is a constant t such that $(x - x_1, y - y_1) = t(a, b)$. Write it in a component form, we have the parametric equation of the line:

$$\begin{cases} x = x_1 + at, & -\infty < t < \infty \\ y = y_1 + bt, & -\infty < t < \infty. \end{cases} \tag{6.17}$$

We can also use vector function to represent the above linear equation:

$$\vec{f}(t) = (x_1 + at, y_1 + bt), \quad t \in (-\infty, \infty). \tag{6.18}$$

Now, we consider the equation of the circle with the center at the origin and radius r. Let t be the angle between the positive x-axis and the line segment between the origin and a given point (x, y) on the circle, — the range of t is from 0 to 2π. Then, from the definition of the trigonometric functions, we can directly obtain the parametric equations of the circle:

$$\begin{cases} x = r\cos t, & 0 \le t \le 2\pi \\ y = r\sin t, & 0 \le t \le 2\pi. \end{cases} \tag{6.19}$$

Similarly, we can also use a vector function to represent the above parametric equations for the circle:

$$\vec{g}(t) = (r\cos t, r\sin t), \quad t \in [0, 2\pi). \tag{6.20}$$

Example 6.22 If a and b are positive, show that the vector function:

$$\vec{g}(t) = (x_0 + a\cos t, y_0 + b\sin t), \quad t \in [0, 2\pi) \tag{6.21}$$

represents an elliptic equation with the center at (x_0, y_0).

Solution: We can write the vector equation as parametric equations:
$$\begin{cases} x = x_0 + a\cos t, & 0 \le t < 2\pi \\ y = y_0 + b\sin t, & 0 \le t < 2\pi. \end{cases}$$

Then (x, y) satisfies
$$\frac{(x-x_0)^2}{a^2} + \frac{(y-y_0)^2}{b^2} = 1.$$

That is, the point (x, y) satisfies an elliptic equation with the center at (x_0, y_0).

□

As can be seen from the above elliptic parametric equations, a remarkable advantage of parametric equations is that they can be drawn by the machine. One obvious drawback of parametric equations is that it is not easy to see the parameter from the graph (can you see what is the parameter t in the parametric equations for an ellipse ?).

6.6.2 Parametric equations in three-dimensional space

Consider a line equation that passes through the point $P_1(x_1, y_1, z_1)$ and is parallel to vector $\vec{v} = (a, b, c)$ (we often call this vector a directional vector of the line)). If $P(x, y, z)$ is a point on the line, then vector $\vec{P_1P} = (x - x_1, y - y_1, z - z_1)$ is parallel to vector (a, b, c). That is, there is a constant t, so that $(x - x_1, y - y_1, z - z_1) = t(a, b, c)$. Writing it in the component form, we obtain the parametric equations for the line:

$$\begin{cases} x = x_1 + at, & -\infty < t < \infty \\ y = y_1 + bt, & -\infty < t < \infty \\ z = z_1 + ct, & -\infty < t < \infty. \end{cases} \quad (6.22)$$

We can also use a vector function to express the above linear equations:

$$\vec{f}(t) = (x_1 + at, y_1 + bt, z_1 + ct), \quad t \in (-\infty, \infty). \quad (6.23)$$

Example 6.23 Find the parameter equations of the intersection line of two planes $x + 2y + 3z - 1 = 0$ and $3x + 6y = 9z + 1$.

Solution: It is known from Example 6.21 that the unit vector $\vec{v} = (-\frac{2\sqrt{5}}{5}, \frac{\sqrt{5}}{5}, 0)$ is parallel to the line. It is easy to verify that point $P_0(\frac{2}{3}, 0, \frac{1}{9})$ satisfies both plane equations, thus is on the line. Therefore, the parameter equations of the intersection line are:

$$\begin{cases} x = \frac{2}{3} - \frac{2\sqrt{5}}{5}t, & -\infty < t < \infty \\ y = \frac{\sqrt{5}}{5}t, & -\infty < t < \infty \\ z = \frac{1}{9}, & -\infty < t < \infty. \end{cases}$$

□

Exercise 6.5 Find the parameter equations for the ellipsoid
$$\frac{x^2}{a^2} + \frac{y^2}{b^2} + \frac{z^2}{c^2} = 1.$$

Finally, we look at the plane equation that contains point $P_1(x_1, y_1, z_1)$ and is perpendicular to vector $\vec{v} = (a, b, c)$ (recall: this is the normal vector of the plane).

If $P(x, y, z)$ is a point in the plane, then vector $\overrightarrow{P_1P} = (x - x_1, y - y_1, z - z_1)$ is perpendicular to vector (a, b, c). That is,
$$(x - x_1, y - y_1, z - z_1) \cdot (a, b, c) = 0.$$

Simplifying it, we have
$$ax + by + cz = ax_1 + by_1 + cz_1.$$

For simplicity, we write $ax_1 + by_1 + cz_1 = d$, then the plane equation with the normal vector of (a, b, c) is given by
$$ax + by + cz = d.$$

For a non-degenerate plane, the normal vector (a, b, c) is not a zero vector. That is, at least one of the three components is not zero. Without loss of generality, we assume that $c \neq 0$. Then we can write the plane equation as parametric equations with two parameters (u, v):

$$\begin{cases} x = u, & -\infty < u < \infty \\ y = v, & -\infty < v < \infty \\ z = \frac{d-ax-by}{c}, & -\infty < u, v < \infty. \end{cases} \quad (6.24)$$

Generally, we need two parameters to express a two-dimensional surface. In the future, we will gradually understand the geometric meaning of parametric equations and their relations to general Riemannian manifolds.

6.6.3 Exercises after class

1. **Basic skills**.

    (1) Drawing: $x = 3t + 1$; $y = 2t + 3$, $-\infty < t < +\infty$.

    (2) Drawing: $x = \sin t + 1$; $y = \cos t - 1$, $-\infty < t < +\infty$.

    (3) Convert the parametric equations into a rectangular coordinate equation: $x = \frac{1}{4}\sin t + 1$; $y = \frac{1}{9}\cos t - 1$, $-\infty < t < +\infty$.

2. **Comprehensive questions**.

    (1) Find the parameter equations of the intersection line of two planes $2x + 3y + z - 6 = 0$ and $3x - y + 2z = 4$.

    (2) Find the parametric equations of the intersection curve of two surfaces $x^2 + y^2 + z^2 = 9$ and $x + y + z = 0$. Can you tell what kind of curve this is?

6.7 Chapter review and exercises

In Chapter 6, we studied the graphs of functions in two-dimensional or three-dimensional coordinates.

1. 2D and 3D coordinate systems.

Basic concepts of two-dimensional and three-dimensional coordinate systems: axis and coordinates;

Verifying whether a point is on the curve or surface or not;

The distance formula between two points $A(x_1, y_1, z_1)$ and $B(x_2, y_2, z_2)$ is

$$dist(A, B) = \sqrt{(x_2 - x_1)^2 + (y_2 - y_1)^2 + (z_1 - z_2)^2}.$$

2. Graphs and properties of functions in rectangular coordinate system.

2.1. Graphs of linear functions: straight line (in a two-dimensional coordinate system), plane (in a three-dimensional coordinate system), and their normal vectors.

2.2. Graphs of quadratic functions: ellipse, hyperbola, and parabola.

2.3. Convexity and concavity of functions and Jensen's inequality: assuming that a function $y = f(x)$ on an interval I is convex, then for k points on the interval I: $x_1, x_2, \cdots, x_k$, we have:

$$f(\frac{x_1 + x_2 + \cdots + x_k}{k}) \leq \frac{f(x_1) + f(x_2) + \cdots + f(x_k)}{k}.$$

3. Polar coordinate system and diagram of polar equations * - selective reading.

3.1. Transformation of rectangular coordinate (x, y) to polar coordinate (r, θ).

Converting polar coordinate (r, θ) to rectangular coordinate (x, y):

$$x = r \cos \theta$$
$$y = r \sin \theta.$$

Converting rectangular coordinate (x, y) to polar coordinate (r, θ):

$$r^2 = x^2 + y^2$$
$$\tan \theta = \frac{y}{x}.$$

3.2. Polar equation:

Line: $\theta = $ Constant; Circle: $r = $ Constant; More general circle: $r = a \cos \theta + b \sin \theta$.

4. Operations on vectors.

4.1. Addition, scalar product, algebraic operation and their geometric meaning, general linear combination: for any given two vectors $\vec{u} = (u_1, u_2, u_3)$ and $\vec{v} = (v_1, v_2, v_3)$,

and any two real numbers a, b,
$$a\vec{u} + b\vec{v} = (au_1 + bv_1, au_2 + bv_2, au_3 + bv_3).$$

4.2. The inner product of two vectors. For any given two vectors $\vec{u} = (u_1, u_2, ..., u_n)$ and $\vec{v} = (v_1, v_2, ..., v_n)$, their standard inner product is defined as
$$\vec{u} \cdot \vec{v} = u_1 v_1 + u_2 v_2 + \cdots + u_n v_n.$$

The angle of two vectors: Suppose the angle between two nontrivial vectors $\vec{u}$ and $\vec{v}$ is $\theta \in [0, \pi]$, then
$$\cos \theta = \frac{\vec{u} \cdot \vec{v}}{|\vec{u}||\vec{v}|}.$$

4.3. Definition of a general inner product $[*, *]$ and its property* —— selective reading.

Cauchy-Schwarz inequality *: For any two vectors $\vec{u}$ and $\vec{v}$, we always have
$$|[\vec{u}, \vec{v}]| \le |\vec{u}||\vec{v}|,$$
or
$$-|\vec{u}||\vec{v}| \le [\vec{u}, \vec{v}] \le |\vec{u}||\vec{v}|.$$

4.4. Definition and properties of cross product *.

Cross product formula: For any two vectors $\vec{u} = (u_1, u_2, u_3)$ and $\vec{v} = (v_1, v_2, v_3)$,
$$\vec{u} \times \vec{v} = \begin{bmatrix} \mathbf{i} & \mathbf{j} & \mathbf{k} \\ u_1 & u_2 & u_3 \\ v_1 & v_2 & v_3 \end{bmatrix}.$$

4.5. Definition and properties of mixed product*

For three nontrivial vectors $\vec{u} = (u_1, u_2, u_3)$, $\vec{v} = (v_1, v_2, v_3)$ and $\vec{w} = (w_1, w_2, w_3)$, we define their mixed product as:
$$\vec{u} \cdot \vec{v} \times \vec{w} = \begin{bmatrix} u_1 & u_2 & u_3 \\ v_1 & v_2 & v_3 \\ w_1 & w_2 & w_3 \end{bmatrix}.$$

5. Line and plane equations.

5.1 Line equation on the plane

Two-point formula: A straight line passing through two points (x_1, y_1) and (x_2, y_2),
$$\frac{y - y_1}{x - x_1} = \frac{y - y_2}{x - x_2}.$$

Point slope formula: a straight line passing through point (a, b), and has a slope of m,
$$y - b = m(x - a).$$

Vector representation of line equation:

Normal vector and a point: a line passing through point (x_1, x_2), and has the normal vector $\vec{n} = (a, b)$,

$$a(x - x_1) + b(y - y_1) = 0.$$

The directional vector and a point: a line passing through point (x_1, y_1), and has the directional vector $\vec{v} = (v_1, v_2)$,

$$(x, y) = (x_1, y_1) + t(v_1, v_2), \quad -\infty < t < +\infty.$$

Two lines (which are not parallel to axes) are perpendicular to each other if and only if $m_1 \cdot m_2 = -1$.

5.2 Plane equations in three-dimensional space

Linear functions represent planes in 3D space: $ax + by + cz = d$.

Vector representations of plane equations:

Normal vector and a point: a plane passing through point (x_1, y_1, z_1), and with the normal vector $\vec{n} = (a, b, c)$,

$$a(x - x_1) + b(y - y_1) + c(z - z_1) = 0.$$

Angle θ between two planes $a_1 x + b_1 + c_1 z = d_1$ and $a_2 x + b_2 y + c_2 z = d_2$:

$$\theta = \arccos \frac{a_1 a_2 + b_1 b_2 + c_1 c_2}{\sqrt{a_1^2 + b_1^2 + c_1^2} \cdot \sqrt{a_2^2 + b_2^2 + c_2^2}}.$$

The intersection of two non-parallel planes $a_1 x + b_1 + c_1 z = d_1$ and $a_2 x + b_2 y + c_2 z = d_2$ yields a line equation in space:

$$\begin{cases} a_1 x + b_1 + c_1 z = d_1 \\ a_2 x + b_2 y + c_2 z = d_2. \end{cases}$$

The vector representations of line equations in space: a line passing through point (x_1, y_1, z_1), and with the directional vector $\vec{v} = (v_1, v_2, v_3)$,

$$(x, y, z) = (x_1, y_1, z_1) + t(v_1, v_2, v_3), \quad -\infty < t < +\infty.$$

6. Parametric equations and vector functions.

6.1. Parametric equations in two-dimensional plane

Parametric equations of a straight line:

$$\begin{cases} x = x_1 + at, & -\infty < t < \infty \\ y = y_1 + bt, & -\infty < t < \infty. \end{cases}$$

Parametric equations of an ellipse:
$$\begin{cases} x = x_0 + a\cos t, & 0 \le t \le 2\pi \\ y = y_0 + b\sin t, & 0 \le t \le 2\pi. \end{cases}$$

6.2. Parametric equations in three-dimensional space

Parametric equations of a straight line:
$$\begin{cases} x = x_1 + at, & -\infty < t < \infty \\ y = y_1 + bt, & -\infty < t < \infty \\ z = z_1 + ct, & -\infty < t < \infty. \end{cases}$$

Parametric equations of an ellipsoid
$$\frac{(x-x_0)^2}{a^2} + \frac{(y-y_0)^2}{b^2} + \frac{(z-z_0)^2}{c^2} = 1$$
$$\begin{cases} x = x_0 + a\cos t, & 0 \le t \le 2\pi \\ y = y_0 + b\sin t, & 0 \le t \le 2\pi \\ z = z_0 + c\sin t, & 0 \le t \le 2\pi. \end{cases}$$

6.7.1 Chapter 6 test

1. **Comprehensive question-1**. The relationship between points and lines.

   (1) Verify that the point $A(1, 3)$ is not on the line $3x + 4y = 7$.

   (2) Find a point B on the line $3x + 4y = 7$, so that $\vec{BA}$ is perpendicular to the line.

   (3) Find the distance from point $A(1, 3)$ to line $3x + 4y = 7$

   (4) Prove: the distance d from the point $P(x_0, y_0)$ to the line $ax + by - c = 0$ is
   $$d = \frac{|ax_0 + by_0 - c|}{\sqrt{a^2 + b^2}}.$$

2. **Comprehensive question-2**. The relationship between point and plane.

   (1) Verify that the point $A(1, 3, 2)$ is not on the plane $x + 2y + 2z = 7$.

   (2) Find a point B on the plane $x + 2y + 2z = 7$, so that $\vec{BA}$ is perpendicular to the plane.

   (3) Find the distance from point $A(1, 3, 2)$ to plane $x + 2y + 2z = 7$

   (4) Prove: The distance formula from point $P(x_0, y_0, z_0)$ to plane $ax+by+cz-d = 0$ is
   $$\text{distance} = \frac{|ax_0 + by_0 + cz_0 - d|}{\sqrt{a^2 + b^2 + c^2}}.$$

3. **Comprehensive question-3**. Use two methods to prove that: for any k positive numbers,
   $$\frac{x_1 + x_2 + \cdots + x_k}{k} \leq \sqrt{\frac{x_1^2 + x_2^2 + \cdots + x_k^2}{k}}.$$

4. **Comprehensive question-4**.

   (1) Verify whether four points $A(0, 0, 0)$, $B(1, 2, 1)$, $C(0, 5, 2)$, and $D(2, -3, 1)$ are on the same plane or not. If they are on the same plane, what is the plane equation?

   (2) Verify whether four points $A(0, 0, 0)$, $B(1, 1, 1)$, $C(0, 5, 2)$, and $D(2, -3, 0)$ are on the same plane or not. If they are on the same plane, what is the plane equation?

5. **Comprehensive question-5**. Convert the following polar coordinate equations into rectangular coordinate equations

   (1) $r = \frac{1}{2+\cos \theta}$

   (2) $r = \frac{1}{5-5 \sin \theta}$

   (3) $r = \frac{1}{1-4 \sin \theta}$

6. **Comprehensive question-6**. Non-coplanar lines. Consider two lines in three dimensional space. The first one $\mathcal{L}_1$: that passes through point $P_1(0, 0, 1)$ and is parallel to vector $\vec{u} = (1, 0, 0)$; The second one $\mathcal{L}_2$: that passes through point $P_2(0, 1, 0)$ and is parallel to vector $\vec{v} = (1, -1, 0)$.

(1) Write down the equations (could be parametric equations or vector equations) for these two lines.

(2) Find the equation for plane $\mathcal{P}$ that contains line $\mathcal{L}_2$ and is parallel to line $\mathcal{L}_1$.

(3) Find the distance from one point on line $\mathcal{L}_1$ to plane $\mathcal{P}$. **This is in fact the distance between two non-coplanar lines.**

Chapter 7 Sequence and series

Introduction

- Sequence
- General term
- Arithmetic sequence
- Geometric sequence
- Summation of sequence
- Fibonacci sequence
- Limit of sequence
- Limit of function
- Continuous function
- Series
- Partial sum of series
- Convergence or divergence of series

In this chapter, we will study sequences, the limits of sequences, and summations of sequences, series, and the convergence or divergence of series, etc. These topics will help us to understand the existence of irrational numbers, the precise meaning of repeating decimals, as well as how to derive Euler's formula for complex numbers in the next chapter.

7.1 Sequence

To be more precise, in this book, we only consider sequences of numbers. A sequence of numbers (we use sequence for abbreviations later) is an enumerated collection of numbers. If the sequence has finite elements, we call it a finite sequence; If it has infinite elements, we call it an infinite sequence.

Example 7.1
$$\{a_1, a_2, \cdots, a_{10}\}$$
is a finite sequence of 10 elements.

Example 7.2
$$\{0.1, 0.01, 0.001, 0.0001, \cdots, \}$$
is an infinite sequence.

In the second example, we may "guess" the 5th term and the following terms. But we can not be certain.

To express an infinite sequence precisely, we should give the formula for its general term - let us rigorously define the infinite sequence in the following.

7.1.1 Definition of sequence

> **Definition 7.1. Definition of sequence**
>
> We call a mapping from the natural number set (or a subset with infinite numbers) to the real number set $\mathbb{R}$
>
> $$f : n \in \{0, 1, 2, 3, \cdots\} \to \mathbb{R}$$
>
> an infinite sequence. We call $f(n)$ the general term formula for the sequence. Traditionally, we use $\{a_n\}_{n=0}^{\infty}$ to represent an infinite sequence.

If we know that the general term of the sequence in the above second example is

$$a_n = f(n) = 10^{-n}, \quad \text{for} \quad n = 1, 2, \cdots,$$

then we can verify that the fifth term of this sequence is 0.00001.

Once we have the general term of an infinite sequence, we can write any term of the sequence.

Example 7.3 If the general term of a sequence is

$$a_n = f(n) = 2(n - 1), \quad \text{for} \quad n = 1, 2, \cdots,$$

we know that this is the sequence of all even natural numbers

$$\{0, 2, 4, 6, 8, \cdots\}.$$

Its first term is 0, and its third term is 4.

□

Example 7.4 Suppose the general term of a sequence is

$$a_n = f(n) = 3(n - 1), \quad \text{for} \quad n = 1, 2, \cdots.$$

We know that this is the sequence of all natural numbers that are multiples of 3

$$\{0, 3, 6, 9, 12, \cdots\}.$$

Its first term is 0, and its third term is 6.

□

Another kind of sequence is given by the recurrence formula.

Example 7.5 Suppose a sequence is given by the following recurrence formula:

$$a_{n+1} = a_n + 2, \quad \text{for} \quad n = 1, 2, 3, \cdots ; \quad \text{and} \quad a_1 = 1.$$

Then from $a_1 = 1$ and the recurrence formula (take $n = 1$), we can get

$$a_2 = a_1 + 2 = 3.$$

By analogy, we can get $a_n = 2n+1$. That is, the above sequence is actually a sequence of all odd natural numbers. We will come back later to discuss general arithmetic sequences more.

□

Exercise 7.1 Suppose a sequence is given by the following recurrence formula:

$$a_{n+1} = a_n + 2, \text{ for } n = 1, 2, 3, \cdots; \text{ and } a_1 = 2.$$

Find the general term formula of the sequence.

Example 7.6 Suppose a sequence is given by the following formula:

$$a_{n+1} = 2a_n, \text{ for } n = 1, 2, 3, \cdots; \text{ and } a_1 = 2.$$

Then from $a_1 = 2$ and the recurrence formula (take $n = 1$), we get

$$a_2 = 2a_1 = 2^2.$$

By analogy, we obtain $a_n = 2^n$. We will also come back later to discuss general geometric sequences more.

□

Now let us study a slightly complex recursive relationship.

Example 7.7 (**Branch bifurcation problem**) Botanists found that some plants only fork after one year. Suppose the first year starts with a new branch; The second year it forks only one new branch, thus, we have a total of two branches; The third year we have has $1+2$ branches (one new forked branch+ two old branches); And the fourth year we have has $2+3$ branches (two new forked branches+ three old branches)... From this we obtain Fibonacci sequence:

$$a_{n+2} = a_{n+1} + a_n, \text{ for } n = 1, 2, 3, \cdots; \text{ and } a_1 = 1, \ a_2 = 2.$$

Not particularly difficult, we can list the first few terms of the Fibonacci series:

$$\{1, 2, 3, 5, 8, 13, 21, ...\}$$

However, it is not easy to see what is its 100th term. The main reason is that it is not easy to find the formula of its general term. At the end of this section, we will discuss how to calculate the general term formula for the Fibonacci sequence.

Now we study two special sequences.

7.1.2 Arithmetic sequence

> **Definition 7.2. Definition of arithmetic sequence**
>
> We call the following sequence
>
> $$a_1, a_1 + d, a_1 + 2d, \cdots$$
>
> an arithmetic sequence. a_1 is called the first item, and d is called the common difference. The general term for the above arithmetic sequence is: for $n = 1, 2, \cdots$,
>
> $$a_n = a_1 + (n-1)d.$$

Example 7.8 Suppose a sequence is given by the following formula:

$$a_{n+1} = a_n + 2, \quad \text{for} \quad n = 1, 2, 3, \cdots ; \quad \text{and} \quad a_1 = 1.$$

Then this is an arithmetic sequence. Its first term is $a_1 = 1$, its common difference is 2. We can also get its general term: for $n = 1, 2, 3, \cdots$,

$$a_n = 2(n-1) + 1.$$

□

For an arithmetic sequence, besides the feature that two consecutive terms differ by the common difference, it also has the following property: for $n = 1, 2, 3, \cdots$

$$a_{n+1} = \frac{a_{n+2} + a_n}{2}.$$

That is: one term in an arithmetic sequence is the arithmetic mean of its previous and next terms [1]

Exercise 7.2 Prove: if $\{a_n\}_{n=1}^{\infty}$ is an infinite arithmetic sequence, then for $n = 1, 2, 3, \cdots$

$$a_{n+1} = \frac{a_{n+2} + a_n}{2}.$$

For a given arithmetic sequence, if its general term is:

$$a_n = a_1 + (n-1)d, \quad \text{for} \quad n = 1, 2, \cdots,$$

then we can easily calculate the sum of its first k terms. The idea of the calculation is similar to finding the area of a trapezoid.

First of all, list the first k terms of this arithmetic sequence as follows:

$$a_1, \quad a_1 + d, \quad a_1 + 2d, \quad \cdots, \quad a_1 + (k-2)d, \quad a_1 + (k-1)d.$$

Then reverse the list:

---

[1] Given any two numbers B and C, their arithmetic mean value A is defined as $A = \frac{B+C}{2}$.

$$a_1 + (k-1)d, \quad a_1 + (k-2)d, \quad a_1 + (k-3)d, \quad \cdots, \quad a_1 + d, \quad a_1.$$

Add the corresponding terms of these two columns (the first term plus the first term) to get a constant sequence:

$$2a_1 + (k-1)d, \quad 2a_1 + (k-1)d, \quad 2a_1 + (k-1)d, \quad \cdots, \quad 2a_1 + (k-1)d, \quad 2a_1 + (k-1)d.$$

The sum of them is obviously $(2a_1 + (k-1)d) \cdot k = [a_1 + a_1 + (k-1)d] \cdot k$. The latter form is written in convenience for memorizing: first term + last term of the sequence. Therefore, we obtain a formula for the sum of the first k terms of an arithmetic sequence:

> **Proposition 7.1. Sum of the first k terms of an arithmetic sequence**
>
> *The formula for the sum of the first k terms of an arithmetic sequence is*
>
> $$\text{Sum of the first } k \text{ terms of an arithmetic sequence} = \frac{\text{First term} + \text{Last term}}{2} \cdot k.$$

Now we introduce some mathematical notations so that we can express the formula in a more concise and consensual way (Human beings have various languages, but a mathematical formula that is close to a certain machine language usually will be the best).

First, we use

$$\sum_{n=1}^{k} a_n = a_1 + a_2 + \cdots + a_k$$

as the sum of k terms. The subscript n in a_n is an indicator. It tells us that the added terms are from $n = 1$ (that is, a_1) to $n = k$ (that is, a_k). In this way, we can write the formula for the sum of the first k terms of an arithmetic sequence as follows:

$$\sum_{n=1}^{k} a_n = \frac{a_1 + a_k}{2} \cdot k. \tag{7.1}$$

Example 7.9 Calculate the sum of the first 100 natural numbers. That is, calculate

$$0 + 1 + 2 + \cdots + 98 + 99.$$

Solution:

$$\text{Sum} = \sum_{n=0}^{99} n = \frac{0 + 99}{2} \cdot 100 = 4950.$$

□

7.1.3 Geometric sequence

> **Definition 7.3. Definition of geometric sequence**
>
> We call the sequence in the following form a geometric sequence:
>
> $$g, gr, gr^2, \cdots.$$
>
> Here, g is called the first term, and the non-zero number r is called the common ratio of the sequence. Geometric sequence $\{g_n\}_{n=1}^{\infty}$ has the following general term formula:
>
> $$g_n = gr^{n-1} \quad \text{for } n = 1, 2, \cdots.$$

Example 7.10 Suppose a sequence is given by

$$g_{n+1} = \frac{g_n}{10}, \quad \text{for } n = 1, 2, 3, \cdots; \quad \text{and } g_1 = \frac{3}{10}.$$

then it is a geometric sequence. Its first term is $g_1 = \frac{3}{10}$, and its common ratio is $\frac{1}{10}$. We can also derive its general term formula as:

$$g_n = \frac{3}{10^n} \quad \text{for } n = 1, 2, 3, \cdots.$$

□

For a geometric sequence, besides the feature that the ratio between two consecutive terms is constant, it also has the following property: for $n = 1, 2, 3, \cdots$,

$$g_{n+1}^2 = g_{n+2} \cdot g_n.$$

That is: the absolute value of a term in the geometric sequence is the geometric mean of the previous term and the next term[2].

Exercise 7.3 Prove: if $\{g_n\}_{n=1}^{\infty}$ is an infinite geometric sequence, then: for $n = 1, 2, 3, \cdots$,

$$g_{n+1}^2 = g_{n+2} \cdot g_n.$$

For a geometric sequence, whose general term is:

$$g_n = g_1 r^{n-1} \quad \text{for } n = 1, 2, \cdots,$$

we can also calculate the sum of its first k terms.

Let the sum of the first k terms of this geometric sequence be S:

$$S = g_1 + g_1 r + g_1 r^2 + \cdots + g_1 r^{k-2} + g_1 r^{k-1}.$$

First, we assume $r \neq 1$. Multiply both sides of the above equation by r to get:

$$rS = g_1 r + g_1 r^2 + g_1 r^3 + \cdots + g_1 r^{k-1} + g_1 r^k.$$

---

[2] For any two given positive numbers B and C, their geometric mean value A is defined as $A = \sqrt{BC}$.

Subtract the above two equations to get:

$$(1-r)S = g_1 - g_1 r^k.$$

Therefore, we obtain the formula for the sum of the first k terms of a geometric sequence in the following.

> **Proposition 7.2. Sum of the first k terms of a geometric sequence**
>
> Consider a geometric sequence, whose general term is:
>
> $$g_n = g_1 r^{n-1} \quad \text{for } n = 1, 2, \cdots.$$
>
> Assume $r \neq 1$, then the sum of its first k terms is
>
> $$\sum_{n=1}^{k} g_n = \sum_{n=1}^{k} g_1 r^{n-1} = \frac{g_1(1-r^k)}{1-r}.$$

In fact, the above formula is the same as the factorization we have learned before: for $n = 2, 3, \cdots$,

$$x^n - 1 = (x-1)(x^{n-1} + x^{n-2} + \cdots x + 1).$$

Example 7.11 For a geometric sequence, whose general term formula is:

$$g_n = \frac{3}{10^n} \quad \text{for } n = 1, 2, \cdot.$$

Find

$$\sum_{n=1}^{5} g_n$$

Solution:

$$\sum_{n=1}^{5} g_n = \frac{\frac{3}{10} \cdot [1 - (\frac{1}{10})^5]}{1 - \frac{1}{10}} = \frac{33333}{100000}.$$

$\square$

If $r = 1$, a geometric sequence is a constant sequence, and the sum of its first k terms is simply its first term multiplied by k.

On the other hand, when we are deriving the sum of the first k terms for an arithmetic sequence and for a geometric sequence, we can not write out completely all middle terms. Some students may have doubts about the formula. To make certain compensations, we shall use mathematical induction to strictly prove one of the formulas.

Example 7.12 For $r \neq 1$, prove: for all natural numbers $n \in \mathbb{N}$,

$$1 + r + r^2 + \cdots + r^n = \frac{1 - r^{n+1}}{1 - r}. \tag{7.2}$$

Proof: For $n = 0$, the left and right sides of identity (7.2) are 1. The equation holds.

Suppose that identity (7.2) holds for $n = k$, that is
$$1 + r + r^2 + \cdots + r^k = \frac{1 - r^{k+1}}{1 - r}. \tag{7.3}$$
Under this assumption, we will prove that identity (7.2) also holds for $n = k + 1$.

In fact
$$\begin{aligned}
1 + r + r^2 + \cdots + r^k + r^{k+1} &= (1 + r + r^2 + \cdots + r^k) + r^{k+1} \\
&= \frac{1 - r^{k+1}}{1 - r} + r^{k+1} && \text{(Use (7.3))} \\
&= \frac{1 - r^{(k+1)+1}}{1 - r}. && \text{(Simplifying)}
\end{aligned}$$
So identity (7.2) holds for $n = k + 1$. The proof is completed. □

Finally, we look at a slightly more complicated example.

Example 7.13 (**Mixture of arithmetic sequence and geometric sequence**) Let $\{a_n\}_{n=1}^{\infty}$ be an arithmetic sequence and $\{g_n\}_{n=1}^{\infty}$ be a geometric sequence: for $n = 1, 2, \cdots$,
$$a_n = a_1 + (n - 1)d, \qquad g_n = g_1 r^{n-1}, \qquad r \neq 1.$$
We will study how to calculate the sum of the first k terms of the new sequence $\{a_n g_n\}_{n=1}^{\infty}$.

First, we have
$$\begin{aligned}
\sum_{n=1}^{k} a_n g_n &= \sum_{n=1}^{k} (a_1 + (n - 1)d) \cdot g_1 r^{n-1} \\
&= \sum_{n=1}^{k} a_1 g_1 r^{n-1} + \sum_{n=1}^{k} (n - 1) d g_1 r^{n-1}.
\end{aligned} \tag{7.4}$$

Let
$$S_1 = \sum_{n=1}^{k} (n - 1) r^{n-1} = r + 2r^2 + \cdots + (k - 1) r^{k-1}.$$

Then
$$r S_1 = r^2 + \cdots + (k - 2) r^{k-1} + (k - 1) r^k.$$

So
$$(1 - r) S_1 = r + r^2 + \cdots + r^{k-1} - (k - 1) r^k = \frac{r - r^k}{1 - r} - (k - 1) r^k.$$

That is
$$S_1 = \frac{r - r^k}{(1 - r)^2} - \frac{(k - 1) r^k}{1 - r}.$$

Bringing the value of S_1 into (7.4), we have
$$\sum_{n=1}^{k} a_n g_n = \frac{(1 - r^k) a_1 g_1}{(1 - r)} - \frac{(k - 1) r^k d g_1}{1 - r} + \frac{(r - r^k) d g_1}{(1 - r)^2}.$$

□

7.1.4 The connection between summation and difference*

Here we want to introduce the connection between summation and difference. This relation has certain similarities to the Fundamental Theorem of Calculus, which will be covered in the next chapter.

Example 7.14 We first choose another method to calculate:

$$1 + 2 + \cdots + k.$$

Observe: For $n \in \mathbb{N}$,

$$(n+1)^2 - n^2 = 2n + 1.$$

So we have

$$2^2 - 1^2 = 2 \cdot 1 + 1$$
$$3^2 - 2^2 = 2 \cdot 2 + 1$$
$$\cdots$$
$$(k+1)^2 - k^2 = 2 \cdot k + 1$$

We add up the above identities altogether to get:

$$(k+1)^2 - 1^2 = 2 \cdot (1 + 2 + \cdots + k) + k.$$

Thus, we have:

$$1 + 2 + \cdots + k = \frac{(k+1)^2 - 1 - k}{2} = \frac{(1+k)k}{2}.$$

$\square$

Some students may think that the above method is not as simple as using the summation formula for an arithmetic sequence. Yes, we should pay more attention to the connection between the summation and the difference. Motivated by the above approach, one can look at the following example.

Example 7.15 Calculate:

$$1^2 + 2^2 + \cdots + k^2$$

Obviously, $\{n^2\}_{n=1}^{\infty}$ is neither an arithmetic sequence nor a geometric sequence. But we observe: for $n \in \mathbb{N}$,

$$(n+1)^3 - n^3 = 3n^2 + 3n + 1.$$

So we have

$$2^3 - 1^3 = 3 \cdot 1^2 + 3 \cdot 1 + 1$$
$$3^3 - 2^3 = 23 \cdot 2^2 + 3 \cdot 2 + 1$$
$$\cdots$$
$$(k+1)^3 - k^3 = 3 \cdot k^2 + 3 \cdot k + 1$$

We add up the above to get:

$$(k+1)^3 - 1^3 = 3 \cdot (1^2 + 2^2 + \cdots + k^2) + 3 \cdot (1 + 2 + \cdots + k) + k$$
$$= 3 \cdot (1^2 + 2^2 + \cdots + k^2) + \frac{3k(1+k)}{2} + k.$$

Thus, we have:

$$1^2 + 2^2 + \cdots + k^2 = \frac{(k+1)^3 - 1 - k}{3} - \frac{3k(1+k)}{6} = \frac{k(k+1)(2k+1)}{6}.$$

□

It is good for students, using the same strategy, to calculate

$$1^3 + 2^3 + \cdots + k^3.$$

And you shall be able to, at the end of your calculation, the following wonderful formula

$$1^3 + 2^3 + \cdots + k^3 = (1 + 2 + \cdots + k)^2.$$

Of course, you can also use mathematical induction to prove the above formula (see Question 3 in "Exercises after class" at the end of this section).

7.1.5 General term for Fibonacci sequence*

Recall that the Fibonacci sequence is given by the following recurrence relation:

$$a_{n+2} = a_{n+1} + a_n, \text{ for } n = 1, 2, 3, \cdots; \text{ and } a_1 = 1, a_2 = 2.$$

We first wand to find two numbers α and β, such that

$$a_{n+2} - \alpha a_{n+1} = \beta[a_{n+1} - \alpha a_n].$$

Comparing this identity with the recurrence relation, we know α, β shall satisfy

$$\alpha\beta = -1$$
$$\alpha + \beta = 1.$$

There are two sets of solutions:

$$\alpha_1 = \frac{1 + \sqrt{5}}{2}, \quad \beta_1 = \frac{1 - \sqrt{5}}{2};$$

or
$$\alpha_2 = \frac{1-\sqrt{5}}{2}, \quad \beta_2 = \frac{1+\sqrt{5}}{2}.$$

For $n \geq 2$, we write
$$A_n = a_n - \alpha_1 a_{n-1}, \quad B_n = a_n - \beta_1 a_{n-1}.$$

then
$$A_{n+2} = \beta_1 A_{n+1} = \alpha_2 A_{n+1}, \quad A_2 = \frac{3-\sqrt{5}}{2} = \beta_1^2$$
$$B_{n+2} = \beta_2 B_{n+1} = \alpha_1 B_{n+1}, \quad B_2 = \frac{3+\sqrt{5}}{2} = \beta_2^2.$$

We thus concluded that: for $n = 1, 2, 3, \cdots$,
$$A_n = \beta_1^n, \quad B_n = \beta_2^n.$$

That is
$$a_n - \alpha_1 a_{n-1} = \beta_1^n \tag{7.5}$$
$$a_n - \beta_1 a_{n-1} = \beta_2^n. \tag{7.6}$$

From $(7.5) \times \beta_1 - (7.6) \times \beta_2$, we have, for $n = 1, 2, \cdots$,
$$a_n = \frac{\beta_1^{n+1} - \beta_2^{n+1}}{\beta_1 - \beta_2} = \frac{1}{\sqrt{5}} \cdot \left\{ \left(\frac{1+\sqrt{5}}{2}\right)^{n+1} - \left(\frac{1-\sqrt{5}}{2}\right)^{n+1} \right\}.$$

7.1.6 Exercises after class

1. **Basic skills**.

    (1) Find the general term for the following sequence: $a_0 = 1$, $a_{n+1} = a_n - 2$ for $n = 0, 1, 2, \cdots$.

    (2) Find the general term for the following sequence: $a_1 = 3$, $a_{n+1} = \frac{a_n}{5}$ for $n = 1, 2, \cdots$.

    (3) If $a_n = 3 + 2n$ for $n = 1, 2, \cdots$, find
    $$\sum_{i=1}^{4} a_i$$

    (4) If $a_n = 5 \cdot 10^{-n+1}$ for $n = 1, 2, \cdots$, find
    $$\sum_{k=1}^{6} a_k.$$

2. **Comprehensive questions**. Find the summation (without using the formula from Example 7.15):
$$\sum_{k=1}^{100} k(k+1).$$

3. **Comprehensive questions**. Prove: for $n = 1, 2, 3, \cdots$,
$$\sum_{k=1}^{n} k^3 = \left(\sum_{k=1}^{n} k\right)^2.$$

4. **Comprehensive questions**.

    (1) For $n = 1, 2, 3, \cdots$, $a_{n+1} = -2a_n - 3$.
    (i). Prove that $b_n = a_n + 1$ is a geometric sequence.
    (ii). Find the general term for a_n.

    (2) For $n = 1, 2, 3, \cdots$, $a_{n+1} = 3a_n - 1$, find the general term for a_n.

    (3)* For $n = 0, 1, 2, 3, \cdots$, $a_{n+2} - 3a_{n+1} + 2a_n = 0$ and $a_0 = 0, a_1 = 1$. Find the general term for a_n.

7.2 Limit

The introduction of the concept of the limit usually is viewed as the starting point of Advanced Mathematics (compared with Elementary Mathematics). Here, we need to answer some questions left behind while we started systematically studying elementary mathematics.

7.2.1 Motivation for introducing the limit and applications of the limit

From the viewpoint of the expansion of the number fields, we need to: (I) Deepen the understanding of rational numbers, especially the understanding of infinite repeating decimals; (II) Deepen the understanding of irrational numbers and understand that irrational numbers can always be approximated by rational numbers.

The introduction of the limit enables us to continue our study of the new concepts of differentiation and integration in the following chapters. From the perspective of understanding functions, we also need to answer the reason that we have the "fundamental assumptions on elementary mathematics". We also want to understand the relation between general functions and polynomials — This requires us to use the concept of differentiation.

From the perspective of real-life applications, the study of differentiation and derivative is indispensable for instantaneous speed, the tangent of the curve, etc; The area of an irregular region can only be calculated by using the integration.

7.2.2 Definition and operation

The limit (trend) of a sequence is sometimes easy to see, such as

Example 7.16 Can we see the trend of sequence $\{\frac{1}{n}\}_{n=1}^{\infty}$?

Answer: We can see that this is a positive sequence (all elements are positive). As n is getting larger and larger, the sequence $\frac{1}{n}$ is getting smaller and smaller with the trend to 0

□

Although we can see that the trend of the above series is 0, the argument is very imprecise. Some students will argue that the sequence $\{\frac{n+1}{n}\}$ is also a positive sequence (all positive numbers). At the same time, numbers in the sequence are getting smaller and smaller as n is getting larger and larger. Is its limit also 0? We will see later that

7.2 Limit

the limit actually is 1 instead of 0. Therefore, a rigorous mathematical definition is necessary.

Next, we use the general debating strategy to define the limit of a sequence.

> **Definition 7.4. Definition of the limit for sequence**
>
> *Consider an infinite sequence $\{a_n\}_{n=1}^{\infty}$. We say the sequence $\{a_n\}$ converges to L: if there is a finite number L, such that, for any positive number ϵ, there is a corresponding natural number N_0, we always have*
>
> $$|a_n - L| < \epsilon \quad \text{for } n > N_0.$$
>
> *We also say that the sequence has a limit of L. We use the notation:*
>
> $$\lim_{n \to \infty} a_n = L.$$

The above definition is the notorious "$\epsilon - N$ definition", which gives many students a headache. Perhaps there will be more suitable definitions in the future that can be easily mastered by more people.

Here we do some simple exercises (**warning:** which may not be simple at all for some students).

Example 7.17 Prove:
$$\lim_{n \to \infty} \frac{1}{n} = 0.$$

Proof: For any given positive number ϵ, we take the natural number N_0 satisfying
$$N_0 > \frac{1}{\epsilon}.$$

Then, for any natural number $n > N_0$, we always have $\frac{1}{n} < \frac{1}{N_0} < \epsilon$. Therefore, **for any natural number $n > N_0$, it holds**
$$\left|\frac{1}{n} - 0\right| = \frac{1}{n} < \frac{1}{N_0} < \epsilon.$$

By the definition (comparing the sentences in boldface with the definition of the limit), we know that
$$\lim_{n \to \infty} \frac{1}{n} = 0.$$

□

Exercise 7.4 Find
$$\lim_{n \to \infty} \frac{n+1}{n},$$
and prove your conclusion.

7.2.3 Algebraic properties for the limit operation

We now derive the algebraic properties for the limit operation. These "proofs" are not required to be fully understood by students in their first reading (we add * as a reminder: these are selective contents for reading/self-study).

> **Proposition 7.3. Algebraic properties of limit**
>
> *Suppose*
> $$\lim_{n \to \infty} a_n = L, \quad \lim_{n \to \infty} b_n = M.$$
>
> *(1) For any given two numbers α and β, sequence $\alpha a_n + \beta b_n$ is convergent, and*
> $$\lim_{n \to \infty} (\alpha a_n + \beta b_n) = \alpha L + \beta M.$$
>
> *(2) Sequence $a_n b_n$ is convergent, and*
> $$\lim_{n \to \infty} a_n b_n = LM.$$
>
> *(3) We further assume that for large n, $b_n \neq 0$, and $M \neq 0$. Then, sequence $\frac{a_n}{b_n}$ is convergent, and*
> $$\lim_{n \to \infty} \frac{a_n}{b_n} = \frac{L}{M}.$$

Prove*: Here, we only prove the above rule (1) for the case of $\alpha = \beta = 1$, that is, to prove :
$$\lim_{n \to \infty} (a_n + b_n) = L + M. \tag{7.7}$$

Other proofs are left for students to practice.

From
$$\lim_{n \to \infty} a_n = L$$
we know that for any positive number ϵ, there is a natural number N_1, so that for $n > N_1$,
$$|a_n - L| < \frac{\epsilon}{2}. \tag{7.8}$$

Similarly, from
$$\lim_{n \to \infty} b_n = M$$
we know: for any positive number ϵ, there is a natural number of N_2, so that for $n > N_2$,
$$|b_n - M| < \frac{\epsilon}{2}. \tag{7.9}$$

7.2 Limit

Choose a natural number $N_0 > N_1 + N_2$. Then, for $n > N_0$, we have:

$$\begin{aligned}
|a_n + b_n - (L + M)| &= |(a_n - L) + (b_n - M)| \\
&\leq |a_n - L| + |b_n - M| \quad &\text{(Triangle inequality)} \\
&< \frac{\epsilon}{2} + \frac{\epsilon}{2} \quad &\text{(Using (7.8) and (7.9))} \\
&= \epsilon.
\end{aligned}$$

Therefore, we know from the definition of the limit

$$\lim_{n \to \infty} (a_n + b_n) = L + M.$$

$\square$

The above arguments about the algebraic operation properties of limits are not easy to grasp, but they are very useful. Students can grasp these properties only by doing more exercises.

Example 7.18 Find

(a)
$$\lim_{n \to \infty} \frac{1}{n^2};$$

(b)
$$\lim_{n \to \infty} \frac{n^2 + n + 1}{n^2 + 1}.$$

(c)
$$\lim_{n \to \infty} \frac{n^2 - 1}{\sqrt{n^4 - n^2 + 1}}.$$

Solution:

(a)
$$\begin{aligned}
\lim_{n \to \infty} \frac{1}{n^2} &= \lim_{n \to \infty} \frac{1}{n} \cdot \lim_{n \to \infty} \frac{1}{n} \quad &\text{(Proposition (7.3))} \\
&= 0 \cdot 0 \quad &\text{(Results of Example 7.15)} \\
&= 0.
\end{aligned}$$

(b) We first manipulate the rational polynomial by dividing the numerator and the denominator simultaneously by the highest order monomial (here, it is n^2):

$$\lim_{n\to\infty} \frac{n^2+n+1}{n^2+1} = \lim_{n\to\infty} \frac{n^2+n+1}{n^2+1} \cdot \frac{1/n^2}{1/n^2}$$

$$= \lim_{n\to\infty} \frac{1+n^{-1}+n^{-2}}{1+n^{-2}} \quad \text{(Simplify)}$$

$$= \frac{\lim_{n\to\infty}(1+n^{-1}+n^{-2})}{\lim_{n\to\infty}(1+n^{-2})} \quad \text{(Proposition (7.3))}$$

$$= 1.$$

(c) We again manipulate the irrational polynomial by dividing the numerator and the denominator simultaneously by the highest order monomial (here, it is n^2):

$$\lim_{n\to\infty} \frac{n^2-1}{\sqrt{n^4-n^2+1}} = \lim_{n\to\infty} \frac{n^2-1}{\sqrt{n^4-n^2+1}} \cdot \frac{1/n^2}{1/n^2}$$

$$= \lim_{n\to\infty} \frac{1-1/n^2}{\sqrt{1-1/n^2+1/n^4}} \quad \text{(Simplifying)}$$

$$= \frac{\lim_{n\to\infty}(1-1/n^2)}{\lim_{n\to\infty}\sqrt{1-1/n^2+1/n^4}} \quad \text{(Proposition (7.3) and the comment below)}$$

$$= 1.$$

□

Remark In solving Example 7.18 part (c), we also use the fact: if $\lim_{n\to\infty} a_n = L \geq 0$, then $\lim_{n\to\infty} \sqrt{a_n} = \sqrt{L}$. This can be proved via the definition of the limit, or via the continuity of $y = \sqrt{x}$ for $x > 0$.

Often, it is difficult to determine whether a sequence converges or not. To this end, we shall learn more about the properties of sequences.

Definition 7.5. Definition of bounded sequence

Consider an infinite sequence $\{a_n\}_{n=1}^{\infty}$. If there is a finite number M, such that, for all $n \in \mathbb{N}$, it holds

$$|a_n| < M,$$

then we say that sequence a_n is bounded.

Note that $|a_n| < M$ is equivalent to $-M < a_n < M$.

If $a_n > L$ for all n, we say that a_n has a lower bound L; If $a_n < M$ for all n, we say that a_n has an upper bound M. We call the largest lower bound of the sequence (if it exists) the infimum; Call the smallest upper bound of the sequence (if it exists) the supremum[3].

---

[3] The existence of the largest lower bound and the smallest upper bound can be obtained from the com-

7.2 Limit

> **Proposition 7.4. Boundedness of convergence limit**
>
> If $\lim_{n\to\infty} a_n = L$, then sequence $\{a_n\}_{n=0}^{\infty}$ is bounded.

Proof: From the definition of $\lim_{n\to\infty} a_n = L$, we know that for a given constant of 1, there is a natural number N_0, such that, for $n > N_0$,

$$|a_n - L| < 1.$$

That is:

$$L - 1 < a_n < L + 1 \quad \text{for any } n > N_0.$$

Therefore, for all natural numbers n, a_n is not less than the smallest number in set $\{a_0, a_1, \cdots, a_{N_0}, L - 1\}$, and a_n is not greater than the maximum number in set $\{a_0, a_1, \cdots, a_{N_0}, L + 1\}$. So $\{a_n\}_{n=0}^{\infty}$ is bounded.

□

A sequence a_n is bounded, but may not converge.

In fact, by the definition of the limit, we can prove (we skip the proof here)

> **Proposition 7.5. Convergence of subsequences of a convergent sequence**
>
> If $\lim_{n\to\infty} a_n = L$, Then any of its subsequence a_{i_n} must converge to L.

Example 7.19 Sequence $a_n = (-1)^n$, $n = 1, 2, \cdots$ is a bounded sequence. However, by proposition 7.5, we know this is not a convergent sequence.

Usually, it is not easy to prove that a sequence is convergent. We give two more propositions here (Proofs will be given in a Calculus course).

> **Proposition 7.6. Monotone bounded sequence must converge**
>
> If sequence $\{a_n\}_{n=1}^{\infty}$ is monotone in n and bounded, then it is a convergent sequence.

Through some complicated calculations, we can show that for all $n \in \mathbb{N}$,

$$(1 + \frac{1}{n})^n < (1 + \frac{1}{n+1})^{n+1}, \quad \text{and} \quad (1 + \frac{1}{n})^n < (1 + 1)^2.$$

So the limit

$$\lim_{n\to\infty} (1 + \frac{1}{n})^n$$

exist. We denote the limit as e – the natural constant as we mentioned earlier (see Section 3.3), also known as the Euler constant. After introducing the concept and basic properties of derivatives, we will come back to see why this is a monotonically increasing and bounded sequence. (See the comprehensive exercises in Section 8.2.4.)

---

pleteness of real number field.

Next, we introduce

> **Proposition 7.7. Squeeze Theorem**
> Consider three sequences $\{a_n\}_{n=1}^{\infty}$, $\{b_n\}_{n=1}^{\infty}$, $\{c_n\}_{n=1}^{\infty}$, which satisfy
> (1) There is a number N_0, such that for all $n > N_0$,
> $$a_n \leq b_n \leq c_n.$$
> (2)
> $$\lim_{n \to \infty} a_n = \lim_{n \to \infty} c_n = L.$$
> Then $\{b_n\}$ is also a convergent sequence, and $\lim_{n \to \infty} b_n = L$.

We use the following two examples to show how useful the squeeze theorem is.

Example 7.20 Prove:
$$\lim_{n \to \infty} \frac{n+1}{\sqrt{n^2+1}} = 1.$$

proof: Observe: for positive natural number n,
$$n < \sqrt{n^2+1} < n+1.$$

Let
$$a_n = \frac{n+1}{n}, \quad b_n = \frac{n+1}{\sqrt{n^2+1}}, \quad c_n = \frac{n+1}{n+1},$$

so we have
$$a_n \geq b_n \geq c_n.$$

Easy to verify

$$\lim_{n \to \infty} a_n = \lim_{n \to \infty} \frac{n+1}{n}$$
$$= \lim_{n \to \infty} (1 + \frac{1}{n})$$
$$= 1;$$

and

$$\lim_{n \to \infty} c_n = \lim_{n \to \infty} \frac{n+1}{n+1}$$
$$= \lim_{n \to \infty} 1$$
$$= 1.$$

Therefore, from the squeeze theorem, we have:
$$\lim_{n \to \infty} b_n = 1.$$

□

Example 7.21 Prove:
$$\lim_{n \to \infty} \frac{n}{2^n} = 0$$

Proof: Claim: for $n \geq 4$, $2^n \geq n^2$. We use mathematical induction to prove this inequality.

For $n = 4$,
$$2^4 = 16 = 4^2.$$

Inequality holds.

Suppose that for $n = k \geq 4$, the inequality holds, that is:
$$2^k \geq k^2.$$

Then for $n = k + 1$,
$$\begin{aligned}
2^{k+1} &= 2 \cdot 2^k \\
&\geq 2k^2 & \text{(Inductive hypothesis)} \\
&> k^2 + 2k + 1 & \text{(Because } k \geq 4\text{)} \\
&= (k+1)^2
\end{aligned}$$

So the inequality also holds for $n = k + 1$. The claim is proved.

Let $a_n = 0$, $b_n = \frac{n}{2^n}$, $c_n = \frac{n}{n^2}$. Then
$$a_n \leq b_n \leq c_n.$$

Also
$$\lim_{n \to \infty} 0 = 0, \quad \lim_{n \to \infty} \frac{n}{n^2} = \lim_{n \to \infty} \frac{1}{n} = 0.$$

Therefore, from the squeeze theorem, we have
$$\lim_{n \to \infty} \frac{n}{2^n} = 0.$$

□

Exercise 7.5 (a). Suppose that $\lim_{n \to \infty} |a_n| = 0$. Use the squeeze theorem to prove:
$$\lim_{n \to \infty} a_n = 0.$$

(b). Suppose that $\lim_{n \to \infty} a_n = 0$. Can you prove that
$$\lim_{n \to \infty} |a_n| = 0?$$

Roughly speaking, it is not an easy job to find limits with rigorous reasons based on the definition of limits and some of the above theorems. For example, we may easily see that
$$\lim_{n \to \infty} \frac{1}{\sqrt{n} - 2} = 0.$$

But it is not easy to prove it rigorously. We shall continue to learn the limits of functions and hope to have some relatively easy ways to find the limits.

7.2.4 Limit for function

Consider function $y = f(x)$ defined on an interval I. We define its limit at a point $x_0 \in I$.

> **Definition 7.6. Limit for function**
> We say for $x \to x_0$, function $f(x)$ converges to L: if for any $\epsilon > 0$, there is a positive number δ, such that, for all $0 < |x - x_0| < \delta$, we have
> $$|f(x) - L| < \epsilon.$$
> We also say for $x \to x_0$, $f(x)$ has a limit L. We denote it as: $\lim_{x \to x_0} f(x) = L$.

The definition is difficult to understand. It even has a hidden meaning: The limit (trend) of the function $f(x)$ at point x_0 is not necessarily the same as the function value at point x_0.

Example 7.22 Prove:
$$\lim_{x \to a} x = a.$$

Proof: For any $\epsilon > 0$, let $\delta = \epsilon$. Then, for $0 < |x - a| < \delta = \epsilon$,
$$|x - a| < \epsilon.$$
We obtain the result by the definition of the limit.

□

We check the following similar but somewhat confusing example.

Example 7.23 Define function
$$f(x) = \begin{cases} x & \text{for } x \neq a, \\ 0 & \text{for } x = a. \end{cases}$$

Prove:
$$\lim_{x \to a} f(x) = a.$$

Proof: Students who do understand the definition of the limit for functions already know that the answer to this question is the same as that to the previous example. If you are not certain, then we just repeat it again.

For any $\epsilon > 0$, let $\delta = \epsilon$. Then, for $0 < |x - a| < \delta = \epsilon$,
$$|f(x) - a| = |x - a| < \epsilon.$$

We obtain $\lim_{x \to a} f(x) = a$ by the definition of the limit.

□

Obviously, in the above example, for $a \neq 0$, $\lim_{x \to a} f(x) \neq f(a)$.

Similar to limits of sequences, limits for functions also have the following algebraic properties (we leave the proofs in Calculus books).

Proposition 7.8. Algebraic properties of function limits

Suppose that
$$\lim_{x \to x_0} f(x) = L, \quad \lim_{x \to x_0} g(x) = M.$$

(1) Then, for any two numbers given, α and β, function $\alpha f(x) + \beta g(x)$ converges as $x \to x_0$, and
$$\lim_{x \to x_0} (\alpha f(x) + \beta g(x)) = \alpha L + \beta M.$$

(2) Function $f(x)g(x)$ converges as $x \to x_0$, and
$$\lim_{x \to x_0} f(x)g(x) = LM.$$

(3) Further assume that $M \neq 0$. Then, function $\frac{f(x)}{g(x)}$ converges as $x \to x_0$, and
$$\lim_{x \to x_0} \frac{f(x)}{g(x)} = \frac{L}{M}.$$

Example 7.24 Find
$$\lim_{x \to a} x^2$$

Solution:

$$\lim_{x \to a} x^2 = \lim_{x \to a} x \cdot \lim_{x \to a} x \qquad \text{(Proposition (7.8))}$$
$$= a \cdot a \qquad \text{(Results of Example 7.23)}$$
$$= a^2.$$

□

We also have the following squeeze convergence theorem for the limits of functions.

Proposition 7.9. Squeeze theorem

Suppose three functions $f(x)$, $g(x)$, $h(x)$ satisfy

(1) There is a number ϵ_1, such that: for any $0 < |x - x_0| < \epsilon_1$,
$$f(x) \leq g(x) \leq h(x).$$

(2)
$$\lim_{x \to x_0} f(x) = \lim_{x \to x_0} h(x) = L.$$

Then $\lim_{x \to x_0} g(x) = L$.

It is easy to get from the definition: if $\lim_{x \to a} f(x) = 0$, then $\lim_{x \to a} |f(x)| = 0$. In turn, we have the following example.

Example 7.25 Assume $\lim_{x \to a} |k(x)| = 0$. Prove: $\lim_{x \to a} k(x) = 0$.

Proof: Let $f(x) = -|k(x)|$, $h(x) = |f(x)|$. So
$$f(x) \le k(x) \le h(x),$$
and
$$\lim_{x \to a} f(x) = 0, \quad \text{and} \quad \lim_{x \to a} h(x) = 0.$$

From the squeezing convergence theorem, we immediately have
$$\lim_{x \to a} k(x) = 0.$$

□

Generally, it is pretty complicated to calculate (with proof) the limit of a function by the definition. We will do more calculations for the function limit after learning more about the concepts of the limit and continuity of a function.

The limits of composite functions have the following properties.

Proposition 7.10. Limit theorem of composite function

If two functions $f(x)$, $g(x)$ satisfy

(1)
$$\lim_{x \to x_0} f(x) = L;$$

(2)
$$\lim_{y \to L} g(y) = M,$$

then
$$\lim_{x \to x_0} g[f(x)] = \lim_{y \to L} g(y) = M.$$

Example 7.26 Find the limit $\lim_{x \to 2} e^{x^2}$.

Solution: Observe
$$\lim_{x \to 2} x^2 = 2^2 = 4,$$
and we will learn how to find the limits for elementary functions in the following section (we use the result ahead here):
$$\lim_{x \to 4} e^x = e^4.$$

From Proposition 7.10 we have
$$\lim_{x \to 2} e^{x^2} = e^4.$$

□

There is another limit of the function: the limit as the variable x goes to infinite.

Definition 7.7. Limit of function

We say that for $x \to \infty$, the function $f(x)$ converges to L: if for any $\epsilon > 0$, there is a number N, such that for any $x > N$, it holds
$$|f(x) - L| < \epsilon.$$
We also say for $x \to \infty$, $f(x)$ has the limit L. We use the notation: $\lim_{x \to \infty} f(x) = L$.

♣

Exercise 7.6 Try to give the limit definition of the function as variable x becomes very negative: $\lim_{x \to -\infty} f(x) = L$.

This limit also satisfies the algebraic operation rules and the squeezing convergence rule.

Proposition 7.11. Algebraic properties of function limits

Suppose that
$$\lim_{x \to \infty} f(x) = L, \quad \lim_{x \to \infty} g(x) = M.$$

Then

(1) For any two numbers α and β, function $\alpha f(x) + \beta g(x)$ converges, and
$$\lim_{x \to \infty} (\alpha f(x) + \beta g(x)) = \alpha L + \beta M.$$

(2) Function $f(x)g(x)$ converges, and
$$\lim_{x \to \infty} f(x)g(x) = LM.$$

(3) If further assume $M \neq 0$, then function $\frac{f(x)}{g(x)}$ converges, and
$$\lim_{x \to \infty} \frac{f(x)}{g(x)} = \frac{L}{M}.$$

♠

Proposition 7.12. Squeeze theorem

If three functions $f(x)$, $g(x)$, $h(x)$ satisfy:

(1) There exists a number N, such that for any $x > N$,
$$f(x) \leq g(x) \leq h(x).$$

(2)
$$\lim_{x \to \infty} f(x) = \lim_{x \to \infty} h(x) = L.$$

> Then $\lim_{x \to \infty} g(x) = L$.

Proposition 7.12 also holds for $L = \infty$ or $-\infty$.

Here we give an application of the limit in finding constant coefficients for two identical polynomials.

Example 7.27 Suppose
$$ax^3 + bx^2 + cx + d \equiv x^3 + 2x^2 + 3x + 4.$$
("$\equiv$" means that the above two polynomials take the same value for all variables x). Prove $a = 1$, $b = 2$, $c = 3$, $d = 4$.

Proof: In an algebra course, we learned that we can obtain four linear equations about a, b, c, d by choosing four different x values, and then solve the resulting 4×4 linear system to get the values of a, b, c, d.

Here we use the concept of the limit, which leads to an easier way to solve the problem. We present two methods (both methods are based on the limiting argument).

Method 1: First choosing $x = 0$, we get
$$d = 4.$$
So
$$ax^3 + bx^2 + cx \equiv x^3 + 2x^2 + 3x.$$
for $x \neq 0$, we have
$$ax^2 + bx + c \equiv x^2 + 2x + 3.$$
Sending $x \to 0$ at both sides of the above formula, we obtain $c = 3$.

So now, we obtain
$$ax^3 + bx^2 \equiv x^3 + 2x^2.$$
For $x \neq 0$, we have
$$ax + b \equiv x + 2.$$
Sending $x \to 0$ at both sides of the above formula we have $b = 3$.

In the same way, we finally get $a = 1$.

Method 2: Consider $x \neq 0$. Dividing both sides of the equation by x^3, we have
$$a + bx^{-1} + cx^{-2} + dx^{-3} \equiv 1 + 2x^{-1} + 3x^{-2} + 4x^{-3}.$$
Sending $x \to \infty$. We obtain
$$a = 1.$$

Subtracting x^3 from both sides of the original equation, we have
$$bx^2 + cx + d \equiv 2x^2 + 3x + 4,$$
Again consider $x \neq 0$. Dividing both sides of the above equation by x^2, and we have
$$b + cx^{-1} + dx^{-2} \equiv 2 + 3x^{-1} + 4x^{-2}.$$
Sending $x \to \infty$. We get
$$b = 2.$$
By analogy, we can also get $c = 3$, $d = 4$.

$\square$

✍ **Exercise 7.7** Suppose
$$a(x+2)^3 + b(x+2)^2 + c(x+2) + d \equiv 4(x+2)^3 + 3(x+2)^2 + 2(x+2) + 1,$$
Prove $a = 4$, $b = 3$, $c = 2$, $d = 1$.

7.2.5 Continuity of elementary functions

We list all typical elementary functions in the following figure: (1) Polynomial $y = c_n x^n + c_{n-1} x^{n-1} + \cdots + c_1 x + c_0$; (2) Power function $y = x^a$; (3) Exponential function $y = a^x$; (4) Logarithmic function $y = \ln x$; (5) Trigonometric function $y = \sin x$. We can intuitively see that these curves are not "broken". We use the continuity of functions to rigorously describe the meaning for non- "broken" graphs.

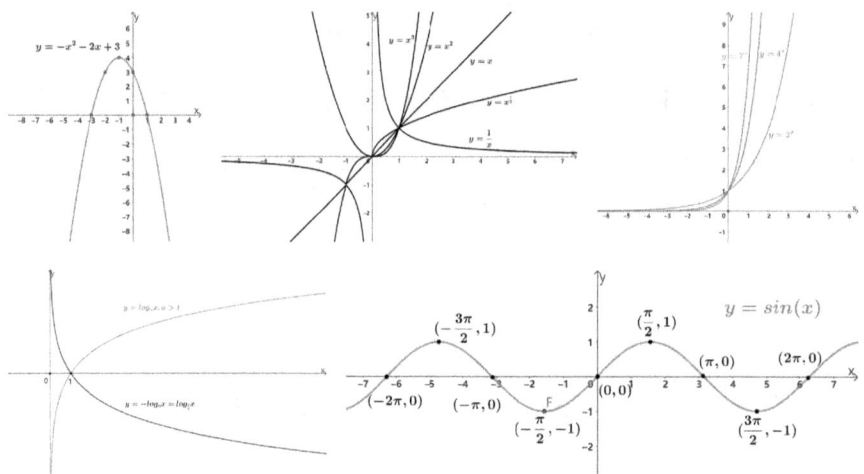

Figure 7.1: Image of elementary function

> **Definition 7.8. Continuity of function at one point**
>
> *Consider function $y = f(x)$ defined on the interval $I = (a, b)$. If at point $x_0 \in I$, the function satisfies the following two conditions*
>
> *(1) $\lim_{x \to x_0} f(x)$ exists;*

(2) $\lim_{x \to x_0} f(x) = f(x_0)$,

then we say function $y = f(x)$ is continuous at point x_0.

Example 7.28 (a) (**Examples of discontinuous points of functions**). Consider function

$$f(x) = \begin{cases} 1 & \text{for } x \geq a, \\ 0 & \text{for } x < a. \end{cases}$$

Because $\lim_{x \to a} f(x)$ does not exist, $x = a$ is a discontinuous point for the function $y = f(x)$.

(b) (**Examples of functions that have removable discontinuous points**). Define function

$$g(x) = \begin{cases} x & \text{for } x \neq 1, \\ 0 & \text{for } x = 1. \end{cases}$$

We can show $\lim_{x \to a} f(x) = 1$, but it is different from the function value at $x = 1$. So $x = a$ is also a discontinuous point for the function $y = f(x)$. Different from (a), $\lim_{x \to a} f(x)$ exists. Such a discontinuous point is called a removable discontinuous point.

Definition 7.9. Continuous function
Consider function $y = f(x)$ defined on the interval $I = (a, b)$. If it is continuous at any point on I, We call the function $y = f(x)$ a continuous function on the interval I.

We list some propositions about the continuity of functions (their proofs will be carefully studied in a Calculus course later).

Proposition 7.13. Algebraic properties of continuous functions
If the functions $f(x), g(x)$ are continuous functions on the interval I, then
(1). For any given number α and β, $\alpha f(x) + \beta g(x)$ is also a continuous function on the interval I.
(2). $f(x)g(x)$ is also a continuous function on the interval I.
(3). If $g(x) \neq 0$, $\frac{f(x)}{g(x)}$ is also a continuous function on the interval I.

Proposition 7.14. Continuous theorem of composite function
If the function $f(x)$ is a continuous function on the interval I with a range in the interval J, $g(x)$ is a continuous function on the interval J, then the composite function $g[f(x)]$ is a continuous function on the interval I.

Finally, we state the following theorem without providing proof.

> **Theorem 7.1. Continuity of elementary functions**
>
> Elementary function: (1a) Polynomial function
> $$y = c_n x^n + c_{n-1} x^{n-1} + \cdots + c_1 x + c_0$$
> (1b) Exponential function (a is any positive real number that is not 1)
> $$y = a^x$$
> (1c) Trigonometric function
> $$y = \sin x \quad \text{and} \quad y = \cos x$$
> They are all continuous functions in the real number set.
>
> Elementary function: (2a) Power function (r is any non-zero real number)
> $$y = x^r$$
> (2b) Logarithmic function (b is any positive real number that is not 1)
> $$y = \log_b x$$
> They are continuous functions in the positive real number set.

Example 7.29 Function $y = \ln(x^2 + 1)$ is a continuous function on $\mathbb{R}$.

□

Example 7.30 Find $\lim_{x \to 1} \ln(x^2 + 3)$.

Solution: Use the continuity of the function to get the result directly:
$$\lim_{x \to 1} \ln(x^2 + 3) = \ln(1 + 3) = \ln 4.$$

□

Finally, we use examples to show why we have Fundamental Assumptions of Elementary Mathematics. Here we choose the area of a rectangle as an example.

Consider a rectangle $ABCD$ with the length $AB = L$ and the width $BC = W$. Then its area $\mathcal{A}$ is a function that depends on L and W. As we know from "Introduction to Introductory Geometry and Proofs", if both L, W are fractions,
$$\mathcal{A} = LW.$$

If L is a fraction and W is an irrational number, we consider the linear function $f(x) = Lx$: it is a continuous function. Choosing a sequence of rational numbers W_i tending to W. The rectangular area with a length of L and a width of W_i is LW_i. As W_i tends to W, we can reasonably use the limit of area value LW_i to define the rectangular area with a length of L and width of W. From the continuity of a linear function, we know that the limit is LW! Therefore, we have: assuming that L is a fraction and W is an irrational number, then the rectangular area with a length of L and width of W is LW.

Similarly, we can also have: assuming that L and W are irrational numbers, the rectangular area with length of L and width of W is LW.

For continuous functions, we have the following theorem.

> **Theorem 7.2. Intermediate value value theorem of continuous function**
> Suppose that $y = f(x)$ is a continuous function on interval $[a, b]$ and $f(a)f(b) < 0$. Then there is a point $x_0 \in (a, b)$ such that $f(x_0) = 0$.

Example 7.31 * As an application of the mean value theorem, let's prove: the following cubic equation
$$x^3 + px^2 + qx + r = 0$$
has at least one real number solution.

proof: First of all, we make
$$f(x) = x^3 + px^2 + qx + r = x^3(1 + \frac{p}{x} + \frac{q}{x^2} + \frac{r}{x^3}).$$

Be aware that
$$\lim_{x \to \pm\infty} (1 + \frac{p}{x} + \frac{q}{x^2} + \frac{r}{x^3}) = 1,$$

so
$$\lim_{x \to +\infty} f(x) \to +\infty, \quad \lim_{x \to -\infty} f(x) \to -\infty.$$

Therefore, there is a large number b, such that $f(b) > 0$; and a very negative number a, such that $f(a) < 0$. According to the mean value theorem (all polynomials are continuous): There exists a x_0 so that $f(x_0) = 0$. That is, equation $x^3 + px^2 + qx + r = 0$ has at least one real number solution.

□

Finally, we present a direct application of function limits to determine the limits of sequences.

> **Proposition 7.15. Sequence limit via function limit**
> Let $a_n = f(n)$, and $\lim_{n \to \infty} f(n) = L$. Then
> $$\lim_{n \to \infty} a_n = L.$$

Example 7.32 Show that
$$\lim_{n \to \infty} \sqrt{1 + \frac{1}{n}} = 1.$$

proof:

Method 1: We can use the squeeze theorem to prove this statement.

Note:
$$1 < \frac{n+1}{n} < \frac{n+1}{n} \cdot \frac{n+1}{n}.$$

So,
$$1 < \sqrt{1 + \frac{1}{n}} = \sqrt{\frac{n+1}{n}} < \frac{n+1}{n}.$$

On the other hand,
$$\lim_{n \to \infty} 1 = 1, \quad \lim_{n \to \infty} \frac{n+1}{n} = \lim_{n \to \infty} \left(1 + \frac{1}{n}\right) = 1.$$

By the squeeze theorem we conclude
$$\lim_{n \to \infty} \sqrt{1 + \frac{1}{n}} = 1.$$

Method 2: Let $f(x) = \sqrt{1 + \frac{1}{x}}$. Since
$$\lim_{x \to \infty} \frac{1}{x} = 0,$$
and
$$\lim_{y \to 0} \sqrt{1 + y} = 1. \quad \text{(Continuity of the function)}.$$

Thus
$$\lim_{n \to \infty} \sqrt{1 + \frac{1}{n}} = \lim_{n \to \infty} f(n) = 1.$$

□

7.2.6 Exercises after class

1. **Basic skills**

   (1) Using definition to prove
   $$\lim_{n\to\infty} \frac{1}{\sqrt{n}} = 0.$$

   (2) Using definition to prove
   $$\lim_{x\to 4} \sqrt{x} = 2.$$

2. **Basic skills**

   (1) Find the limit of the sequence

   (i)
   $$\lim_{n\to\infty} \frac{\frac{1}{n}+4}{2-\frac{1}{n^2}}.$$

   (ii) Find the limit of the sequence
   $$\lim_{n\to\infty} \frac{n+1}{\sqrt{9n^2+3}}.$$

   (2) Find the limit for functions.

   (i)
   $$\lim_{x\to\infty} \frac{2+\frac{1}{\sqrt{x}}}{3-\frac{2}{x^2}}.$$

   (ii)
   $$\lim_{x\to\infty} \frac{2x^2+\frac{1}{\sqrt{x}}}{3x^2-\frac{2}{x^2}}.$$

3. **Comprehensive questions**. (1). Prove: for $n = 4, 5, 6, \cdots$,
   $$2^n \geq n^2.$$

   (2). Prove:
   $$\lim_{n\to\infty} \frac{n}{2^n} = 0.$$

4. **Comprehensive questions**.

   (1) Prove:
   $$\lim_{n\to\infty} \frac{n}{e^n} = 0.$$

   (2) Prove:
   $$\lim_{n\to\infty} \frac{\ln n}{n} = 0.$$

5. **Comprehensive questions**. Suppose $\lim_{x\to\infty} f(x) = L$. Let $a_n = f(n)$. Using definition to prove
   $$\lim_{n\to\infty} a_n = L.$$

6. **Comprehensive questions***.

   (1) (**Theorem:**) Suppose $\lim_{n\to\infty} a_n = L$. If $f(x)$ is a monotone decreasing function, and $a_n = f(n)$, use the definition to prove
   $$\lim_{x\to\infty} f(x) = L.$$

   (2) Use the above theorem in part (1) to prove that
   $$\lim_{x\to\infty} \frac{x}{e^x} = 0,$$
   and
   $$\lim_{x\to\infty} \frac{\ln x}{x} = 0.$$

7.3 Series

For a given sequence $\{a_n\}_{n=1}^{k}$ (k may be infinite), we will discuss the summation of each term in this section.

7.3.1 Finite series

First, consider the sum of a finite sequence $\{a_n\}_{n=1}^{k}$ (k is a finite number):

Example 7.33 Consider the first k terms of an arithmetic sequence. We list it as follows:
$$a_1, \ a_1 + d, \ a_1 + 2d, \ \cdots, \ a_1 + (k-2)d, \ a_1 + (k-1)d.$$

Their sum is
$$\sum_{n=1}^{k}(a_1 + (n-1)d) = \frac{2a_1 + (k-1)d}{2} \cdot k.$$

$\square$

Example 7.34 We have also calculated the sum of the first k terms of a geometric sequence. For $r \neq 1$,
$$\sum_{n=1}^{k} r^{n-1} = 1 + r + r^2 + \cdots + r^{k-1} = \frac{1 - r^k}{1 - r}.$$

Generally speaking, the addition of finite series is not too complicated.

7.3.2 Infinite series

We now consider the sum of an infinite sequence $\{a_n\}_{n=1}^{\infty}$. We can use the notation
$$\sum_{n=1}^{\infty} a_n = a_1 + a_2 + a_3 + \cdots.$$

The addition seems to go on forever. But its specific meaning must be clear. In view of this, we first define the partial sum of this infinite series.

For positive natural numbers $k = 1, 2, \cdots$, define
$$S_k = a_1 + a_2 + \cdots + a_{k-1} + a_k = \sum_{n=1}^{k} a_n.$$

In this way, we get a new sequence.

Definition 7.10. Convergence of series

Consider a series $\sum_{n=1}^{\infty} a_n$.

(1) If $\lim_{k \to \infty} S_k = L$, we say that the series $\sum_{n=1}^{\infty} a_n$ converges to L, and denote

it as
$$\sum_{n=1}^{\infty} a_n = L.$$

(2) If $\lim_{k \to \infty} S_k$ does not exist, then we call the series $\sum_{n=1}^{\infty} a_n$ divergent.

Example 7.35 (**Infinite recurring decimals**) Consider the series:
$$\sum_{n=0}^{\infty} 9 \times 10^{-n}.$$

This is exactly a recurring decimal that we talked about before:
$$9.\dot{9} = 9.999...$$

Its partial sum is
$$S_k = \sum_{n=0}^{k} 9 \times 10^{-n} = \frac{9 \times (1 - 10^{-k-1})}{1 - 10^{-1}}.$$

Because
$$\lim_{k \to \infty} S_k = \lim_{k \to \infty} \frac{9 \times (1 - 10^{-k-1})}{1 - 10^{-1}} = \frac{9}{0.9} = 10,$$

So this series converges to 10. This is the reason for
$$9.\dot{9} = 9.999... = 10$$

Example 7.36 Consider the series:
$$\sum_{n=1}^{\infty} (-1)^{-n}.$$

The partial sum of its even terms is
$$S_{2k} = \sum_{n=1}^{2k} (-1)^{-n} = 0;$$

And the partial sum of its odd terms is
$$S_{2k+1} = \sum_{n=1}^{2k+1} = -1.$$

The partial sum does not converge, thus the series is divergent.

We define the Remainder term R_{k+1} of the series as
$$R_{k+1} = \sum_{n=1}^{\infty} a_n - S_k = \sum_{n=k+1}^{\infty} a_n.$$

In a Calculus course, we will also prove (by definition)

> **Proposition 7.16. Remainder term of convergent series**
>
> If the series $\sum_{n=1}^{\infty} a_n$ converges, then
> $$\lim_{k \to \infty} R_{k+1} = 0.$$

Once we understand series and the convergence of series, we look back at those algebraic operations involving repeating decimals. For example, we verify
$$3 \times 0.\dot{3} = 1.$$

Recall: We pointed out that it is wrong if one uses the distribution rule to get
$$3 \times 0.\dot{3} = 0.\dot{9}$$

But the following calculation is correct. Let
$$S_k = \sum_{n=1}^{k} 3 \times 10^{-n}.$$

So that
$$3 \times 0.\dot{3} = 3 \times \lim_{k \to \infty} S_k$$
$$= \lim_{k \to \infty} (3 \times \sum_{n=1}^{k} 3 \times 10^{-n})$$
$$= \lim_{k \to \infty} (\sum_{n=1}^{k} 9 \times 10^{-n}) \quad \text{(Here we use the distributive rule)}$$
$$= \lim_{k \to \infty} \frac{0.9 \times (1 - 10^{-k})}{1 - 10^{-1}}$$
$$= 1.$$

We will learn more about series in a Calculus course in the future.

At the end of this chapter, we briefly introduce the power series — natural generalization of polynomials.

> **Definition 7.11. Definition of power series**
>
> We call
> $$\sum_{n=0}^{\infty} c_n (x - x_0)^n = c_0 + c_1(x - x_0) + c_2(x - x_0)^2 + \cdots.$$
> an infinite power series centered at $x = x_0$, where c_n (for $n = 0, 1, \cdots$, they are all real numbers) are called the coefficients.

If the power series always converges for $|x - x_0| < R$ (or for $x \in (x_0 - R, x_0 + R)$), but never converge for $|x - x_0| > R$, we call R the radius of convergence of the power series.

The following power series is very useful

$$\sum_{n=0}^{\infty} \frac{x^n}{n!} = 1 + x + \frac{x^2}{2!} + \cdots.$$

In a Calculus course, we will prove that its radius of convergence is $R = \infty$. In other words, the above series converges for any given real number x (even for any complex number). We will explain in the next chapter that the above power series actually converges to the natural exponential function.

$$e^x = \sum_{n=0}^{\infty} \frac{x^n}{n!}.$$

7.3.3 Exercises after class

1. **Basic skills**. Find the value of the following finite series

    (1)
    $$\sum_{n=1}^{20} 3n;$$

    (2)
    $$\sum_{k=1}^{10} 2^{-k};$$

    (3)
    $$\sum_{n=1}^{2022} \frac{1}{n(n+1)};$$

    (4)
    $$\sum_{n=1}^{200} (-1)^n.$$

2. **Comprehensive questions**. Check whether the following infinite series is convergent, and why?

    (1)
    $$\sum_{n=1}^{\infty} (\frac{1}{2})^n;$$

    (2)
    $$\sum_{n=1}^{\infty} (-1)^n;$$

    (3)
    $$\sum_{n=1}^{\infty} \frac{1}{n(n+1)};$$

    (4*)
    $$\sum_{n=1}^{\infty} \frac{1}{n}.$$

7.4 Chapter review and exercises

In Chapter 7, we learned sequences, the limits for sequences and functions, and series.

1. Sequence.

1.1 Several basic sequences

Arithmetic sequence: First term: a_1, common difference: d, General term formula:
$$a_n = a_1 + (n-1)d.$$

Geometric sequence: First term: a_1, Common ratio: r, General term formula:
$$a_n = a_1 r^{n-1}.$$

Fibonacci sequence *: The first two terms: $a_1 = 1$, $a_2 = 2$, recurrence formula : $a_{n+2} = a_{n+1} + a_n$. General term formula:
$$a_n = \frac{1}{\sqrt{5}} \cdot \left\{ \left(\frac{1+\sqrt{5}}{2}\right)^{n+1} - \left(\frac{1-\sqrt{5}}{2}\right)^{n+1} \right\}.$$

1.2 The concept of partial sum*

2. The concept of limits.

2.1. Limits for sequences

$\epsilon - N$ Definition*

Algebraic operation formulas of sequence limit: suppose $\lim_{n\to\infty} a_n$, $\lim_{n\to\infty} b_n$ exist, then

$$\lim_{n\to\infty} (\alpha a_n + \beta b_n) = \alpha \lim_{n\to\infty} a_n + \beta \lim_{n\to\infty} b_n;$$

$$\lim_{n\to\infty} a_n b_n = \lim_{n\to\infty} a_n \cdot \lim_{n\to\infty} b_n;$$

and (suppose $\lim_{n\to\infty} b_n \neq 0$)

$$\lim_{n\to\infty} \frac{a_n}{b_n} = \frac{\lim_{n\to\infty} a_n}{\lim_{n\to\infty} b_n}.$$

The limit exists for the monotone bounded sequence: If sequence $\{a_n\}_{n=1}^{\infty}$ is monotone and bounded, then it is a convergent sequence.

The convergence of subsequences for a convergent sequence: If $\lim_{n\to\infty} a_n = L$, Then its subsequence a_{i_n} must converge to L.

The squeeze theorem of for sequence limit: If three series a_n, b_n, c_n satisfy

(1) There is a number N_0, such that all $n > N_0$,

$$a_n \leq b_n \leq c_n.$$

(2)

$$\lim_{n \to \infty} a_n = \lim_{n \to \infty} c_n = L.$$

Then b_n is also a convergent sequence, and $\lim_{n \to \infty} b_n = L$.

2.2. Limit for Function

$\epsilon - \delta$ Definition*

Algebraic operation rules for function limits: Suppose that $\lim_{x \to x_0} f(x)$, $\lim_{x \to x_0} g(x)$ exist. Then

$$\lim_{x \to x_0} (\alpha a_n + \beta b_n) = \alpha \lim_{x \to x_0} f(x) + \beta \lim_{x \to x_0} g(x);$$

$$\lim_{x \to x_0} f(x)g(x) = \lim_{x \to x_0} f(x) \cdot \lim_{x \to x_0} g(x);$$

and (Suppose that $\lim_{x \to x_0} g(x) \neq 0$)

$$\lim_{x \to x_0} \frac{f(x)}{g(x)} = \frac{\lim_{x \to x_0} f(x)}{\lim_{x \to x_0} g(x)}.$$

The squeeze theorem for function limits: If three functions $f(x)$, $g(x)$, $h(x)$ satisfy

(1) There exists a number ϵ_1, such that for all $|x - x_0| < \epsilon_1$,

$$f(x) \leq g(x) \leq h(x).$$

(2)

$$\lim_{x \to x_0} f(x) = \lim_{x \to x_0} h(x) = L,$$

then $\lim_{x \to x_0} g(x) = L$.

Continuity for functions

Continuity of elementary functions: Polynomial functions, exponential functions, trigonometric functions, and logarithmic functions are continuous functions in their domain.

3. Series.

3.1. Finite series $\sum_{n=1}^{k} a_n$:

Summation of arithmetic sequence:

$$\sum_{n=1}^{k} (a_1 + (n-1)d) = \frac{2a_1 + (k-1)d}{2} \cdot k.$$

Summation of geometric series:

$$\sum_{n=1}^{k} r^{n-1} = 1 + r + r^2 + \cdots + r^{k-1} = \frac{1-r^k}{1-r}.$$

3.2. Infinite series $\sum_{n=1}^{\infty} a_n$:

Partial Sum: $S_k = \sum_{n=1}^{k} a_n$

Series converge $\Leftrightarrow \lim_{k \to \infty} S_k$ exists. And we use notation $\sum_{n=1}^{\infty} a_n = \lim_{k \to \infty} S_k$.

Series diverge $\Leftrightarrow \lim_{k \to \infty} S_k$ does not exist.

7.4.1 Comprehensive exercise questions

1. **Comprehensive questions-1**. Find the sum of the sequence:
$$\sum_{n=1}^{k} n3^{n-1}.$$

2. **Comprehensive questions-2**.

    (1) Find
    $$\lim_{n\to\infty} (\sqrt{n+1} - \sqrt{n}).$$

    (2) Find
    $$\lim_{x\to 0} \frac{\sqrt{1+4x} - 1}{x}.$$

    (3) Consider a sequence a_n given by the following recurrence formula:
    $$a_1 = 1, \quad a_{n+1} = \frac{1}{2}(a_n + 6) \quad \text{when } n = 1, 2, 3 \cdots.$$
    (i). Prove that a_n is a monotone and bounded sequence.
    (ii). Find $\lim_{n\to\infty} a_n$.

3. **Comprehensive questions-3***. Prove the convergence or divergence for the following series.

    (1) $\sum_{n=1}^{\infty} \frac{n!}{2^n}$;

    (2) $\sum_{n=1}^{\infty} \frac{2^n}{n!}$;

    (3) $\sum_{n=1}^{\infty} \frac{1}{n(n+1)}$;

    (4) $\sum_{n=1}^{\infty} \frac{1}{n^2}$;

    (5) $\sum_{n=1}^{\infty} \frac{1}{n^3}$.

Chapter 8 Differentiation and Integration

Introduction

- Tangent line
- Formula for motions
- Definition of derivative
- Derivative operations
- Applications of derivatives
- Derivative for a vector function
- Indefinite integral
- Substitution for integral
- Integration by part
- Definite integral
- Fundamental Theorem of Calculus
- Area
- Volume

8.1 Tangent and instantaneous speed— concept of differential

In Example 3.4, in order to find the tangent line we looked for a suitable straight line so that it intersects the curve $y = x^2$ only at a point $(1, 1)$. But for other curves, such as $y = x^3$, the above method does not work.

We examine the following diagram. When point x_1 is close to x_0, the line connecting point $T(x_0, y_0)$ and point $G(x_1, y_1)$ is getting closer and closer to the tangent line of the curve passing through (x_0, y_0). Therefore, we define:

Definition 8.1. Slope of tangent line

The slope of the tangent line for function $y = f(x)$ at point $(a, f(a))$ is defined as
$$m_a = \lim_{x \to a} \frac{f(x) - f(a)}{x - a}.$$

The slope of the tangent is a limit, so it may not exist (we will learn more about the differentiable of functions in the future).

Similarly, we study the problems about the distance and speed encountered in physics.

Let $y = s(t)$ be a distance function (for example, with the unit of kilometers per hour). From $t = t_0$ to $t = t_1$, the average speed is
$$v(t) = \frac{s(t_1) - s(t_0)}{t_1 - t_0}.$$

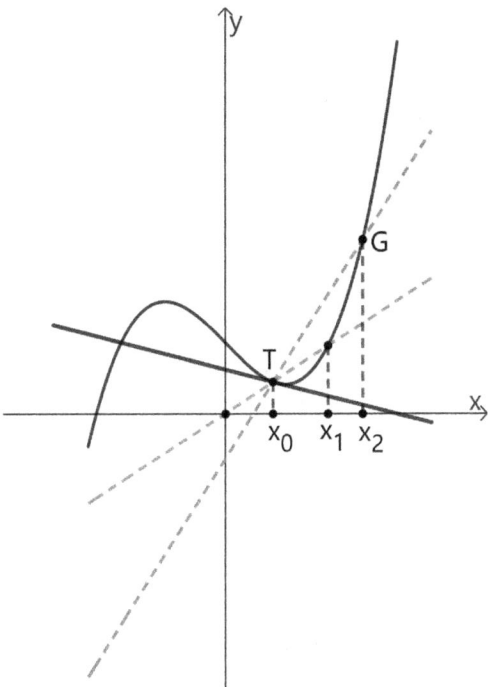

Figure 8.1: Tangent of function

This is again the slope of the line connecting two points $(t_0, s(t_0))$, $(t_1, s(t_1))$. At $t = t_0$, the instantaneous speed is defined as:

$$v(t_0) = \lim_{t \to t_0} \frac{s(t) - s(t_0)}{t - t_0}.$$

It is also natural to introduce the second derivative while studying physics problems. For example, the instantaneous acceleration at time $t = t_0$ is the derivative of the speed function:

$$a(t_0) = \lim_{t \to t_0} \frac{v(t) - v(t_0)}{t - t_0}.$$

8.1.1 Exercises After Class

1. **Basic skills.** Find the values of the following limits

   (1)
   $$\lim_{x \to 0} \frac{(1+x)^3 - 1}{x}$$

   (2)
   $$\lim_{h \to 0} \frac{\sqrt{1+h} - 1}{h}$$

   (3)
   $$\lim_{h \to 0} \frac{\frac{1}{2}a(t+h)^2 - \frac{1}{2}at^2}{h}$$

2. **Comprehensive Word Problems.**

   (1) Find the tangent line equation of the following curve at the given point.

   (i) The tangent line of $y = x^2$ at point $(2, 4)$.

   (ii) The tangent line of $y = 2\sqrt{x}$ at point $(1, 2)$.

   (2) The distance formula for a given vehicle is $S(t) = \frac{1}{2}at^2$, where t is the time variable and a is a given parameter.

   (i) Find the velocity formula.

   (ii) Find the acceleration formula.

8.2 Derivative and applications

> **Definition 8.2. Definition of derivative**
> Consider the function $y = f(x)$ defined on the interval I. At point $a \in I$, if the following limit exists:
> $$\lim_{x \to a} \frac{f(x) - f(a)}{x - a} = D_a,$$
> we say that function $y = f(x)$ has a derivative at the point $x = a$, and call D_a the derivative of the function at a, denoted as as $f'(a)$.

If $f(x)$ has a derivative at every point on the interval I, we obtain the derivative function $f'(x)$ of $f(x)$. The derivative of the derivative function $f'(x)$ is the second-order derivative of the original function, denoted as $f''(x)$. We iterate the above procedure to obtain nth-order derivative $f^{(n)}(x) = (f^{(n-1)}(x))'$. We use the notation:
$$f^{(n)}(x) = \frac{d^n f(x)}{dx^n} = \frac{d^n}{dx^n}\{f(x)\}.$$

Example 8.1 Find the derivative of the constant value function $y = f(x) = C$.

Solution: For any point a, using the definition we have
$$f'(a) = \lim_{x \to a} \frac{C - C}{x - a} = 0.$$
That is: the derivative of a constant function is a zero function.

$\square$

We also use other notations $\frac{df(x)}{dx}$ or $\frac{d}{dx}f(x)$ for $f'(x)$. The advantage of using the notation $\frac{df(x)}{dx}$ is that: it specifies which variable the function takes its derivative with respect to.

Example 8.2 Find
$$\frac{d}{dy}e^x.$$

Solution: For any point a, using the definition we have
$$\frac{de^x}{dy}(a) = \lim_{y \to a} \frac{e^x - e^x}{y - a} = 0.$$
That is, the derivative of function $f(x) = e^x$ with respect to variable **y** is a zero function (the change of **y** has no effect on the change of function value).

$\square$

8.2.1 Derivative of Elementary Function

Using the definition of the derivative and the algebraic properties of limits we have the following theorem (all proofs are left in a calculus course in the future):

> **Theorem 8.1. Algebraic Properties of Derivatives**
> Consider two functions $f(x), g(x)$ which have derivatives. Their linear combination, that is, for any two numbers α and β, $\alpha f(x) + \beta g(x)$ also has derivative, and
> $$(\alpha f(x) + \beta g(x))' = \alpha f'(x) + \beta g'(x).$$

Here we list the derivatives for all typical elementary functions. We will learn how to derive these formulas in a calculus course in the future.

> **Theorem 8.2. Derivatives of Elementary Functions**
> $$\frac{d}{dx}(x^n) = nx^{n-1} \tag{8.1}$$
>
> $$\frac{d}{dx}(e^x) = e^x \tag{8.2}$$
>
> $$\frac{d}{dx}(\ln x) = \frac{1}{x} \tag{8.3}$$
>
> $$\frac{d}{dx}(\sin x) = \cos x \tag{8.4}$$
>
> $$\frac{d}{dx}(\cos x) = -\sin x \tag{8.5}$$
>
> $$(f(x)g(x))' = f'(x)g(x) + f(x)g'(x) \tag{8.6}$$

The following chain rule allows us to calculate the derivatives of more functions obtained from the compositions of elementary functions.

> **Theorem 8.3. Chain Rule of Derivative of Compound Function**
> The derivative of the composite functions is given by
> $$\{f[g(x)]\}' = f'[g(x)] \cdot g'(x). \tag{8.7}$$
> If we write $f[g(x)] = f(u)$, where $u = u(x)$ is a function of x, then
> $$\frac{df(u)}{dx} = \frac{df(u)}{du} \cdot \frac{du}{dx}. \tag{8.8}$$

Example 8.3 Find
$$\frac{de^{x^2}}{dx}.$$

Solution: Let $u = x^2$. We have
$$\begin{aligned}\frac{de^{x^2}}{dx} &= \frac{de^u}{dx}\\ &= \frac{de^u}{du} \cdot \frac{du}{dx} &\text{(Chain rule)}\\ &= e^u \cdot (2x) &\text{(Using (8.2) and (8.1))}\\ &= 2xe^{x^2}.\end{aligned}$$

$\square$

Example 8.4 Using the chain rule, we derive the derivative of $\ln u$ with respect to u from $\frac{d}{dx}e^x = e^x$. (That is, we derive formula (8.3) from formula (8.2).)

Solution: First of all, we know
$$\ln(e^x) = x.$$

We have
$$\frac{d}{dx}(\ln e^x) = \frac{d}{dx}(x) = 1.$$

Let $u = e^x$. Using the chain rule, we have
$$\begin{aligned}1 &= \frac{d \ln e^x}{dx}\\ &= \frac{d \ln u}{du} \cdot \frac{du}{dx} &\text{(Chain rule)}\\ &= \frac{d \ln u}{du} \cdot \frac{de^x}{dx} &(u = e^x)\\ &= \frac{d \ln u}{du} \cdot e^x\\ &= \frac{d \ln u}{du} \cdot u.\end{aligned}$$

So
$$\frac{d \ln u}{du} = \frac{1}{u}.$$

$\square$

Example 8.5 We calculate the derivative of the division of two functions. We compute
$$\frac{d}{dx}\left(\frac{f(x)}{g(x)}\right).$$

Solution: We provide two methods to calculate it.

Method I. Let $u = f(x)/g(x)$. Then $f(x) = g(x)u(x)$,

$$f' = g'u + gu'$$
$$= g' \cdot \frac{f}{g} + gu'.$$

So
$$f' - g' \cdot \frac{f}{g} = g \cdot u'.$$

Thus
$$u' = \frac{f'g - fg'}{g^2}.$$

That is:
$$(\frac{f}{g})' = \frac{f'g - fg'}{g^2}. \tag{8.9}$$

Method II. Let $v = g(x)$, then
$$\frac{d}{dx}(\frac{f(x)}{g(x)}) = \frac{d}{dx}(f \cdot v^{-1})$$
$$= f' \cdot v^{-1} + f \cdot \frac{d(v^{-1})}{dx}$$
$$= f' \cdot \frac{1}{g} - f \cdot v^{-2}v'$$
$$= \frac{f'g - fg'}{g^2}.$$

$\square$

Example 8.6 Assume $a > 0$. Find
$$\frac{da^x}{dx}.$$

Solution: Recall:
$$a^x = e^{x \ln a}.$$

Let $u = x \ln a$, we have
$$\frac{da^x}{dx} = \frac{de^{x \ln a}}{dx}$$
$$= \frac{de^u}{dx}$$
$$= \frac{de^u}{du} \cdot \frac{du}{dx}$$
$$= e^u \cdot \ln a$$
$$= a^x \ln a.$$

$\square$

Example 8.7 For $x > 0$, find
$$\frac{dx^p}{dx}.$$

Solution: Recall:
$$x^p = e^{p \ln x}.$$

Let $u = p \ln x$, we have
$$\begin{aligned}\frac{dx^p}{dx} &= \frac{de^{p \ln x}}{dx} \\ &= \frac{de^u}{dx} \\ &= \frac{de^u}{du} \cdot \frac{du}{dx} \\ &= e^u \cdot \frac{p}{x} \\ &= px^{p-1}.\end{aligned}$$

□

8.2.2 Application of Derivative

We study the applications of derivatives.

From the definition of the derivative, we can get the following theorem.

Theorem 8.4. Derivative Properties of Increasing Functions
If the function $y = f(x)$ has a derivative on the interval (a, b) and its derivative is not less than zero, then $f(x)$ is an increasing function on the interval (a, b).

We use the above theorem to prove the inequality in the following example that cannot be proved before.

Example 8.8 Proof: for $x \geq 0$,
$$e^x \geq x + 1.$$

Proof: Let $f(x) = e^x - x - 1$, then
$$f'(x) = e^x - 1 \geq 0 \quad \text{for} \quad x \geq 0.$$
So $f(x)$ is an increasing function on the interval $[0, \infty)$. Thus
$$f(x) \geq f(0) = 0 \quad \text{for} \quad x \geq 0.$$
That is, for $x \geq 0$,
$$e^x \geq x + 1.$$

□

We can also prove the following mean value theorem (Detailed proof will be given in a calculus course).

8.2 Derivative and applications

> **Theorem 8.5. Mean value theorem**
>
> Suppose the function $y = f(x)$ has a derivative on the interval (a, b). If $f(a) = f(b)$, then there is a point $c \in (a, b)$, such that $f'(c) = 0$.

From the above, we can derive a more general mean value theorem (see Question 3 in the exercise after class at the end of this section).

> **Theorem 8.6. Mean value theorem-2**
>
> Suppose the function $y = f(x)$ has a derivative on the interval (a, b), then there is a point $c \in (a, b)$, such that
> $$f'(c) = \frac{f(b) - f(a)}{b - a}.$$

> **Definition 8.3. Extreme Value of Function**
>
> Suppose the function $y = f(x)$ has a derivative on the interval (a, b). If at $x_0 \in (a, b)$, $f'(x_0) = 0$, we call that x_0 is a critical point.

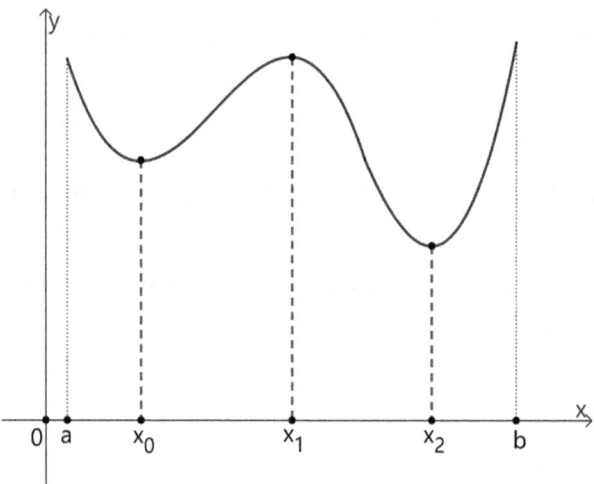

Figure 8.2: Extreme point (x_0, x_1, x_2): a valley or a peak

> **Definition 8.4. Local maximum and minimum of function**
>
> Consider functions $y = f(x)$ defined on interval (a, b), $x_0 \in (a, b)$ is a point on the interval. If there is a small interval $(x_0 - \epsilon, x_0 + \epsilon) \subset (a, b)$, such that for any $x \in (x_0 - \epsilon, x_0 + \epsilon)$,
> $$f(x) \leq f(x_0),$$
> we say that $x_0 \in (a, b)$ is a local maximum point of $f(x)$, and $f(x_0)$ is a local maximum.
>
> Similarly, we can define a local minimum.

We have the following theorem.

> **Theorem 8.7. Extreme Value of Function**
> Consider a function $f(x)$ which has a continuous derivative on the interval (a, b). Assume $x_0 \in (a, b)$ is a local maximum or local minimum of $f(x)$, then $x_0 \in (a, b)$ is a critical point of $f(x)$, that is, $f'(x_0) = 0$.

Example 8.9 Find the critical points of the function $f(x) = \frac{1}{3}x^3 - 4x + 2$.

Solution: $f'(x) = x^2 - 4$. Solving equation

$$x^2 - 4 = 0$$

yields the critical points of the function: $x = 2$ and $x = -2$.

$\square$

The introduction of derivatives makes it easy to calculate the tangent line of a general curve.

Example 8.10 Find the tangent line equation of the following curve at a given point.

(a) Tangent line of $y = x^3 - 8$ at point $(2, 0)$.

(b) Tangent line of $y = 2\sin x$ at point $(\frac{\pi}{6}, 1)$.

Solution: (a) The derivative of function $f(x) = x^3 - 8$ is: $f'(x) = 3x^2$. At point $(2, 0)$, $f'(2) = 12$. This is also the slope of the tangent. So the tangent line equation is:

$$y = 12(x - 2).$$

After simplifying, we have $y = 12x - 24$.

(b) The derivative of function $f(x) = 2\sin x$ is: $f'(x) = 2\cos x$. At point $(\frac{\pi}{6}, 1)$, $f'(\frac{\pi}{6}) = \sqrt{3}$. This is also the slope of the tangent. So the tangent line equation is:

$$y - 1 = \sqrt{3}(x - \frac{\pi}{6}).$$

After simplifying, we have $y = \sqrt{3}x - \frac{\sqrt{3}\pi}{6} + 1$.

$\square$

We also list the following applications for the second-order derivatives.

> **Theorem 8.8. Maximum and Minimum Value of Function**
> Consider a function $f(x)$ which has a second-order derivative on the interval (a, b). Let $x_0 \in (a, b)$ be a critical point of $f(x)$.
> (a) If $f''(x_0) > 0$, then x_0 is a local minimum of $f(x)$;
> (b) If $f''(x_0) < 0$, then x_0 is a local maximum of $f(x)$.

Theorem 8.9. Convexity and Concavity of Functions

Consider a function $f(x)$ which has a second-order derivative on the interval (a, b).

(a) If for all $x \in (a, b)$, $f''(x) > 0$, then $f(x)$ is a convex function on (a, b);

(b) If for all $x \in (a, b)$, $f''(x) < 0$, then $f(x)$ is a concave function on (a, b).

Example 8.11 For any two real numbers x_1, x_2, proof:
$$4e^{\frac{1}{4}x_1+\frac{3}{4}x_2} \leq e^{x_1} + 3e^{x_2}.$$

Proof: Because $(e^x)'' = e^x > 0$, so $y = e^x$ is a convex function over the real number set. Therefore, for any $\lambda \in [0, 1]$,
$$e^{\lambda x_1+(1-\lambda)x_2} \leq \lambda e^{x_1} + (1-\lambda)e^{x_2}.$$

Especially for $\lambda = 1/4$, we have
$$e^{\frac{1}{4}x_1+\frac{3}{4}x_2} \leq \frac{1}{4}e^{x_1} + \frac{3}{4}e^{x_2}.$$

Multiplying both sides by 4 we have
$$4e^{\frac{1}{4}x_1+\frac{3}{4}x_2} \leq e^{x_1} + 3e^{x_2}.$$

$\square$

Finally, we discuss how to approximate general functions by polynomials. Polynomial functions are natural generalizations of decimal numbers, and they are also the functions that we are most familiar with. For any given functions, people naturally want to compare them with polynomials.

Consider a function $y = f(x)$ that has infinite order continuous derivative (For example, the exponential function $y = e^x$, or trigonometric function $y = \sin x$, etc.). If it can be expressed as an infinite power series
$$f(x) = \sum_{n=0}^{\infty} c_n(x-x_0)^n,$$

that is
$$f(x) = c_0 + c_1(x-x_0) + c_2(x-x_0)^2 + c_3(x-x_0)^3 + \cdots,$$

then we can calculate the values of the coefficients c_n in the power series. For example, if we choose $x = x_0$, you get
$$c_0 = f(x_0).$$

Therefore
$$f(x) - f(x_0) = c_1(x-x_0) + c_2(x-x_0)^2 + c_3(x-x_0)^3 \cdots.$$

(Rigorously speaking, we use the partial sum and a negligible remainder term to represent the infinite power series. So we can use the three rules for algebraic operation on the partial sum (thus it is a finite summation).)

It is more tricky to find other coefficients.

Let us assume that we can take the derivative of the power series term by term[1]. We have
$$f'(x) = c_1 + 2c_2(x - x_0) + 3c_3(x - x_0)^2 \cdots.$$

Now, we choose $x = x_0$, we obtain
$$c_1 = f'(x_0)$$

We can repeat the above steps. Taking another derivative on both sides again, we have
$$f''(x) = 2c_2 + 3 \cdot 2c_3(x - x_0)^2 + \cdots$$
$$= 2!c_2 + 3!c_3(x - x_0) + \cdots.$$

Now, we choose $x = x_0$, we obtain
$$c_2 = \frac{f''(x_0)}{2!}.$$

Then repeat the above steps.

We summarize the above discussion as the following definitions.

> **Definition 8.5. Definition of Taylor series**
>
> *Consider a function $y = f(x)$ that has infinite order continuous derivative on interval I. For all natural numbers $n \in \mathbb{N}$, let*
> $$c_n = \frac{f^n(x_0)}{n!}.$$
> *We call*
> $$\sum_{n=0}^{\infty} c_n(x - x_0)^n = f(x_0) + f'(x_0)(x - x_0) + \frac{f''(x_0)}{2}(x - x_0)^2 + \cdots$$
> *Taylor series[a] of function $f(x)$. For $x_0 = 0$, that is*
> $$f(0) + f'(0)x + \frac{f''(0)}{2!}x^2 + \cdots$$
> *We call it Maclaurin[b] series.*
>
> [a]Brook Taylor (1685.8.18—1731.12.29), English mathematician
>
> [b]Colin Maclaurin (1686.2—1746.6.14), Scottish mathematician. Despite both parents passed away when he was young, he entered university at the age of 11 and became a math professor at the age of 19.

Example 8.12 Find the Maclaurin series for function $f(x) = e^x$.

---

[1] In a calculus course, we will discuss in detail when we can take the derivative of the power series term by term.

Solution: We observe
$$\frac{d(e^x)}{dx} = e^x, \quad \frac{d^n(e^x)}{dx^n} = e^x.$$
So
$$\frac{d^n(e^x)}{dx^n}(0) = 1.$$
Therefore its Maclaurin series is
$$\sum_{k=0}^{\infty} \frac{x^k}{k!} = 1 + x + \frac{x^2}{2!} + \frac{x^3}{3!} + \dots.$$

□

In a calculus course, we will prove the following formulas.

Theorem 8.10. Maclaurin Expansion of Elementary Functions

$$e^x = \sum_{k=0}^{\infty} \frac{x^k}{k!} = 1 + x + \frac{x^2}{2!} + \frac{x^3}{3!} + \dots \tag{8.10}$$

$$\sin x = \sum_{k=0}^{\infty} (-1)^k \frac{x^{2k+1}}{(2k+1)!} = x - \frac{x^3}{3!} + \frac{x^5}{5!} + \dots \tag{8.11}$$

$$\cos x = \sum_{k=0}^{\infty} (-1)^k \frac{x^{2k}}{(2k)!} = 1 - \frac{x^2}{2!} + \frac{x^4}{4!} + \dots \tag{8.12}$$

By formula (8.10), (8.11) and (8.12), we can obtain Euler's formula.

8.2.3 Derivative of Vector Function and Circular Motion*

For a given vector function (For convenience, we will only discuss two-dimensional vector functions here) $\vec{F}(x) = (f(x), g(x))$, we naturally define

$$\lim_{x \to x_0} \vec{F}(x) = (\lim_{x \to x_0} f(x), \lim_{x \to x_0} g(x)),$$

and define $\vec{F}(x)$ as a continuous function if and only if $f(x)$ and $g(x)$ are continuous.

We also define derivative
$$\frac{d\vec{F}(x)}{dx} = \vec{F}'(x) = (f'(x), g'(x)).$$

It can be verified that those formulas of the limits and derivatives for functions are also valid for vector functions. We list these formulas but skip the proofs.

> **Proposition 8.1. Algebraic Properties of Limit of Vector Function**
>
> Assume
> $$\lim_{x \to x_0} \vec{F}(x) = \vec{L}, \quad \lim_{x \to x_0} \vec{G}(x) = \vec{M}.$$
>
> Then, for the two numbers α and β, function $\alpha \vec{F}(x) + \beta \vec{G}(x)$ converges, and
> $$\lim_{x \to x_0} (\alpha \vec{F}(x) + \beta \vec{G}(x)) = \alpha \vec{L} + \beta \vec{M}.$$

> **Theorem 8.11. Algebraic Properties of Derivatives for Vector Functions**
>
> Consider two functions $\vec{F}(x), \vec{G}(x)$ which have derivatives. Their linear combination, that is, for any two numbers α and β, $\alpha \vec{F}(x) + \beta \vec{G}(x)$, also has derivative, and
> $$(\alpha \vec{F}(x) + \beta \vec{G}(x))' = \alpha \vec{F}'(x) + \beta \vec{G}'(x).$$

We also have the formula for the derivative of an inner product:

$$(\vec{F}(x) \cdot \vec{G}(x))' = \vec{F}'(x) \cdot \vec{G}(x) + \vec{F}(x) \cdot \vec{G}'(x). \tag{8.13}$$

Consider a curve given by the parametric equation: $\vec{F}(x) = (f(x), g(x))$. Its derivative $\vec{F}'(x_0)$ at point x_0, is the tangent vector of the curve at this point.

Example 8.13 Find the tangent line equation of ellipse $\vec{F}(x) = (4\cos x, 9\sin x)$ at $x = \frac{\pi}{4}$.

Solution: First
$$\vec{F}'(x) = (-4\sin x, \ 9\cos x).$$

Therefore
$$\vec{F}'(\frac{\pi}{4}) = (-4\sin\frac{\pi}{4}, \ 9\cos\frac{\pi}{4}) = (-2\sqrt{2}, \frac{9\sqrt{2}}{2}).$$

We also know $\vec{F}(\frac{\pi}{4}) = (2\sqrt{2}, \frac{9\sqrt{2}}{2})$. Thus, the parameter equation of the tangent line is:
For $t \in (-\infty, +\infty)$,
$$\begin{cases} x &= 2\sqrt{2} - 2\sqrt{2}t \\ y &= \frac{9\sqrt{2}}{2} + \frac{9\sqrt{2}}{2}t. \end{cases}$$

$\square$

The derivative of the vector function also makes it easy for us to understand that the acceleration of a circular motion with a constant speed is perpendicular to its velocity (thus pointing to the center of the circle).

Assume that the velocity vector function of the circular motion is $\vec{V}(t)$, then

$$|\vec{V}(t)| = \text{constant} \Leftrightarrow <\vec{V}, \vec{V}> = \text{constant}.$$

Thus:
$$<\vec{V}', \vec{V}> = 0.$$

That is, acceleration vector function $\vec{a}(t) = \vec{V}'$ satisfies:
$$\vec{a}(t) \perp \vec{V}(t).$$

8.2.4 Exercises After Class

1. **Basic skills**. Calculate the derivative of the following functions

   (1) (i) $f(x) = e^{2x}$, (ii) $f(x) = e^{x^2}$.

   (2) (i) $f(x) = \cos 2x$, (ii) $f(x) = \sin x^2$.

   (3) (i) $f(x) = \cos^2 x$, (ii) $f(x) = \sin^2 x^2$.

   (4) (i) $f(x) = \tan x$, (ii) $f(x) = \frac{e^x}{x}$.

2. **Comprehensive Word Problems**.

   (1) Find the tangent line equation of the following curve at the given point.

   (i) Tangent line of $y = x \cos x$ passing through point $(0, 0)$.

   (ii) Tangent line of $y = (x + 1)e^x$ at point $(0, 1)$.

   (2) Prove the following inequalities:

   (i) For $x \geq 0$, $e^x \geq 1 + x$.

   (ii) For $x \geq 0$, $x \geq \sin x$

   (3) For $x \in [0, \pi]$, prove:
   $$\sin x - \frac{x}{2} \geq \frac{\sqrt{2}}{2} - \frac{\pi}{6}.$$

   (4) For $x \in (0, \infty)$, prove:
   $$\ln x + \frac{1}{x} \geq 1.$$

   (5) Assume $\vec{F}(x) = (f(x), g(x))$, $\vec{G}(x) = (r(x), s(x))$. Prove
   $$(\vec{F}(x) \cdot \vec{G}(x))' = \vec{F}'(x) \cdot \vec{G}(x) + \vec{F}(x) \cdot \vec{G}'(x).$$

3. **Comprehensive Word Problems-1**. Using Theorem 8.5 to prove Theorem 8.6.
   (hint: Applying Theorem 8.5 for a test function
   $$g(x) = f(x) - \{\frac{f(b) - f(a)}{b - a}(x - a) + f(a)\}$$
   Try to think about the geometric figure of the test function.)

4. **Comprehensive Word Problems-2**.

   (1) Prove : Function $y = (1 + \frac{1}{x})^{x+1}$ is a decreasing function on interval $(0, \infty)$.

   (2) Prove: For $x \geq 5$,
   $$(1 + \frac{1}{x})^{x+1} < 3.$$

   (3) Prove: Function $y = (1 + \frac{1}{x})^x$ is an increasing function on interval $(0, \infty)$.

   (4) Prove:
   $$\lim_{n \to \infty} (1 + \frac{1}{n})^n \in (2, 3).$$

8.3 Anti-derivative——indefinite integral

Like many mathematical operations, people naturally ask: can you find a function, whose derivative is a given function?

Example 8.14 Find a function $f(x)$ such that $f'(x) = x^2$.

Solution: Since
$$\frac{d}{dx}x^3 = 3x^2$$

We know:
$$\frac{d}{dx}(\frac{1}{3}x^3) = x^2,$$

That is, the derivative of function $\frac{1}{3}x^3$ is x^2. Is there any other functions whose derivative is also x^2? Yes. Easy to verify, for any constant C, the derivative of function $\frac{1}{3}x^3 + C$ is x^2. We will see below that this representation gives all functions whose derivatives are x^2.

□

> **Definition 8.6. Indefinite Integral**
>
> If $g'(x) = f(x)$, we call $g(x)$ the anti-derivative, or indefinite integral of the function $f(x)$. We use the notation
> $$g(x) = \int f(x)dx.$$

It is not difficult to verify from the definition of derivative: a differentiable function whose derivative is zero everywhere if and only if it is a constant function. So we have

> **Proposition 8.2. General Representation of Indefinite Integral**
>
> If $f'(x) = g'(x)$, then there is a constant C, such that
> $$f(x) = g(x) + C.$$

It is easy to prove by the definition of the integral

> **Theorem 8.12. Algebraic Properties of Integral**
>
> For any two numbers α and β,
> $$\int (\alpha f(x) + \beta g(x))dx = \alpha \int f(x)dx + \beta \int g(x)dx.$$

From Proposition 8.2 and Theorem 8.2 in the previous section, we can obtain the following indefinite integral formulas and the formula of integration by parts (formula (8.19) below).

Theorem 8.13. Integral Formulas of Elementary Function

Indefinite integral formulas

$$\int x^n dx = \frac{x^{n+1}}{n+1} + C, \quad \text{for } n \neq -1. \tag{8.14}$$

$$\int \frac{1}{x} dx = \ln|x| + C \tag{8.15}$$

$$\int e^x dx = e^x + C. \tag{8.16}$$

$$\int \cos x \, dx = \sin x + C. \tag{8.17}$$

$$\int \sin x \, dx = -\cos x + C. \tag{8.18}$$

And the formula of integration by parts

$$\int f(x)g'(x)dx = f(x)g(x) - \int f'(x)g(x)dx. \tag{8.19}$$

From the chain rule, we can obtain the following substitution formula for the integral.

Theorem 8.14. Substitution Formula for the Integral

$$\int f(g(x))g'(x)dx = \int f(u)du.$$

Let $u = g(x)$. With the help of the differential forms $du = dg(x) = g'(x)dx$, we can easily remember the above substitution formula as

$$\int f(g(x))g'(x)dx = \int f(u)g'(x)dx \qquad \text{(Let } u = g(x)\text{)}$$

$$= \int f(u)du. \qquad \text{(Because } du = dg(x) = g'(x)dx\text{)}$$

The integration by parts formula can also be given in the following more concise form:

$$\int u \, dv = uv - \int v \, du$$

where $u = u(x), v = v(x)$ are functions of x.

Using substitution, we can calculate more complicated indefinite integrals.

Example 8.15 Find: (a) $\int e^{2x} dx$, (b) $\int (x+1)^2 dx$

Solution: (a)

$$\int e^{2x}\,dx = \int e^u\,dx \qquad \text{(Let } u = 2x\text{)}$$
$$= \int e^u \frac{du}{2} \qquad \text{(So } du = 2dx\text{)}$$
$$= \frac{e^u}{2} + C$$
$$= \frac{e^{2x}}{2} + C.$$

(b)

$$\int (x+1)^2\,dx = \int u^2\,dx \qquad \text{(Let } u = x+1\text{)}$$
$$= \int u^2\,du \qquad \text{(So } du = dx\text{)}$$
$$= \frac{u^3}{3} + C$$
$$= \frac{(x+1)^3}{3} + C.$$

Integration by parts can also help us find more "difficult" indefinite integrals.

Example 8.16 Find: (a) $\int xe^x\,dx$, (b) $\int x\sin x\,dx$

Solution: (a)

$$\int xe^x\,dx = \int x\,de^x$$
$$= \int u\,dv \qquad \text{(Let } u = x,\ v = e^x\text{)}$$
$$= uv - \int v\,du \qquad \text{(Integration by parts)}$$
$$= xe^x - \int e^x\,dx$$
$$= xe^x - e^x + C.$$

(b)

$$\int x\sin x\,dx = \int x\,d(-\cos x)$$
$$= \int u\,dv \qquad \text{(Let } u = x,\ v = -\cos x\text{)}$$
$$= uv - \int v\,du \qquad \text{(Integration by parts)}$$
$$= -x\cos x + \int \cos x\,dx$$
$$= -x\cos x + \sin x + C.$$

8.3.1 Exercises after class

1. **Basic skills**. Find the following indefinite integral.

   (1) (i) $\int x^2 dx$;   (ii) $\int \frac{x^{\frac{1}{2}}}{2} dx$.

   (2) (i) $\int e^{2x} dx$;   (ii) $\int \frac{\sin 2x}{2} dx$.

   (3) (i) $\int (1+2x) e^{x^2+x} dx$;   (ii) $\int (x-2) \cos(x^2-4x) dx;$.

   (4) (i) $\int x(x^2+1)^{\frac{2}{3}} dx$;   (ii) $\int \frac{x+1}{x-1} dx$.

2. **Comprehensive Problems-1**. Find the indefinite integral

   (1) (i) $\int \frac{2x-1}{x^2-x} dx$;   (ii) $\int \frac{3x-1}{x^2-x} dx$.

   (2) (i) $\int \frac{1}{x^2-x} dx$;   (ii) $\int \frac{1}{x^2-3x-4} dx$.

   (3) (i) $\int x e^{x^2} dx$;   (ii) $\int x e^x dx$.

   (4) (i) $\int x \sin x dx$;   (ii) $\int e^x \cos x dx$.

3. **Comprehensive Problem-2**. Prove Proposition 8.2.

8.4 Definite integral

The motivation for inventing definite integral is inherent. How to measure the area of irregular areas? How to calculate the volume of irregularly shaped objects? These are all very natural problems.

8.4.1 An Example of Finding Area

Let us see how to find the area of the domain that is between $x = 0$ and $x = 6$, is below curve $y = f(x) = \frac{x^2}{4}$ but above x-axis. We write this area as

$$\int_0^6 \frac{1}{4}x^2\,dx.$$

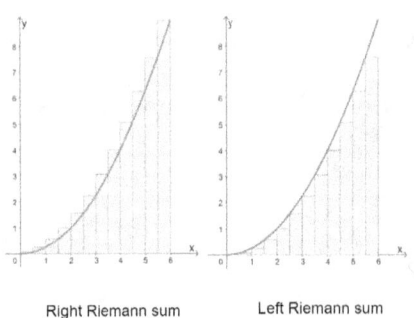

Right Riemann sum   Left Riemann sum

Figure 8.3: Right Riemann sum and left Riemann sum of $\int_0^6 \frac{1}{4}x^2\,dx$.

First, we divide the interval $[0, 6]$ into n equal length sub-intervals:

$$I_1 = [0, \frac{6}{n}),\ I_2 = [\frac{6}{n}, \frac{12}{n}),\cdots,\ I_k = [\frac{(k-1)6}{n}, \frac{6k}{n}),\cdots,\ I_n = [\frac{6(n-1)}{n}, 6).$$

We first calculate the areas of two regions M_n and m_n: One of them covers the domain (right Riemann sum), and the other is covered by this domain (left Riemann sum).

$$M_n = \sum_{k=1}^{n} f(\frac{6k}{n}) \cdot \frac{6}{n} = 54 \sum_{k=1}^{n} \frac{k^2}{n^3},$$

and

$$m_n = \sum_{k=1}^{n} f(\frac{6(k-1)}{n}) \cdot \frac{6}{n} = 54 \sum_{k=1}^{n} \frac{(k-1)^2}{n^3},$$

If $\lim_{n\to\infty} M_n = \lim_{n\to\infty} m_n$, it is more natural to define

$$\int_0^6 \frac{1}{4}x^2\,dx = \lim_{n\to\infty} M_n = \lim_{n\to\infty} m_n.$$

In fact, this could be the original idea of Riemann when he introduced definite integrals.

> **Definition 8.7. Definition of Riemann integral**
>
> *Consider continuous function $y = f(x)$ on interval (a,b). Divide (a,b) into n equal length sub-intervals: $x_0 = a < x_1 < x_2 < \cdots < x_n$. The right Riemann Sum is defined as:*
>
> $$M_n = \sum_{k=1}^{n} f(x_k) \cdot \frac{b-a}{n}.$$
>
> *The left Riemann Sum is defined as:*
>
> $$M_n = \sum_{k=1}^{n} f(x_{k-1}) \cdot \frac{b-a}{n}.$$
>
> *Then $\lim_{n \to \infty} M_n = \lim_{n \to \infty} m_n$, and we define the definite integral*
>
> $$\int_a^b f(x)\,dx = \lim_{n \to \infty} M_n.$$
>
> *Here we call a the lower bound of the integral, b the upper bound of the integral, and $f(x)$ the integrand.*

Example 8.17 Evaluate
$$\int_0^6 \frac{1}{4} x^2 \, dx$$

Solution: From the above discussion, we only need to calculate
$$\lim_{n \to \infty} 54 \sum_{k=1}^{n} \frac{k^2}{n^3}.$$

As shown in Example 7.15,
$$\sum_{k=1}^{n} k^2 = \frac{n(n+1)(2n+1)}{6}.$$

So
$$\lim_{n \to \infty} 54 \sum_{k=1}^{n} \frac{k^2}{n^3} = \lim_{n \to \infty} \frac{54}{n^3} \sum_{k=1}^{n} k^2$$
$$= \lim_{n \to \infty} \{\frac{54}{n^3} \cdot \frac{n(n+1)(2n+1)}{6}\}$$
$$= 18,$$

that is
$$\int_0^6 \frac{1}{4} x^2 \, dx = 18.$$

□

From Example 8.16, we also obtained the area of the domain asked at the beginning of this section. Obviously, it is too complicated to calculate the integral by finding the limit for the Riemann sum. The integration theory was widely accepted by the public later thanks to the discovery of the Fundamental Theorem of Calculus (also known as the Leibniz-Newton formula). We discuss this in next subsection.

8.4.2 Fundamental Theorem of Calculus

We can all agree on that seeking an anti-derivative of a function is regarded as an intelligence game without practical applications. But we will see below that these primitive and natural questions inadvertently become the driving force for the development of science and technology.

We first consider the derivative of a definite integral whose upper bound is the variable.

> **Theorem 8.15. Fundamental Theorem of Calculus Part I**
> Consider the continuous function $f(x)$ defined on the interval $I = (a, b)$. For all $a \leq x \leq b$,
> $$\frac{d}{dx}\int_a^x f(t)dt = f(x).$$

From the above theorem, we can obtain (details will be given in a Calculus course):

> **Theorem 8.16. Fundamental Theorem of Calculus Part II**
> Consider the differentiable function $f(x)$ defined on interval I. If its derivative is continuous on I, then for any two points $a < b \in I$,
> $$\int_a^b f'(t)dt = f(b) - f(a).$$
> The above is also called the "Leibniz-Newton Formula".

Example 8.18 Using the Leibniz-Newton formula to compute
$$\int_0^6 \frac{1}{4}x^2 dx$$

Solution: First observe
$$\int \frac{1}{4}x^2 dx = \frac{x^3}{12} + C.$$

That is
$$(\frac{x^3}{12} + C)' = \frac{1}{4}x^2.$$

So
$$\int_0^6 \frac{1}{4}x^2 dx = (\frac{6^3}{12} + C) - (0 + C) = 18.$$

□

As applications, we give more examples of calculating definite integrals.

1. Distance, speed, and acceleration

Consider a motion in a straight line. If we know the distance function $y = s(t)$ with respect to time variable t, then, its velocity function $v(t)$ and acceleration function $a(t)$ are given by the following:

$$v(t) = s'(t), \quad a(t) = v'(t) = s''(t).$$

Conversely, we consider a motion on the line with a constant acceleration, that is, its acceleration function $a(t) = a$ is a constant. Then, from $v'(t) = a$, we have

$$\begin{aligned} v(t) &= \int_0^t v'(x)dx + v(0) \\ &= v(0) + \int_0^t a\,dx \\ &= v(0) + at \end{aligned} \tag{8.20}$$

Then, using $s'(t) = v(t)$, we get

$$\begin{aligned} s(t) &= \int_0^t s'(x)dx + s(0) \\ &= s(0) + \int_0^t (v(0) + at)dx \\ &= s(0) + v(0)t + \frac{1}{2}at^2. \end{aligned} \tag{8.21}$$

2. Area formula

If we know the length $h(x)$ of all cross sections of a region from $x = a$ to $x = b$, then we can calculate the area of the region by

$$\textbf{Area} = \int_a^b h(x)dx.$$

Example 8.19 Find the area enclosed by the ellipse

$$\frac{x^2}{a^2} + \frac{y^2}{b^2} = 1.$$

Solution: We first calculate the area A of the region above the x-axis, it is half of the total area B: $B = 2A$.

$$A = \int_{-a}^{a} y\,dx.$$

The parameter equations for the ellipse are:

$$\begin{cases} x = a\cos t, & 0 \le t \le 2\pi \\ y = b\sin t, & 0 \le t \le 2\pi. \end{cases}$$

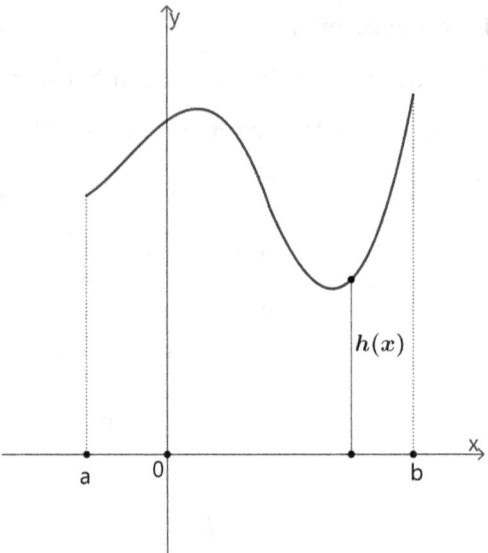

Figure 8.4: Integral: area diagram

We can calculate

$$A = \int_{-a}^{a} y\,dx$$
$$= \int_{\pi}^{0} b\sin t\, d(a\cos t)$$
$$= ab\int_{0}^{\pi} \sin^2 t\,dt$$
$$= ab\int_{0}^{\pi} \frac{1+\cos 2t}{2}\,dt$$
$$= \frac{1}{2}\pi ab.$$

So the whole area $B = \pi ab$.

□

3. Volume formula (ball, cone)

If we know the area $A(x)$ of all cross sections of a solid from $x = a$ to $x = b$, then we can calculate the volume of the solid by

$$\textbf{volume} = \int_{a}^{b} A(x)\,dx.$$

Example 8.20 Calculate the volume for the ball

$$x^2 + y^2 + z^2 \le r^2.$$

Solution: For $x = t$ ($-r \le t \le r$), the area of the section is

$$A(t) = \pi(r^2 - t^2).$$

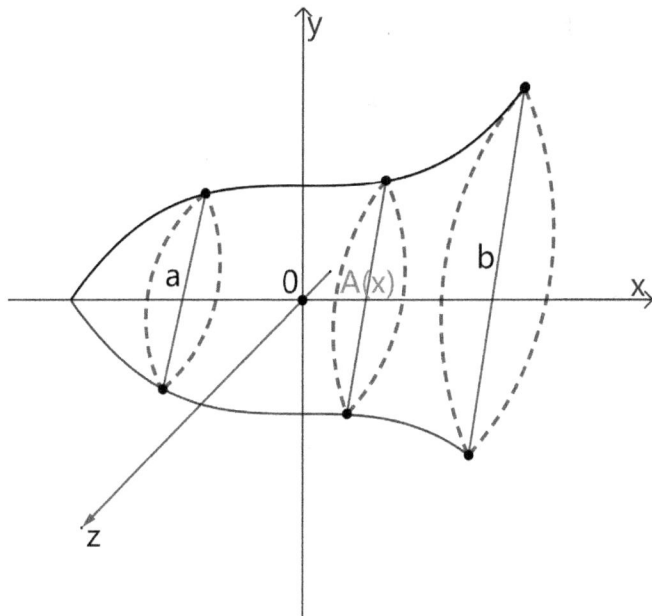

Figure 8.5: Integral: volume

So the volume of the ball is

$$V = \int_{-r}^{r} A(t)\,dt$$
$$= \int_{-r}^{r} \pi(r^2 - t^2)\,dt$$
$$= \frac{4}{3}\pi r^3.$$

□

Example 8.21 Calculate the volume for the cone with the bottom area of B and height of h.

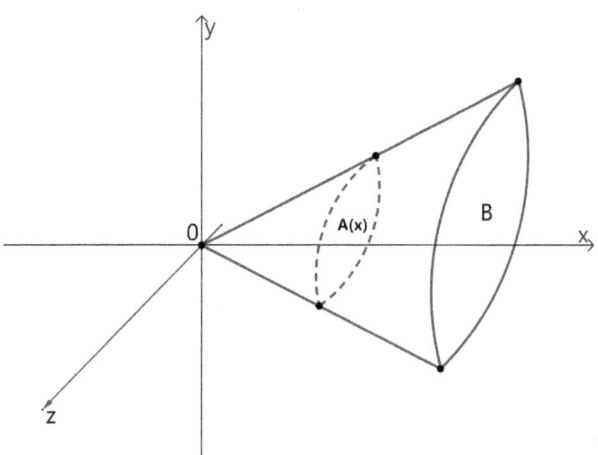

Figure 8.6: Cone volume

Solution: We place the vertex of the cone at the origin, and the base on the plane of $x = h$. Then, for $x = t$ ($0 \leq t \leq h$), using the similarity properties we know that its

sectional area $A(t)$ satisfies
$$\frac{A(t)}{B} = \frac{t^2}{h^2}.$$
thus
$$A(t) = \frac{t^2 B}{h^2}.$$
So, the volume of the cone
$$\begin{aligned} V &= \int_0^h A(t)\,dt \\ &= \int_0^h \frac{t^2 B}{h^2} \\ &= \frac{1}{3} hB. \end{aligned}$$

□

8.4.3 Exercises After Class

1. **Basic skills.** Find the following definite integral.

   (1) (i) $\int_{-1}^{1} 4x^2 dx$; (ii) $\int_{-1}^{1} x^3 dx$

   (2) (i) $\int_{-\frac{\pi}{2}}^{\frac{\pi}{2}} \cos x dx$; (ii) $\int_{-\frac{\pi}{2}}^{\frac{\pi}{2}} \sin x dx$.

   (3) (i) $\int_{1}^{2}(1+2x)e^{x^2+x} dx$; (ii) $\int_{0}^{2} x(x^2+1)^{\frac{2}{3}} dx$;.

   (4) (i) $\int_{-1}^{1} \sin x^3 dx$; (ii) $\int_{1}^{e} \ln x dx$.

2. **Comprehensive Problems.** The properties and the applications of definite integrals.

   (1) Suppose that $y = f(x)$ is an odd function on the interval $[-2, 2]$. Prove
   $$\int_{-1}^{1} f(x) dx = 0.$$

   (2) Consider a motion on the line with an acceleration of $a(t) = t$. Assume that the initial velocity is zero and the initial distance is zero. Find:

   (i) $v(10)$ (speed at $t = 10$);

   (ii) $S(10)$ (distance at $t = 10$).

   (3) The volume of a sphere with a known radius of r is $\frac{4}{3}\pi r^3$. Calculate the surface area of the sphere with a radius of r by using the differentiation approach.

   (4) The surface area of a sphere with a known radius of r is $4\pi r^2$. Use the integration approach to find the volume of a sphere with a radius of r.

8.5 Chapter review and exercises

In Chapter 8, we learned the derivatives and integrals of functions and their applications.

1. Tangent line and equation of motion. Through the discussion of the tangent line and equation of motion, understand the necessity of introducing the concepts of differentiation and derivative.

The slope of the tangent. The slope of the tangent line of function $y = f(x)$ at point $(a, f(a))$ is defined as

$$m_a = \lim_{x \to a} \frac{f(x) - f(a)}{x - a}.$$

The instantaneous speed of movement. The instantaneous speed at $t = t_0$ is defined as:

$$v(t_0) = \lim_{t \to t_0} \frac{s(t) - s(t_0)}{t - t_0}.$$

2. Concept and application of derivative.

2.1. Definition of the derivative function

First derivative:

$$f'(a) = \frac{df}{dx}(a) = \lim_{x \to a} \frac{f(x) - f(a)}{x - a}.$$

Higher order derivative:

$$\frac{d^n f}{dx^n}(x) = \frac{d}{dx}\left(\frac{d^{n-1} f}{dx^{n-1}}(x)\right).$$

Derivative formulas

$$\frac{d}{dx}(x^n) = nx^{n-1}$$

$$\frac{d}{dx}(e^x) = e^x$$

$$\frac{d}{dx}(\ln x) = \frac{1}{x}$$

$$\frac{d}{dx}(\sin x) = \cos x$$

$$\frac{d}{dx}(\cos x) = -\sin x$$

$$(f(x)g(x))' = f'(x)g(x) + f(x)g'(x)$$

2.2. Applications of derivatives

Function monotonicity: Function $y = f(x)$ on interval $I = (a, b)$ is monotonically increasing function $\Leftrightarrow$ For all $x \in I$, $f'(x) \geq 0$.

Critical point: The maximum and minimum values of a differentiable function $y = f(x)$ on interval $I = [a, b]$ are critical points or endpoints of the interval.

Tangent of function $y = f(x)$ at point $(a, f(a))$:

$$y = f'(a)(x - a) + f(a).$$

Convexity and concavity of functions: Assume function $f(x)$ has the second-order derivative.

(a) If $f''(x) > 0$ for all $x \in I$, then $f(x)$ is a convex function on I; (b) If $f''(x) < 0$ for all $x \in I$, then $f(x)$ is a concave function on I.

2.3. Differential of the vector function

Consider a two-dimensional vector function $\vec{F}(x) = (f(x), g(x))$.

Define limit

$$\lim_{x \to x_0} \vec{F}(x) = (\lim_{x \to x_0} f(x), \lim_{x \to x_0} g(x)),$$

And define $\vec{F}(x)$ as a continuous function if and only if $f(x)$ and $g(x)$ are continuous.

For derivatives, we also define

$$\frac{d\vec{F}(x)}{dx} = \vec{F}'(x) = (f'(x), g'(x)).$$

3. Indefinite integral.

3.1. Algebraic properties of integrals

$$\int (\alpha f(x) + \beta g(x)) dx = \alpha \int f(x) dx + \beta \int g(x) dx.$$

3.2. Indefinite integral formulas.

$$\int x^n dx = \frac{x^{n+1}}{n+1} + C, \quad \text{for } n \neq 1,$$

$$\int \frac{1}{x} dx = \ln |x| + C,$$

$$\int e^x dx = e^x + C,$$

$$\int \cos x\, dx = \sin x + C,$$

$$\int \sin x\, dx = -\cos x + C.$$

Integration by parts formula

$$\int f(x)g'(x)\, dx = f(x)g(x) - \int f'(x)g(x)\, dx.$$

Substitution formula.

$$\int f(g(x))g'(x)\, dx = \int f(u)\, du.$$

4. Definite integral and application.

4.1. Fundamental Theorem of Calculus

Fundamental Theorem of Calculus Part 1:

$$\frac{d}{dx}\int_a^x f(t)\, dt = f(x).$$

Fundamental Theorem of Calculus Part II (Leibniz-Newton Formula):

$$\int_a^b f'(t)\, dt = f(b) - f(a).$$

4.2. Applications of definite integrals

Motion with constant acceleration:

$$v(t) = v(0) + at$$

$$s(t) = s(0) + v(0)t + \frac{1}{2}at^2.$$

Area formula

$$\textbf{Area of the region} = \int_a^b h(x)\, dx.$$

Volume formula

$$\textbf{Volume of the solid} = \int_a^b A(x)\, dx.$$

8.5.1 Comprehensive Exercises

1. **Comprehensive questions-1.** Consider function $y = \ln(x+1)$.

   (1) Find the tangent line equation at point $(0, 0)$.

   (2) Find $\int_0^1 \ln(x+1)\,dx$.

   (3) Prove: for any two positive numbers x_1, x_2,
   $$\ln(\frac{x_1 + x_2 + 3}{3}) \geq \frac{\ln(x_1 + 1) + \ln(x_2 + 1)}{3}.$$

2. **Comprehensive questions-2.** Application of limit and derivative.

   (1) If two polynomials are identical:
   $$ax^2 - 2x + c \equiv 3x^2 + bx - 1.$$
   Find the values of a, b, c by selecting the appropriate values of x.

   (2) If two polynomials are identical:
   $$ax^6 - bx^5 + cx^4 - dx^3 + ex^2 - fx + g \equiv x^6 + 2x^5 + 3x^4 + 4x^3 + 5x^2 + 6x + 7.$$
   Find the values of a, b, c, d, e, f, g via taking the limits.

   (3) If two polynomials are identical
   $$ax^6 - bx^5 + cx^4 - dx^3 + ex^2 - fx + g = x^6 + 2x^5 + 3x^4 + 4x^3 + 5x^2 + 6x + 7.$$
   Find the values of a, b, c, d, e, f, g by using the derivatives.

   (4) Suppose $y = f(x)$ is a differentiable function on the interval $I = (a, b)$. Prove
   $$f'(x) = 0 \Leftrightarrow f(x) = \textbf{constant}.$$

3. **Comprehensive questions-3.** Understanding integration by parts

   (1) Use the derivative of the products to derive the integration by part formula:
   $$\int f(x)g'(x)\,dx = f(x)g(x) - \int f'(x)g(x)\,dx.$$

   (2) Use (1) to prove
   $$\int_a^b f(x)g'(x)\,dx = f(x)g(x)\vert_a^b - \int_a^b f'(x)g(x)\,dx.$$

4. **Comprehensive questions-4.** Applications and Understanding of Fundamental Theorem of Calculus.

   (1) Assume that $y = f(x)$ has a continuous derivative function on the interval $I = [a, b]$, $[c, d] \subset (a, b)$.

   (i). For any $x \in [c, d]$. Use the definition of a continuous function, to prove that for any positive number $\epsilon > 0$, there is $h > 0$, such that $[x - h, x + h] \subset (a, b)$, and
   $$|f(w) - f(x)| < \epsilon \quad \text{for all } w \in [x - h, x + h].$$

(ii) Prove: for $x \in (c, d]$,
$$\frac{d}{dx}\int_c^x f(t)dt = f(x).$$
(iii) Use comprehensive question-2(4) and comprehensive question-4(ii) to prove
$$\int_c^d f'(x)dx = f(d) - f(c).$$
(2) Use the idea in the Fundamental Theorem of Calculus to understand some algebraic operation skills.

(i) Find the sum
$$1 \times 2 + 2 \times 3 + \cdots + 99 \times 100.$$

Hint: $3k(k+1) = k(k+1)(k+2) - (k-1)k(k+1)$.

(ii) Find the sum
$$1^2 + 2^2 + 3^2 + \cdots + n^2.$$

Chapter 9 Solutions to even numbered questions

Exercise 1.1.5 1(2). 131

1(4). 1254

1(6). −60

1(8). $= 48 \times \frac{1}{14} \times 7 = 24$

1(10). 18174

1(12). 11

1(14). $\frac{5}{21}$

1(16). $1\frac{4}{9}$

1(18). $\frac{1}{16}$

1(20). $1\frac{3}{4}$

2(2). 121

2(4). 13200

2(6). $= 60 \times \frac{1}{3 \times \frac{1}{2}} = 60 \times \frac{2}{3} = 40$

2(8). 96

2(10). 320

2(12). $\frac{23 \times 13 + 46}{15} + \frac{-13 \times 14 + 11 \times 7}{15} = \frac{23 \times 15}{15} + \frac{-15 \times 7}{15} = 16$

3(2). 7999

3(4). 8001

$$\begin{array}{rrrr}
 & 4 & -2 & 1 \\
\times & & 2 & 1 \\
\hline
 & 4 & -2 & 1 \\
+ 8 & -4 & 2 & \\
\hline
8 & 0 & 0 & 1.
\end{array}$$

3(6). 421

$$
\begin{array}{r}
421 \\
2\hat{1}\,\overline{)\,800\hat{1}} \\
-\;8\hat{4} \\
\hline
0\,4\,0 \\
-\;0\,4\,\hat{2} \\
\hline
2\,\hat{1} \\
-\;2\,\hat{1} \\
\hline
0.
\end{array}
$$

4(2). 324_5

4(4). 410_5

4(6). $= 215_5 \times 10_5 = 2150_5$

4(8). 10506_7

5(2). 2, 3, 5, 29

5(4). 7, 11, 13

5(6). 43, 47

6(2). 20

6(4). 12

7(2). $2\frac{1}{11}$

7(4). $\frac{13}{10}$

8(2). $2\frac{1}{7}$

8(4). 9

8(6). 2

9(2). 112_5

Exercise 1.2.3

1(2). $5x^2 - x$

1(4). $7x^2$

1(6). $3x^2 + 5$

1(8). $4x^2 - 6x - 2$

2(2). $x^4 + 3x^2 + 2$

2(4). $x^4 - 3x^2 + 2$

2(6). $x^4 - x^2 - 42$

2(8). $x^4 + x^2 - 42$

2(10). $2x^4 + 5x^2 - 3$

2(12). $2x^4 - 5x^2 - 3$

2(14). $x^6 - 1$

2(16). $x^6 + 1$

2(18). $x^6 + a^3$

3(2). $x^2 + 10x + 25$

3(4). $x^3 + 1$

3(6). $16x^2 + 4x + 1$

$$\begin{array}{r}
16x^2 \quad +4x \quad +1 \\
\hline
4x-1 \overline{\smash{)}\; 64x^3 \quad\;\; 0 \quad\;\; 0 \quad -1} \\
-\quad 64x^3 \;-16x^2 \\
\hline
0 \quad 16x^2 \quad\;\; 0 \\
-\quad 0 \quad 16x^2 \;-4x \\
\hline
4x \quad -1 \\
-\quad 4x \quad -1 \\
\hline
0.
\end{array}$$

3(8). $x^2 - x + 4$

$$\begin{array}{r}
x^2 \quad -x \quad +4 \\
\hline
x+1 \overline{\smash{)}\; x^3 \quad 0 \quad 3x \quad +4} \\
-\quad x^3 \;+x^2 \\
\hline
0 \;-x^2 \;+3x \\
-\quad 0 \;-x^2 \;-x \\
\hline
4x \quad +4 \\
-\quad 4x \quad +4 \\
\hline
0.
\end{array}$$

3(10). $4x^2 - 2x + 4$

4 (1). $\overline{1ab1ab} = \overline{1ab} \times 1001$. $1001 = 7 \times 11 \times 13$ is divisible by 7.

4(2). $\overline{1abcd1abcd} = \overline{1abcd} \times (10^5 + 1)$. $10^5 + 1$ is divisible by 11.

Exercise 1.3.5

1(2). $\frac{1}{5}$

1(4). $\frac{3}{13}$

1(6). $\frac{3}{5}$

1(8). $1\frac{7}{8}$

1(10). $10\frac{1}{8}$

1(12). $4 - \frac{3a}{2}$

1(14). Case 1: if $a = -1$, x can be any number; Case 2: if $a \neq -1$, no solution.

1(16). Case 1: if $a \neq 1$, $x = \frac{b}{1-a}$; Case 2: if $a = -1$ and $b \neq 0$, no solution; Case 3: if $a = -1$ and $b = 0$, x can be any number.

2(2). $x = 3, y = 2$

2(4). $x = \frac{5}{2}, y = \frac{4}{3}$

2(6). $x = -1\frac{8}{47}, y = 1\frac{13}{47}$

3(2). $x = \frac{1}{3}, y = \frac{1}{2}$

4(2). $x^2 + 1)(x^2 + 2)$

4(4). $(x - 1)(x + 1)(x^2 - 2)$

4(6). $(x^2 + 4)(x^2 - 6)$

4(8). $(2x^2 + 1)(x^2 + 3)$

4(10). $(2x - 1)(2x + 1)(x^2 + 3)$

4(12). $(x^2 + 2)(x^4 - 2x^2 + 4)$

4(14). $(x - 2)(x + 2)(x^2 + 4)$

4(16). $(x - 2)(x + 2)(x^2 - 2x + 4)(x^2 + 2x + 4)$

4(18). $(x^2 - x + 1)(x^2 + x + 1)$

5(2). $x = 2,\ or\ x = 3$

5(4). $x_1 = 2, x_2 = -2, x_3 = 3, x_4 = -3$

5(6). $x_1 = 2, x_2 = -2$

5(8). No rational solutions.

6(2). Ann's age is 5.

7(2). $a \neq 1$

8(2). Case 1: $a \neq -1$, $x = \frac{b+1}{a+1}$; Case 2: $a = -1$ and $b \neq -1$, no solution; Case 3: $a = -1$ and $b = -1$, $x = $ any number.

Exercise 1.4.6

1(2). $x > -2$

1(4). $x > 1\frac{1}{8}$

1(6). $x > \frac{6}{a} + 1$

1(8). Case 1: $a > 0$, then $x < 2 + \frac{2}{a}$; Case 2: $a = 0$, $x = $ any number; Case 3: $a < 0$, $x > 2 + \frac{2}{a}$.

1(10). Case 1: $a - 2b > 0$, then $x < \frac{2}{a+2b}$; Case 2: $a - 2b = 0$, $x = $ any number; Case 3: $a - 2b < 0$, $x > \frac{2}{a+2b}$.

2(2). $-7 < x < 1$

2(4). $x < -7$ or $x > -3$

2(6). $x < 1$ or $x > 4$

3(2). Proof: $a^2 < b^2 \rightleftharpoons a^2 - b^2 < 0 \rightleftharpoons (a-b)(a+b) < 0$. Since $a + b > 0$, we know $(a-b)(a+b) < 0 \rightleftharpoons a - b < 0 \rightleftharpoons a < b$.

3(4). Proof: $x^2 > C^2 \rightleftharpoons (|x|)^2 - C^2 > 0 \rightleftharpoons (|x| - C)(|x| + C) > 0$. Since $|x| + C > 0$, we know $(|x| - C)(|x| + C) > 0 \rightleftharpoons |x| - C > 0 \rightleftharpoons |x| > C$.

3(6). Proof: First observe that
$$x^3 + y^3 + z^3 - 3xyz = (x + y + z)(x^2 + y^2 + z^2 - xy - xz - yz),$$
and
$$x^2 + y^2 + z^2 - xy - xz - yz = \frac{1}{2}[(x-y)^2 + (x-z)^2 + (y-z)^2] \geq 0.$$
For positive number x, y, z, we have $x + y + z > 0$. Thus $x^3 + y^3 + z^3 - 3xyz \geq 0$, which yields
$$x^3 + y^3 + z^3 \geq 3xyz.$$

Exercise 1.5.7

1(2). $5^{-\frac{1}{2}}$

1(4). $x^3 y$

1(6). $3\sqrt{2} + \sqrt{3}$

1(8). $-\frac{\sqrt{2}}{2} - \frac{2\sqrt{3}}{3}$

1(10). 0

1(12). $\frac{1}{2}$

1(14). -1

2(2). $-2\sqrt{7} - 4\sqrt{5}$

2(4). $\sqrt{3} + 2\sqrt{2}$

2(6). $2\frac{1}{2}$

3(2). Proof by contradiction. If $\sqrt{p}$ is a rational number, then
$$\sqrt{p} = \frac{m}{n},$$
where m, n are two positive integers and their largest common factor $(m, n) = 1$ (we can assume $\frac{m}{n}$ is a simple fraction).

So we have $m^2 = pn^2$, thus m is divisible by p. We write $m = pk$, where k is another positive integer. Bringing this into the above equality and simplifying, we have
$$n^2 = pk^2.$$
Thus n is divisible by p. So far, we know that both m, n are divisible by p, thus $(m, n) \geq p > 1$. Contradicts to the assumption that $(m, n) = 1$.

Exercise 1.6.4

1(2). $-1 - 5i$

1(4). 41

1(6). $5 + 4i$

1(8). $2i$

2(2). $|z| = 1$, $arg(z) = \frac{5\pi}{3}$

2(4). $|z| = 1$, $arg(z) = \frac{11\pi}{6}$

2(6). $|z| = 1$, $arg(z) = \frac{7\pi}{4}$

2(8). $\sqrt{2} + \sqrt{2}i$

3(2). $x^2 = (\sqrt{5}i)^2$. So $x_1 = \sqrt{5}i$, $x_2 = -\sqrt{5}i$

3(4). $x_1 = \frac{-3+\sqrt{15}i}{2}$, $x_2 = \frac{-3-\sqrt{15}i}{2}$.

3(6). $x_1 = e^{\frac{\pi}{6}} = \frac{\sqrt{3}}{2} + \frac{1}{2}i$, $x_2 = e^{\frac{5\pi}{6}} = -\frac{\sqrt{3}}{2} + \frac{1}{2}i$, $x_1 = e^{\frac{3\pi}{2}} = -i$.

4(2). $x^5 = 32$

Exercise 1.7.2

1(2). $1 - \frac{\sqrt{2}}{4}$

1(4). 31

1(6). b^2

1(8). $8 + i$

1(10). $8 - i$

1(12). $2x - 1$

1(14). $(x+1)^8 = x^8 + C_8^1 x^7 + C_8^2 x^6 + \ldots + C_8^7 x + 1$

2(2a). For $a \neq 0$, $x = \frac{a}{2}$.

2(2b). For $a \neq -1$, $x = 2a + 2$; For $a = -1$, no solution.

2(4a). $x = 4$, or $x = 7$.

2(4b). $x = -\frac{3}{2}$, or $x = \frac{1}{3}$.

2(6a). $x = 3$.

2(6b). $x = 6$.

2(8a). $-1 \leq x \leq 0$.

2(8a). $-2 \leq x < 0$.

3(2). Proof:

$$A > B \Rightarrow A + x > B + x \quad \textbf{(Addition invariant)}$$

$$x > y \Rightarrow B + x > B + y \quad \textbf{(Addition invariant)}$$

Since $A + x > B + x$ and $B + x > B + y$, using **transitive property** we conclude: $A + x > B + y$.

3(4). Proof: For any two real numbers x, y,

$$(x - y)^2 \geq 0 \Rightarrow x^2 + y^2 \geq 2xy.$$

If x, y are positive, we have (since $f(x) = \ln x$ is an increasing function)

$$\ln(x^2 + y^2) \geq \ln(2xy) = \ln 2 + \ln x + \ln y.$$

This yields

$$\ln(x^2 + y^2) - \ln 2 \geq \ln x + \ln y.$$

Exercise 2.1.4

2(1). $A \cap B = \{x \in \mathbb{R} \mid 2 < x < 12\}$.

2(2). $A \cup B = \{x \in \mathbb{R} \mid x > -8\}$

2(3). $A^c \cup B^c = \{x \in \mathbb{R} \mid x < 2, \text{ or } x > 12\}$.

2(4). Since $(A \cap B)^c = \{x \in \mathbb{R} \mid x < 2, \text{ or } x > 12\}$. The proof for general sets is given in the proof of Proposition 2.2.

Exercise 2.2.5

1(2). (a) 51. (b) 33. (c) 16. (d). 51 − 16 = 35. (e). 33 − 16 = 17.

2(1). 10.

$\{A, B\}$, $\{A, C\}$, $\{A, D\}$, $\{A, D\}$, $\{A, E\}$, $\{B, C\}$, $\{B, D\}$, $\{B, E\}$, $\{C, D\}$, $\{C, E\}$, $\{D, E\}$.

2(2). $C_5^2 \cdot C_3^2 / 2 = 15$ pairs.

$\{A, B\}$ and $\{C, D\}$, $\{A, B\}$ and $\{C, E\}$, $\{A, B\}$ and $\{D, E\}$;
$\{A, C\}$ and $\{B, D\}$, $\{A, C\}$ and $\{B, E\}$, $\{A, C\}$ and $\{D, E\}$;
$\{A, D\}$ and $\{B, C\}$, $\{A, D\}$ and $\{B, E\}$, $\{A, D\}$ and $\{C, E\}$;
$\{A, E\}$ and $\{B, C\}$, $\{A, E\}$ and $\{B, D\}$, $\{A, E\}$ and $\{C, D\}$;
$\{B, C\}$ and $\{D, E\}$, $\{B, D\}$ and $\{C, E\}$, $\{B, E\}$ and $\{C, D\}$.

Exercise 2.3.7

2(1). (ii): $D = \{x \in \mathbb{R} \mid x \neq 1\}$, $R = \{y \in \mathbb{R} \mid y \neq 1\}$.

(iv): $D = \{x \in \mathbb{R} \mid x \geq \sqrt{3} \text{ or } x \leq -\sqrt{3}\}$, $R = \{y \in \mathbb{R} \mid y \geq 0\}$.

2(2). (i): $f[g(x)] = e^{x^2-1}$; (ii): $g[f(x)] = e^{2x} - 1$.

3(2). Proof: $g(x)$ is even, so $g(-x) = g(x)$. Thus
$$f[g(-x)] = f[g(x)].$$

That is: $y = f[g(x)]$ is an even function.

3(4). Proof: Since $f(x)$ and $g(x)$ are increasing functions on $\mathbb{R}$, we know that for any $x_1 < x_2$,
$$f(x_1) \leq f(x_2), \text{ and } g(x_1) \leq g(x_2).$$

So,
$$f(x_1) + g(x_1) \leq f(x_2) + g(x_2).$$

Thus $y = f(x) + g(x)$ is an increasing function.

$y = f(x)g(x)$ may not be an increasing function. For example: $f(x) = x$ and $g(x) = x - 1$ are increasing. But $y = x^2 - x$ is not an increasing function.

3(6). Proof: Since $y = g(x)$ is a periodic function, there is a $T > 0$, such that $g(x + T) = g(x)$. So
$$f[g(x + T)] = f[g(x)].$$

Thus $y = f[g(x)]$ is also a periodic function with T as its period.

Exercise 2.4.2

1(2). $A \cap B = B = \{6n \mid n \in \mathbb{N}\}$, $A \cup B = A = \{2n \mid n \in \mathbb{N}\}$, $A \cap (B \cup C) = A = \{2n \mid n \in \mathbb{N}\}$.

1(4). 4.

2(2). $D = \{x \in \mathbb{R} \mid 1 \leq x \leq 3\}$, $R = \{y \in \mathbb{R} \mid y \geq 0\}$.

2(4). $D = \{x \in \mathbb{R} \mid x \neq -3\}$, $R = \{y \in \mathbb{R} \mid y \neq 1\}$.

2(6). $D = \{x \in \mathbb{R} \mid x \neq 1, x \neq 3\}$, $R = \{y \in \mathbb{R} \mid y < 0 \text{ or } y \geq \frac{1}{7}\}$.

2(8). (i). $g \circ g \circ g(x) = \frac{1}{8}x + 3\frac{1}{2}$; (ii). $g^{-1}(x) = 2x - 2$; (iii). $R(f) = \{y \in \mathbb{R} \mid y \geq -3\}$; (iv). $f^{-1}(-3) = -2$.

3(2). If $f(x)$ is an even function, then for $x_0 < 0$,
$$f(-x_0) = f(x_0).$$
Since $-x_0 < x_0$, the above inequality contradicts to the fact that $f(x)$ is an increasing function.

3(4). Since $f(0) = [f(0)]^2$ and $f(0) \neq 0$, we know that $f(0) = 1$. For all natural number b,
$$1 = f(0) = f(0)f(b) = f(b).$$
So $f(2022) = 1$.

Exercise 3.1.5

1(2). $x = \frac{3+\sqrt{5}}{4}$, or $x = \frac{3-\sqrt{5}}{4}$.

1(4). $m < -\frac{9}{2}$.

1(6). $x > 3$ or $x < -3$.

1(8). $-1 < x < 3$.

2(2). $x = -4$, or $x = 4$, or $x = 5$.

2(4). $C_{10}^8 = 45$.

Exercise 3.2.3

1(2). (i). $D = \{x \in \mathbb{R} \mid x \geq -4\}$. $R = \{y \in \mathbb{R} \mid y \geq 0\}$. $f(x) = x^2 - 4$.

1(2). (ii). $D = \mathbb{R}$. $R = \{y \in \mathbb{R} \mid y \geq 3\}$. $f^{-1}(x) = \frac{2+\sqrt{4y-12}}{2}$ or $f^{-1}(x) = \frac{2-\sqrt{4y-12}}{2}$.

2(2). $x = -59$.

2(4). $x = 7$ ($x = 4$ is an extraneous solution).

Exercise 3.3.1

2(2). $D = \mathbb{R}$, $R = \{y \in \mathbb{R} \mid y \geq 2\}$.

3(2). Proof. Step 1, we prove, via the induction that, for any positive number $0 < p < 1$ and positive integer n,
$$(1-p)^n \geq 1 - pn.$$

For $n = 1$, $1 - p = 1 - p$. The statement is true.

Assume that the inequality holds for $n = k$. That is, for $n = k$,
$$(1-p)^k \geq 1 - pk.$$

Then, for $n = k + 1$, using induction assumption, we have
$$(1-p)^{k+1} \geq (1-pk)(1-p)$$
$$= 1 - p(k+1) + p^2 k$$
$$\geq 1 - p(k+1).$$

We thus prove the statement.

Step 2, we show that for all positive integers n,
$$(1 + \frac{1}{n})^n < (1 + \frac{1}{n+1})^{n+1}.$$

First we observe
$$\frac{(1 + \frac{1}{n+1})^{n+1}}{(1 + \frac{1}{n})^n}$$
$$= \frac{(\frac{n+2}{n+1})^{n+1}}{(\frac{n+1}{n})^n}$$
$$= \left(\frac{(n+2)n}{(n+1)^2}\right)^n \cdot \frac{n+2}{n+1}$$
$$= \left(1 - \frac{1}{(n+1)^2}\right)^n \cdot \left(1 + \frac{1}{n+1}\right)$$
$$\geq \left(1 - \frac{n}{(n+1)^2}\right) \cdot \left(1 + \frac{1}{n+1}\right) \quad \text{(Using the inequality obtained in step 1)}$$
$$= \left(1 - \frac{1}{n+1} + \frac{1}{(n+1)^2}\right) \cdot \left(1 + \frac{1}{n+1}\right)$$
$$= 1 + \frac{1}{(n+1)^3}$$
$$> 1.$$

Step 3. So we have, for all positive integers n,
$$2 = (1+1)^1 \leq (1 + \frac{1}{n})^n$$

which yields
$$2^{\frac{1}{n}} \leq 1 + \frac{1}{n}.$$

Exercise 3.4.1

2(2). (a). $D = \mathbb{R}$, $R = \{\in \mathbb{R} \mid y \geq -2\}$; (b). $D = \{x \in \mathbb{R} \mid x \geq \sqrt{3} \text{ or } x \leq -\sqrt{3}\}$, $R = \{y \in \mathbb{R} \mid y \geq 0\}$.

2(4). (a). $f[g(x)] = xe^x$; (b). $g[f(x)] = x^x$.

3(2). (a). $x = 1 + \sqrt{2}$; (b). $x = 1 + \sqrt{2}$ or $x = 1 - \sqrt{2}$.

3(4). (a). $x = 4$; (b). $x = 4$ or $x = -2$.

3(6). Proof. Step 1, we prove that for three positive numbers a, b, c,
$$a^3 + b^3 + c^3 \geq 3abc.$$

In fact
$$a^3 + b^3 + c^3 - 3abc = (a+b+c)(a^2+b^2+c^2 - ab - ac - bc)$$
$$= \frac{1}{2}(a+b+c)[(a-b)^2 + (a-c)^2 + (b-c)^2]$$
$$\geq 0.$$

Step 2. Let $x = a^3$, $y = b^3$, $z = c^3$. We obtain from step 1 that
$$x + y + z \geq 3(xyz)^{\frac{1}{3}}.$$

Thus
$$\frac{x+y+z}{3} \geq (xyz)^{\frac{1}{3}}.$$

Since $f(x) = \ln x$ is an increasing function, we have
$$\ln \frac{x+y+z}{3} \geq \ln(xyz)^{\frac{1}{3}} = \frac{1}{3}\ln(xyz) = \frac{\ln x + y + \ln z}{3}.$$

Exercise 3.5.1

1(2).

(2.2). $D = \mathbb{R}$, $R = \{y \in \mathbb{R} \mid y \geq 2^{-\frac{1}{4}}\}$.

1(2).

(2.4). $D = \mathbb{R}$, $R = \{y \in \mathbb{R} \mid y > 0\}$.

1(3).

(3.2). $x^3 + 3x^2 - 2x + 1$; (3.4). $\frac{2}{3}$; (3.6). 2.

2(1).

(1.2). $x = \frac{5 \pm \sqrt{17}}{2}$; (1.4). $x_1 = x_2 = 1$, $x_3 = 2$; (1.6). $x = 0$ or $x = 3$; (1.8). $x = 1$, or $x = 3$; (1.10). $x = \frac{-3+3\sqrt{17}}{2}$.

2(2).

(2.2). $x \geq \sqrt{2}$ or $x < -\sqrt{5}$; (2.4). $x \geq 2$; (2.6). $x > 3$ or $x < 1$.

3(2). $x = \frac{11 \pm 3\sqrt{17}}{8}$.

3(4). $m < 0 \text{ or } m > 12$.

Exercise 4.1.1

1(2). $-\frac{1}{2}$; 1(4). $\frac{1}{2}$; 1(6). $-\frac{\sqrt{3}}{3}$.

2(1). (ii). $T = \frac{8\pi}{3}$; (iv). $T = 2\pi$.

2(2). (ii). It is an even function; (iv). It is an even function.

3(3). $-\sqrt{2} \leq \sin x + \cos x \leq \sqrt{2}$.

Exercise 4.2.8

1(2). $\frac{1}{\sin \theta} = \csc \theta$.

1(4). $\cos^2 x$.

1(6). 1

2(2). Proof:
$$\frac{1}{1 - \cos^2 x} = \frac{1}{\sin^2 x}.$$

and

$$1 + \cot^2 x = 1 + \frac{\cos^2 x}{\sin^2 x}$$
$$= \frac{\sin^2 x + \cos^2 x}{\sin^2 x}$$
$$= \frac{1}{\sin^2 x}.$$

So, $\frac{1}{1-\cos^2 x} = 1 + \cot^2 x$.

2(4). Proof.
$$\tan(\frac{\pi}{4} + x) = \frac{\tan \frac{\pi}{4} + \tan x}{1 - \tan \frac{\pi}{4} \cdot \tan x}$$
$$= \frac{1 + \tan x}{1 - \tan x}.$$

2(6). Proof.

$$\frac{\cos(x-y)}{\sin x \cos y} = \frac{\cos x \cos y + \sin x \sin y}{\sin x \cos y}$$
$$= \frac{\cos x}{\sin x} + \frac{\sin y}{\cos y}$$
$$= \cot x + \tan y.$$

2(8). Proof.
$$\sin 3\alpha = \sin 2\alpha \cos \alpha + \cos 2\alpha \sin \alpha$$
$$= 2\sin\alpha \cos^2\alpha + (1 - 2\sin^2\alpha)\sin\alpha$$
$$= 2\sin\alpha - 2\sin^3\alpha + \sin\alpha - 2\sin^3\alpha$$
$$= -4\sin^3\alpha + 3\sin\alpha.$$

2(10). Proof.
$$\cos^4 5\alpha - \sin^4 5\alpha = (\cos^2 5\alpha - \sin^2 5\alpha)(\cos^2 5\alpha + \sin^2 5\alpha)$$
$$= \cos^2 5\alpha - \sin^2 5\alpha$$
$$= \cos 10\alpha.$$

2(12). Proof.
$$\frac{\sin 6x}{\sin 5x + \sin x} = \frac{2\sin 3x \cos 3x}{2\sin 3x \cos 2x}$$
$$= \frac{\cos 3x}{\cos 2x}.$$

3(2). (i). $x = 1 \pm \sqrt{2}i$; (ii). $x_1 = \frac{\sqrt{2}}{2} + \frac{\sqrt{2}}{2}i$, $x_2 = -\frac{\sqrt{2}}{2} - \frac{\sqrt{2}}{2}i$, $x_3 = \frac{\sqrt{2}}{2} - \frac{\sqrt{2}}{2}i$, $x_4 = -\frac{\sqrt{2}}{2} + \frac{\sqrt{2}}{2}i$, $x_5 = 2$, $x_6 = -2$, $x_7 = 2i$, $x_8 = -2i$.

3(4). $y = 1 - \frac{1}{2}\sin^2 2x$, $R = \{y \in \mathbb{R} \mid \frac{1}{2} \le y \le 1\}$.

Exercise 4.3.1

1(2). $c = \sqrt{89 - 40\sqrt{3}}$.

1(4). $\sin(\angle A + \angle C) = \sin \angle B = \frac{4}{5}$.

2(2). $a = 10$.

2(4). Proof.
$$\frac{\tan\frac{A}{2} + \tan\frac{B}{2}}{1 - \tan\frac{A}{2}\tan\frac{B}{2}} = \tan(\frac{A}{2} + \frac{B}{2})$$
$$= \tan(\frac{\pi}{2} - \frac{C}{2})$$
$$= \frac{1}{\tan\frac{C}{2}}.$$

This yields
$$\tan\frac{A}{2}\tan\frac{C}{2} + \tan\frac{B}{2}\tan\frac{C}{2} + \tan\frac{A}{2}\tan\frac{B}{2} = 1.$$

Exercise 4.5.1

1(2). Not even, not odd; 2π-periodic function.

1(4). Even function.

2(2). $D = \mathbb{R}; R = \{y \in \mathbb{R} \mid 1 \le y \le 5\}$ ($y = (\sin x + 1)^2 + 1$)

2(4). $D = \mathbb{R}; R = \{y \in \mathbb{R} \mid -3 \le y \le \frac{3}{2}\}$ ($y = (\frac{3}{2} - 2(\sin x - \frac{1}{2})^2)$)

3(2). $r_1 = 2e^{\frac{\pi}{5}}$ (principal root), $r_2 = 2e^{\frac{3\pi}{5}}$, $r_3 = 2e^{\frac{5\pi}{5}} = -2$, $r_4 = 2e^{\frac{7\pi}{5}}$, $r_5 = 2e^{\frac{9\pi}{5}}$.

4(2). Proof.

$$(\cos x - \sin x)^2 = \cos^2 x - 2\cos x \sin x + \sin^2 x$$
$$= 1 - 2\cos x \sin x$$
$$= 1 - \sin 2x.$$

4(4). Proof. First, we note that $x_1 = \cos\frac{\pi}{5} + i\sin\frac{\pi}{5}$ is a solution to $x^5 + 1 = 0$. So (since $x \ne -1$)
$$x_1^4 - x_1^3 + x_1^2 - x_1 + 1 = 0.$$

Thus
$$0 = Re\{x_1^4 - x_1^3 + x_1^2 - x_1 + 1\}$$
$$= \cos\frac{4\pi}{5} - \cos\frac{3\pi}{5} + \cos\frac{2\pi}{5} - \cos\frac{\pi}{5} + 1$$
$$= -\cos\frac{\pi}{5} + \cos\frac{2\pi}{5} + \cos\frac{2\pi}{5} - \cos\frac{\pi}{5} + 1$$
$$= -2\cos\frac{\pi}{5} + 2\cos\frac{2\pi}{5} + 1.$$

This yields
$$\cos\frac{\pi}{5} - \cos\frac{2\pi}{5} = \frac{1}{2}.$$

5(2). (a).
$$x^5 + x^4 + x^3 + x^2 + x + 1 = \frac{x^6 - 1}{x - 1}$$
$$= \frac{(x^3 - 1)(x^3 + 1)}{x - 1}$$
$$= \frac{(x - 1)(x^2 + x + 1)(x + 1)(x^2 - x + 1)}{x - 1}$$
$$= (x + 1)(x^2 + x + 1)(x^2 - x + 1).$$

5(2). (b). $x_1 = -1$, $x_2 = \frac{-1+\sqrt{3}i}{2}$, $x_3 = \frac{-1-\sqrt{3}i}{2}$, $x_4 = \frac{1+\sqrt{3}i}{2}$, $x_2 = \frac{1-\sqrt{3}i}{2}$.

6(2). $x = \frac{\pi}{3} + n\pi$ or $x = -\frac{\pi}{3} + n\pi$, where $n = 0, \pm 1, \pm 2, \cdots$.

6(4). $x = \frac{n\pi}{2}$, or $x = \frac{2\pi}{3} + 2n\pi$, or $x = \frac{4\pi}{3} + 2n\pi$ where $n = 0, \pm 1, \pm 2, \cdots$.

Exercise 5.1.4

1(2). (a). $-\frac{\pi}{6}$; (b). $x = -\frac{\pi}{6} + n\pi$, where $n = 0, \pm 1, \pm 2, \cdots$.

1(4). (a). $\frac{3\sqrt{7}}{8}$; (b). $\frac{\pi}{9}$.

2(2). $\frac{4\sqrt{5}}{9}$.

2(4). $2x\sqrt{1-x^2}$.

Exercise 5.2.3

1(2). $\frac{\sqrt{15}}{60}$.

1(4). $\frac{3\sqrt{21}-1}{16}$.

2(2). $(\cos x + \sin x)(\cos x - \sin x - 1) = 0$. So, $x = \frac{3\pi}{4} + n\pi$, or $x = 2n\pi$, or $x = -\frac{\pi}{2} + 2n\pi$, where $n = 0, \pm 1, \pm 2, \cdots$.

2(4). $-1 \le x < 0$.

Exercise 5.3.1

1(2). Even in the domain $D = \{x \in \mathbb{R} \mid -\frac{\sqrt{2}}{2} \le x \le \frac{\sqrt{2}}{2}\}$.

1(4). Odd and strictly increasing function in the domain $D = \{x \in \mathbb{R} \mid -\infty < x < \infty\}$.

2(2). $-1 \le y \le 1$.

2(4). $\frac{1}{2} < x \le 1$.

Exercise 6.1.1

2(1). $(A, B) = 4$.

2(2). 2.

2(3). $2\sqrt{2}$.

Exercise 6.2.1

1(2). x-intercept: $x = \frac{7}{2}$; y-intercept: $y = -\frac{7}{5}$;

1(4). $(\frac{9}{7}, \frac{2}{7}, -\frac{15}{7})$.

2(2). Proof. Using Example 6.9, we have
$$\frac{a}{b} + \frac{b}{c} + \frac{c}{a} \ge 3\sqrt[3]{\frac{a}{b} \cdot \frac{b}{c} \cdot \frac{c}{a}}$$
$$= 3.$$

2(4). Proof: Step 1: for $k = 2$,
$$f(\frac{x_1 + x_2}{2}) = f(\frac{1}{2}x_1 + \frac{1}{2}x_2)$$
$$\leq \frac{1}{2}f(x_1) + \frac{1}{2}f(x_2) \quad \text{(f(x) is a convex function)}$$
$$= \frac{f(x_1) + f(x_2)}{2}.$$

The statement is correct.

We now assume that for $n = k \geq 2$ the statement is correct. That is, for $n = k \geq 2$,
$$f(\frac{x_1 + x_2 + \cdots + x_k}{k}) \leq \frac{1}{k}f(x_1) + \frac{1}{2}f(x_2) + \cdots + f(x_k).$$

Step 2. Now for $n = k + 1$,
$$f(\frac{x_1 + x_2 + \cdots + x_{k+1}}{k+1})$$
$$= f(\frac{k}{k+1} \cdot (\frac{x_1 + x_2 + \cdots + x_k}{k}) + \frac{x_{k+1}}{k+1})$$
$$\leq \frac{k}{k+1}f(\frac{x_1 + x_2 + \cdots + x_k}{k}) + \frac{1}{k+1} \cdot f(x_{k+1}) \quad \text{(since f(x) is a convex function)}$$
$$\leq \frac{1}{k+1}(f(x_1) + f(x_2) + \cdots + f(x_k)) + \frac{1}{k+1} \cdot f(x_{k+1}) \quad \text{(using the induction assumption)}$$
$$= \frac{f(x_1) + f(x_2) + \cdots + f(x_{k+1})}{k+1}.$$

We prove that the statement holds for $n = k + 1$ thus complete the proof.

Exercise 6.3.3

2(2). $x^2 + y^2 - 3y - 3\sqrt{x^2 + y^2} = 0$.

Exercise 6.4.4

1(2). (a). 0; (b). 5; (c). $(0, 0, -12)$.

2(2). (a). Direct proof. Since
$$x + 16y + 64z \leq |x| + 16|y| + 64|z|,$$
we only need to prove the inequality for non-negative numbers x, y, z. For non-negative numbers x, y, z,
$$x + 16y + 64z \leq 9\sqrt{x^2 + 16y^2 + 64z^2}$$
$$\Leftrightarrow (x + 16y + 64z)^2 \leq 81(x^2 + 16y^2 + 64z^2)$$
$$\Leftrightarrow 0 \leq 81(x^2 + 16y^2 + 64z^2) - (x + 16y + 64z)^2.$$

Notice:

$$81(x^2 + 16y^2 + 64z^2) - (x + 16y + 64z)^2$$
$$= 80x^2 + 65 \cdot 16y^2 + 17 \cdot 64z^2 - 16 \cdot 2xy - 64 \cdot 2xz - 16 \cdot 64 \cdot 2yz$$
$$= 16(x-y)^2 + 64(x-z)^2 + 64 \cdot 16(y-z)^2$$
$$\geq 0.$$

We thus complete the proof of the inequality.

(b). Let $\bar{u} = (1, 4, 8)$ and $\bar{v} = (x, 4y, 8z)$. Notice that

$$|\bar{u}| = \sqrt{1^2 + 4^2 + 8^2} = 9, \quad \text{and} \quad |\bar{v}| = \sqrt{x^2 + 16y^2 + 64z^2}.$$

From Cauchy-Schwartz inequality, we know

$$x + 16y + 64z = \bar{u} \cdot \bar{v}$$
$$\leq |\bar{u}||\bar{v}| = 9\sqrt{x^2 + 16y^2 + 64z^2}.$$

Exercise 6.5.4

1(2).

(i). $-3x - y + 3z = 0$.

(ii). $2x + 3y + z = 13$.

(iii). $2x + 3y - z = \frac{5}{3}$.

2(2). They are not co-planar.

Exercise 6.6.3

2(2).
$$\begin{cases} x = \frac{1}{2}(\sqrt{18 - 3t^2} - t), & -\sqrt{6} \leq t \leq \sqrt{6} \\ y = \frac{1}{2}(-\sqrt{18 - 3t^2} - t), & -\sqrt{6} \leq t \leq \sqrt{6} \\ z = t. & -\sqrt{6} \leq t \leq \sqrt{6} \end{cases}$$

It is a space circle.

Exercise 6.7.1

1(2). $B(\frac{1}{25}, \frac{43}{25})$.

1(4). Choose a point $P_1(x_1, y_1)$ on the line. Thus $ax_1 + by_1 = c$. The unit normal vector of the line is: $\vec{n} = (\frac{a}{\sqrt{a^2+b^2}}, \frac{b}{\sqrt{a^2+b^2}})$. So

$$d = |\vec{P_0P_1} \cdot \vec{n}|$$
$$= |\frac{a(x_1 - x_0) + b(y_1 - y_0)}{\sqrt{a^2 + b^2}}|$$
$$= |\frac{c - ax_0 - by_0}{\sqrt{a^2 + b^2}}|$$
$$= |\frac{ax_0 + by_0 - c}{\sqrt{a^2 + b^2}}|.$$

2(2). $B(\frac{5}{9}, 2\frac{1}{9}, 1\frac{1}{9})$.

2(4). Choose a point $P_1(x_1, y_1, z_1)$ on the plane. Thus $ax_1 + by_1 + cz_1 = d$. The unit normal vector of the plane is: $\vec{n} = (\frac{a}{\sqrt{a^2+b^2+c^2}}, \frac{b}{\sqrt{a^2+b^2+c^2}}, \frac{c}{\sqrt{a^2+b^2+c^2}})$. So

$$d = |\vec{P_0P_1} \cdot \vec{n}|$$
$$= |\frac{a(x_1 - x_0) + b(y_1 - y_0) + c(z_1 - z_0)}{\sqrt{a^2 + b^2 + c^2}}|$$
$$= |\frac{d - ax_0 - by_0 - cz_0}{\sqrt{a^2 + b^2 + c^2}}|$$
$$= |\frac{ax_0 + by_0 + cz_0 - d}{\sqrt{a^2 + b^2 + c^2}}|.$$

4(2). $\vec{AB} \cdot \vec{AC} \times \vec{AD} = 0$. They are coplanar. The plane equation is: $-3x - 2y + 5z = 0$.

5(2). $y = \frac{5}{2}x^2 + \frac{1}{10}$.

6. (1). $\mathcal{L}_1$:
$$\begin{cases} x = t, & -\infty < t < \infty \\ y = 0 \\ z = 1 \end{cases}$$

$\mathcal{L}_2$:
$$\begin{cases} x = t, & -\infty < t < \infty \\ y = 1 - t, & -\infty < t < \infty \\ z = 0 \end{cases}$$

(2). Equation for plane $\mathcal{P}$: $z = 0$.

(3). $d = 1$

Exercise 7.1.6

1(2). $a_n = 3 \cdot 5^{-n+1}$.

1(4). 5.55555

2. Solution:
$$\sum_{k=1}^{100} k(k+1) = \frac{1}{3}\sum_{k=1}^{100}[k(k+1)(k+2) - (k-1)k(k+1)]$$
$$= \frac{1}{3}[100 \cdot 101 \cdot 102 - 0]$$
$$= 343400.$$

4(2). (**Hint:** $a_n - \frac{1}{2}$ is a geometric sequence). $a_n = 3^{n-1}(a_1 - \frac{1}{2}) + \frac{1}{2}$.

Exercise 7.2.6

1(2). Proof. For any $\epsilon > 0$, we choose $\delta = \min(\epsilon, 1)$. Then, for any $0 < |x-4| < \delta$, we know that $x > 0$, and $\sqrt{x} + 2 > 2$. Further,
$$|\sqrt{x} - 2| = \frac{|x-4|}{\sqrt{x}+2} < \frac{\epsilon}{2} < \epsilon.$$
Thus, by the definition of the limit,
$$\lim_{x \to 4} \sqrt{x} = 2.$$

2(2). (i). $\frac{2}{3}$; (ii) $\frac{2}{3}$.

3(1). It can be proved by mathematical induction.

(2). Proof. Let $a_n = 0$, $b_n = \frac{n}{2^n}$ and $c_n = \frac{1}{n}$. Using the inequality in 3(1), we know that for $n \geq 4$,
$$a_n \leq b_n \leq c_n.$$
Also, it is clear that
$$\lim_{n \to \infty} 0 = 0, \quad \text{and} \quad \lim_{n \to \infty} \frac{1}{n} = 0.$$
By the squeeze theorem, we conclude that $\lim_{n \to \infty} \frac{n}{2^n} = 0$.

4(1). Proof. $e > 2$, thus $0 < \frac{n}{e^n} < \frac{n}{2^n}$. Using the squeeze theorem and the result obtained in 3(2), we conclude
$$\lim_{n \to \infty} \frac{n}{e^n} = 0.$$

(2). Proof. Let $m = \ln n$. Then, using the result obtained in 4(1), we know
$$\lim_{n \to \infty} \frac{\ln n}{n} = \lim_{m \to \infty} \frac{m}{e^m} = 0.$$

Exercise 7.3.3

1(2). $1 - 2^{-10} = \frac{1023}{1024}$; (2). 0.

2(2). Divergent; 2(4). Divergent.

Exercise 7.4.1

1. $(\frac{k}{2} - \frac{1}{4}) \cdot 3^k + \frac{1}{4}$.

2(2). 2.

Exercise 8.1.1

1(2). $\frac{1}{2}$.

2(2). (i). $v(t) = at$; (ii). $a(t) = a$.

Exercise 8.2.4

1(2). (i). $-2\sin 2x$; (ii). $2x\cos x^2$

(4). (i). $\sec x = \frac{1}{\cos x}$; (ii). $\frac{e^x}{x} - \frac{e^x}{x^2}$.

2(2). (i). Proof: Since, for $x \geq 0$,
$$\frac{d}{dx}(e^x - 1 - x) = e^x - 1 \geq 0,$$
we know function $f(x) = e^x - 1 - x$ is an increasing function for $x \geq 0$. Thus $f(x) \geq f(0) = 0$. It follows that
$$e^x \geq 1 + x.$$

(ii). Proof. Since, for $x \geq 0$,
$$\frac{d}{dx}(x - \sin x) = 1 - \cos x \geq 0,$$
we know function $f(x) = x - \sin x$ is an increasing function for $x \geq 0$. Thus $f(x) \geq f(0) = 0$. It follows that
$$x \geq \sin x.$$

(4). Proof. Since,
$$\frac{d}{dx}(\ln x + \frac{1}{x} - 1) = \frac{1}{x} - \frac{1}{x^2},$$
we know function $f(x) = \ln x + \frac{1}{x} - 1$ is an increasing function for $x \in (0, 1)$, and is a decreasing function for $x \in (1, \infty)$. Thus $0 = f(1) \leq f(x)$ for $x \in (0, \infty)$. It follows that, for $x \in (0, \infty)$,
$$0 \leq \ln x + \frac{1}{x} - 1,$$
that is:
$$\ln x + \frac{1}{x} \geq 1.$$

4. (1). It is slightly easier to show that $y = (x+1)[\ln(x+1) - \ln x]$ is decreasing.

(2).
$$(1+\frac{1}{x})^{x+1} \leq (1+\frac{1}{5})^6 < 3.$$

(3). It is slightly easier to show that $y = x[\ln(x+1) - \ln x]$ is increasing. In fact, for $x > 0$,
$$g(x) = \frac{d}{dx}x[\ln(x+1) - \ln x] = \ln(x+1) - \ln x - \frac{1}{x+1},$$
and
$$g'(x) = -\frac{1}{x(x+1)} + \frac{1}{(x+1)^2} < 0$$
. Notice that $\lim_{x\to\infty} g(x) = 0$, so, $g(x) > 0$ for all $x \geq 0$. Thus $y = x[\ln(x+1) - \ln x]$ is increasing.

(4). $\lim_{n\to\infty}(1+\frac{1}{n})^n \geq 2$ since $(1+\frac{1}{n})^n$ is increasing in n.

On the other hand, $\lim_{n\to\infty}(1+\frac{1}{n})^n \leq \lim_{n\to\infty}(1+\frac{1}{n})^{n+1} \leq (1+\frac{1}{5})^6 < 3$, since $(1+\frac{1}{n})^{n+1}$ is decreasing in n.

Exercise 8.3.1

1(2). (i). $\frac{e^{2x}}{2} + C$; (ii). $-\frac{\cos 2x}{4} + C$.

(4). (i). $\frac{3}{10}(x^2+1)^{\frac{5}{3}} + C$; (ii). $x + 2\ln|x-1| + C$.

2(2). (i). $\ln|\frac{x-1}{x}| + C$; (ii). $\frac{1}{5}\ln|\frac{x-4}{x+1}| + C$.

(4). (i). $-x\cos x + \sin x + C$; (ii). $\frac{e^x \sin x + e^x \cos x}{2} + C$.

Exercise 8.4.3

1(2). (i). 2; (ii). 0.

(4). (i) 0; (ii). 1.

2(2). (i). $v(10) = 50$; (ii). $s(10) = \frac{1000}{6}$.

(4). $v(r) = \frac{4}{3}\pi r^2$.

Exercise 8.5.1 1(2). $2\ln 2 - 2$.

(3). $y = \ln x$ is a convex function, so
$$\ln(\frac{x_1 + x_2 + 3}{3}) = \ln(\frac{(x_1+1) + (x_2+1) + 1}{3})$$
$$\geq \frac{\ln(x_1+1) + \ln(x_2+1) + \ln 1}{3}$$
$$= \frac{\ln(x_1+1) + \ln(x_2+1)}{3}.$$

2(2). First, we have
$$a - \frac{b}{x} + \frac{c}{x^2} - \frac{d}{x^3} + \frac{e}{x^4} - \frac{f}{x^5} + \frac{g}{x^6} = 1 + \frac{2}{x} + \frac{3}{x^2} + \frac{4}{x^3} + \frac{5}{x^4} + \frac{6}{x^5} + \frac{7}{x^6}.$$
Sending $x \to \infty$, we obtain $a = 1$.

Subtracting 1 in both sides of the above equation and then multiplying x, we have
$$-b + \frac{c}{x} - \frac{d}{x^2} + \frac{e}{x^3} - \frac{f}{x^4} + \frac{g}{x^5} = 2 + \frac{3}{x} + \frac{4}{x^2} + \frac{5}{x^3} + \frac{6}{x^4} + \frac{7}{x^5}.$$
Sending $x \to \infty$, we obtain $b = -2$.

Iterating the above processing, we obtain $c = 3$, $d = -4$, $e = 5$, $f = -6$, $g = 7$.

2(4). Proof. If $f(x) = constant$, we know from the definition of derivative that $f'(x) = 0$.

If $f'(x) = 0$, we can prove that $f = constant$ by a contradiction argument. If not, there are two different points $x_1, x_2 \in I$, such that $f(x_1) \neq f(x_2)$. Then by the Mean Value Theorem (Theorem 8.6) there is a point $c \in (x_1, x_2)$, such that
$$f'(c) = \frac{f(x_2) - f(x_1)}{x_2 - x_1} \neq 0.$$
Contradiction.

4(2). (i).

$1 \times 2 + 2 \times 3 + \cdots + 99 \times 100$

$$= \frac{1}{3} \times \{(1 \times 2 \times 3 - 0 \times 1 \times 2) + \cdots + (99 \times 100 \times 101 - 98 \times 99 \times 100)\}$$

$$= \frac{1}{3} \times (99 \times 100 \times 101)$$

$$= 333300.$$

(ii). First, we have
$$\sum_{k=1}^{n} k(k+1) = \frac{1}{3} \sum_{k=1}^{n} \{k(k+1)(k+2) - (k-1)k(k+1)\}$$
$$= \frac{n(n+1)(n+2)}{3}.$$
And
$$\sum_{k=1}^{n} k = \frac{n(n+1)}{2}.$$
So,
$$\sum_{k=1}^{n} k^2 = \sum_{k=1}^{n} k(k+1) - \sum_{k=1}^{n} k$$
$$= \frac{n(n+1)(2n+1)}{6}.$$

Index

Absolute value, 55
Addition, 2
Addition Principle, 94
Algebraic properties of function limits, 277
Algebraic properties of limit, 270
Angle between two vectors, 227, 229
arccos function, 196
Arccot function, 198
Arcsine function, 194
Arctan function, 198
Area formula, 320
arithmetic mean, 216

Binary, 3
Binary system, 8
Binomial formula, 133
    Pascal/Jian Xian's triangle, 134
borrow number, 12
Bounded sequence, 272

Cardinality, 94
Carry number, 7
Cauchy-Schwarz inequality, 216, 229
Chain rule, 302
Circular motion, 311
Combination, 98
Complex number, 71
    Argument, 74
    de Moivre formula, 176
    Euler formula, 177
    Imaginary part, 71
    Imaginary symbol i, 71
    Modulus, 74
    Principal argument, 74
    Real part, 71
Composite function, 112
Composite number, 21
Concave function, 215
Conjugate number, 64
Continuity of elementary functions, 283
Continuous, 282
Convex function, 215
Cosine law, 184
Cross product, 233
Cross-product form, 48

de Moivre formula, 75, 177
Decimal system, 3
Definite integral, 318
Derivative, 300
    Derivative function, 300
    Derivatives of elementary functions, 301
    N-th order derivative, 300
Derivative of vector functions, 309
Difference of two cubes, 33
Difference of two squares, 33
Discriminant, 125
Distance formula, 208, 209
Division, 14
    dividend, 14
    divisor, 14
    quotient, 14

Ellipse, 213
Equation
    First order/linear equation, 40
    Second degree/quadratic equation, 46

INDEX

Equivalent equations, 37
    Addition invariance, 37
    Multiplication invariance, 37
    Symmetric property, 37
    Transitive property, 39
Euler formula, 74, 187
Exponent, 2
    Negative exponent, 18
    Zero exponent, 18
Exponent rules, 19
    For real number exponents, 65
    For complex exponents, 179
    For fractional exponents, 60
    For integer exponents, 91
    For real number exponents, 91
    The same base operation rule, 19
    The same exponent rule, 19
Extraneous solution, 137, 147, 338
Extreme value, 305

Factor, 20
Fraction
    Complex fraction, 23
    Improper fraction, 23
    Mixed fraction, 23
    Proper fraction, 23
    Simplest fraction, 23
Function, 104
    Domain, 104
    Range, 104
Fundamental Assumption of Elementary Mathematics, 65
Fundamental Assumptions of Elementary Mathematics, 283
Fundamental Theorem of Algebra, 130
Fundamental Theorem of Calculus, 319

General inner product, 228
geometric mean, 216
Greatest common divisor (GCD), 23

Hat notation, 10, 101
Hat notation for negative coefficients, 101, 102
Heron-Qin's formula, 184
Horizontal form, 6
Hyperbola, 213

Image, 104
    Pre-image, 104
Indefinite integral, 313
Integrals of elementary functions, 314
Integration by parts formula, 314
Intermediate value value theorem of continuous function, 284
Interval, 92
Inverse function, 113
irrational number, 62

Jensen's inequality, 55, 216

Least common multiplier (LCM), 23
Leibniz-Newton Formula, 319
Like terms, 28, 63
Limit for function, 276
Limit for sequence, 269
Line equations in space, 241
    Angle between planes, 242
    Normal vector, 241
Linear equations on the plane, 239
    Point-slope formula, 239
    Two-point formula, 239
Linear system, 43
Logarithm, 67
    Operation rules, 68

INDEX

Maclaurin series, 308
Mapping, 103
Maximum, 305
Mean value theorem, 305
Minimum, 305
Mixed product, 236
Monomial, 27, 121
    Coefficient, 121
    Coefficient of the monomial, 27
    Definition, 121
    Degree of the monomial, 27
    Degree/Order, 121
Multiplication, 2
Multiplication Principle, 95

N-th root, 59
Natural number e, 142
Natural numbers, 2
Normal vector of a line, 240
Number line, 55
Number of subsets, 99
Number sets, 87

Operation rules, 3, 90
    Associative rule, 4
    Commutative rule, 3
    Distributive rule, 5
Opposite number, 9
    Negative number, 9
    Subtraction, 10

Parabola, 213
Parameter, 40
Parametric equation, 246
Percentage, 20
Perfect square formula, 33
Permutation, 96

Place value, 3
Polar coordinate system, 219
Polar coordinates
    Complex number, 74
    Point on a plane, 219
Polar form, 74
Polynomial, 28, 123
    Coefficient, 123
    Definition, 123
    Degree of the polynomial, 28
    Degree/Order, 123
Prime number, 21
    Infinitely many prime numbers, 21
    Mersenne prime number, 33
Principal root, 77, 181, 343
Product rule, 301
Properties of function
    Even function, 110
    Monotone functions, 109
    Odd function, 110
    One-to-one function, 111
    Periodic function, 111
Properties of functions, 108
Properties of inequality, 52
    Addition invariance, 52
    Invariance under positive number multiplication, 53
    Reverse inequality under negative number multiplication, 53
    Symmetric property, 53
    Transitive property, 53
Ptolemy addition formula, 167

Rational number, 62
Rationalizing denominators, 63
Reciprocal, 14

INDEX

Division, 14, 16
    Fraction, 14, 16
    Negative power, 14
Relevance and irrelevance, 230
Remainder Theorem, 129
Riemann sum, 317

Scientific notation, 20
Sequence, 257
    Arithmetic sequence, 259
    Fibonacci sequence, 258, 265
    Geometric sequence, 261
Series
    Power series, 290
series, 288
    Convergence, 289
    divergence, 289
    Finite series, 288
    Infinite series, 288
    Partial sum, 288
Set, 86
    Complement set, 87
    Intersection set, 88
    Subset, 87
    Union set, 88
Set operation rules, 89
Sine law, 184
Slope, 239
Solving linear system
    Elimination method, 43
    Matrix method, 45
    Substitution method, 43
Squeeze Theorem, 274
Squeeze theorem, 278, 280
Standard form for a decimal number, 3
Substitution Formula for the Integral, 314

Subtraction
    Minuend, 9
    Subtrahend, 9
Sum of two cubes, 33

Taylor series, 187, 308
Three operation rules holds for all numbers, 11
Translation of graph, 107
    Horizontal translation, 108
    Vertical translation, 107
Trigonometric equations, 202
Trigonometric functions, 158
    Definition, 158
    Signs of trigonometric functions, 163
Trigonometric identities, 165
    Addition and subtraction formulas-1, 167
    Addition and subtraction formulas-2, 169
    Double angle formulas, 174
    General Pythagorean identity, 165
    Half angle formulas, 175
    Product-to-Subtraction formulas, 173
    Product-to-Sum formulas, 173
    Sum-to-Product formulas, 171

Vector, 225
    Inn product, 226
    Scalar product, 225
    Vector addition, 225
Vector function, 246
Venn diagram, 88
Vertex of the parabola, 124
Vertical form , 6
Vieta's formula, 131
Volume formula, 321

Volume of parallelepiped, 236

www.ingramcontent.com/pod-product-compliance
Lightning Source LLC
Chambersburg PA
CBHW051146290426
44108CB00019B/2623